STUDENT'S
SOLUTIONS MANUAL

To Accompany
James T. McClave
and P. George Benson's

STATISTICS
· ·
For Business and Economics

SIXTH EDITION

Nancy S. Boudreau
Bowling Green State University

DELLEN
an imprint of
MACMILLAN COLLEGE PUBLISHING COMPANY
New York

MAXWELL MACMILLAN CA▸

MAXWELL MACMILLA▸
New York Oxford

. . . .

Macmillan Publishing Company
113 Sylvan Avenue, Englewood Cliffs, NJ 07632

ISBN 0-02-312721-X

4 5 6 8 9   Year: 5 6 7

Contents

· ·

Preface

· ·

This solutions manual is designed to accompany the text **Statistics for Business and Economics**, Sixth Edition, by James T. McClave and P. George Benson (Dellen Publishing Company, 1994). It provides answers to most odd-numbered exercises for each chapter in the text. Other methods of solution may also be appropriate; however, the author has presented one that she believes to be most instructive to the beginning statistics student. The student should first attempt to solve the assigned exercises without help from this manual. Then, if unsuccessful, the solution in the manual will clarify points necessary to the solution. The student who successfully solves an exercise should still refer to the manual's solution. Many points are clarified and expanded upon to provide maximum insight into and benefit from each exercise.

Instructors will also benefit from the use of this manual. It will save time in preparing presentations of the solutions and possibly provide another point of view regarding their meaning.

Some of the exercises are subjective in nature and thus omitted from the Answer Key at the end of **Statistics for Business and Economics**, Sixth Edition. The subjective decisions regarding these exercises have been made and are explained by the author. Solutions based on these decisions are presented; the solution to this type of exercise is often most instructive. When an alternative interpretation of an exercise may occur, the author has often addressed it and given justification for the approach taken.

I would like to thank Kelly Evans for creating the art work and Brenda Dobson for her assistance and for typing this work.

Nancy S. Boudreau
Bowling Green State University
Bowling Green, Ohio

CHAPTER ONE

....

What Is Statistics?

1.1   Descriptive statistics utilizes numerical and graphical methods to look for patterns, to summarize, and to present the information in a set of data. Inferential statistics utilizes sample data to make estimates, decisions, predictions, or other generalizations about a larger set of data.

1.3   A population is a set of existing units such as people, objects, transactions, or events. A variable is a characteristic or property of an individual population unit such as height of a person, time of a reflex, amount of a transaction, etc.

1.5   An inference without a measure of reliability is nothing more than a guess. A measure of reliability separates statistical inference from fortune telling or guessing. Reliability gives a measure of how confident one is that the inference is correct.

1.7   a.   The population of interest is all the students in the class. The variable of interest is the GPA of a student in the class.

        b.   Since the population of interest is all the students in the class and you obtained the GPA of every member of the class, this set of data would be a census.

        c.   Assuming the class had more than 10 students in it, the set of 10 GPAs would represent a sample. The set of ten students is only a subset of the entire class.

        d.   This average would have 100% reliability as an "estimate" of the class average, since it is the average of interest.

        e.   The average GPA of 10 members of the class will not necessarily be the same as the average GPA of the entire class. The reliability of the estimate will depend on how large the class is and how representative the sample is of the entire population.

1.9   a.   The population of interest is all citizens of the United States.

        b.   The variable of interest is the view of each citizen as to whether the president is doing a good or bad job. It is nonnumerical.

        c.   The sample is the 2000 individuals selected for the poll.

        d.   The inference of interest is to estimate the proportion of all citizens who believe the president is doing a good job.

1.11   a.   The population of interest is the set of *Fortune 500* companies.

        b.   The variable of interest is the number of new employees the company is likely to hire in the coming year.

....

c. The sample is the set of 50 companies selected from the 500 companies.

d. We could estimate the number of new employees the *Fortune 500* companies are likely to hire in the coming year.

e. We are employing inferential statistics. We do not have information on the number of new employees each of the *Fortune 500* companies is likely to hire in the coming year, only the information from the 50 selected companies. Thus, we are inferring or estimating the total number of new employees all *Fortune 500* companies are likely to hire in the coming year from the sample number of new employees the 50 companies are likely to hire in the coming year.

f. We are employing descriptive statistics. We have information on the net profit of each of the 500 companies from which we will compute the mean. We have information from every member of our population.

1.13 a. The population of interest is the set of all medical doctors and the variable is whether the doctor was involved in one or more malpractice suits.

b. The sample is the collection of 500 doctors selected at random. The insurance company wants to estimate the proportion of all medical doctors who have been involved in one or more malpractice suits.

c. "At random" means that all medical doctors had an equal chance of being selected for the sample.

1.15 a. The population of interest is the collection of all companies that offer job-sharing.

b. The variable of interest is whether the firm's director was satisfied or not with the productivity of workers who shared jobs.

c. The sample is the set of 100 firms selected.

d. The government might want to estimate the proportion of all firms that offer job-sharing that are satisfied with the productivity of their workers with shared jobs.

1.17 a. The population of interest is all RV owners in the United States. The variables of interest are the preferences with respect to the features of a portable generator (e.g., size, manual or electric start, etc.). The sample is the collection of 1052 RV owners who returned the questionnaire. The inference of interest is to generalize the preferences of the 1052 sampled RV owners to the population of all RV owner preferences on the features of a portable generator. Specifically, the sample results will be used to decide what features should be included on a portable generator by estimating the proportion that like each feature suggested on the questionnaires.

b. One factor that may affect the reliability of inferences is the group of RV owners who return the questionnaire. Often, the people who return questionnaires have very strong opinions about the items on the questionnaire and may not be representative of the population in general. Another factor may be those receiving the questionnaire may no longer be RV owners.

CHAPTER TWO

· ·

Methods for Describing Sets of Data

2.1 a. Nominal data are measurements that simply classify the units of the sample or population into categories. These categories cannot be ranked. Ordinal data are measurements that enable the units of the sample or population to be ordered or ranked with respect to the variable of interest.

b. Interval data are measurements that enable the determination of the differential (how much more or less) of the characteristic being measured between one unit of the sample or population and another. Interval data will always be numerical, and the numbers assigned to the two units can be subtracted to determine the difference between the units. However, the zero point is not meaningful for these data. Thus, these data cannot be multiplied or divided. Ratio data are measurements that enable the determination of the multiple (how many times as much) of the characteristic being measured between one unit of the sample or population and another. All the characteristics of interval data are included in ratio data. In addition, the zero point for ratio data is meaningful.

c. Qualitative data have no meaningful numbers associated with them. Qualitative data include nominal and ordinal data. Quantitative data have meaningful numbers associated with them. Quantitative data include interval and ratio data.

2.3 The data consisting of the classifications A, B, C, and D are qualitative. These data are nominal and thus are qualitative. After the data are input as 1, 2, 3, and 4, they are still nominal and thus qualitative. The only difference between the two data sets are the names of the categories. The numbers associated with the four groups are meaningless.

2.5 a. Nominal; possible brands are "Guess," "Levis," "Lee," etc., each of which represents a nonranked category.

b. Ratio; the number of hours of sports programming carried in a typical week is measured on a numerical scale where the zero point has meaning. Ten hours of sports programming is five times as much as two hours of sports programming.

c. Ratio; the percentage of their time spent studying is measured on a numerical scale where the zero point has meaning. Forty percent of time spent studying is twice as much as twenty percent.

d. Ratio; the number of long distance telephone calls is measured on a numerical scale where the zero point has meaning. Twenty phone calls is four times as many as five phone calls.

e. Interval; SAT scores are measured on a numerical scale where the zero point has no meaning. A score of 500 is not twice as good as a score of 250.

· ·

2.7   I.   Nominal; the possible responses are "yes" or "no," which are nonnumerical and are not ranked.

      II.   Ratio; age is measured on a numerical scale where the zero point has meaning. A person who is 30 years old is twice as old as a person who is 15 years old.

     III.   Ordinal; the rating can be from 1 to 10, where the higher the rating the more helpful the *Tutorial* instructions. However, the difference between ratings of 3 and 4 may be different than the difference between ratings 7 and 8.

      IV.   Nominal; the possible responses are "dot-matrix printer" or "another type of printer," which are nonnumerical and cannot be ranked.

       V.   Ordinal; the speeds can be classified as "slower," "unchanged," or "faster." These speeds are ordered.

      VI.   Ratio; the number of people in a household who have used Windows 3.0 at least once is measured on a numerical scale where the zero point has meaning. A household of 4 users has twice as many users as a household of 2 users.

2.9

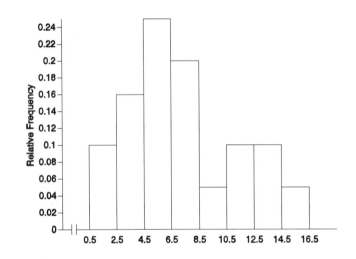

2.11   a.   This is a frequency histogram because the number of observations is graphed for each interval rather than the relative frequency.

       b.   There are 14 measurement classes.

       c.   There are 49 measurements in the data set.

2.13   a.   Most of the observations (12) occur in the interval 97.5 to 102.5. Also, most of the observations are above 97.5 (24). There are only 6 observations less than 97.5. The smallest observation is between 57.5 and 62.5.

b.   There are $12 + 6 + 5 + 1 = 24$ observations greater than 97.5. This is $24/30 = .8$ of the total number of bonds.

c.   The area in the histogram corresponding to this proportion is the area of the bottom four rectangles.

2.15   b.   In all three histograms, most of the observed times were less than 45.5 minutes. There were proportionately more observations in the lower measurement classes in histogram (2) than in histogram (3). This indicates that the time spent when a purchase was made is somewhat shorter than when a purchase is not made.

c.   One possible explanation is that when the purchase is made may not be the first time the customer has been in the store. These customers may have come to the store at an earlier date and spent more time with the salesperson.

2.17   a.   Using MINITAB, the three frequency histograms are as follows (the same starting point and class interval was used for each):

```
Histogram of C1   N = 25

Tenth Performance

Midpoint        Count
    3.50          0
    7.50          0
   11.50          1   *
   15.50          5   *****
   19.50         10   **********
   23.50          6   ******
   27.50          0
   31.50          2   **
   35.50          0
   39.50          1   *

Histogram of C2   N = 25

Thirtieth Performance

Midpoint        Count
    3.50          1   *
    7.50          9   *********
   11.50         12   ************
   15.50          2   **
   19.50          1   *

Histogram of C3   N = 25

Fiftieth Performance

Midpoint        Count
    3.50          3   ***
    7.50         15   ***************
   11.50          4   ****
   15.50          2   **
   19.50          1   *
```

b.   The histogram for the 10th performance shows a much greater spread of observations than do the other two histograms. The histogram for the 30th performance shows a shift to the left—implying shorter completion times than for the 10th performance. In addition, the histogram for the 50th performance shows an additional shift to the left compared to the histogram for the 30th performance. However, the last shift is not as great as the first shift. This agrees with statements made in the problem.

2.19   a.   To construct the stem and leaf displays, the stem will be the tens digit and the leaf will
            be the ones digit.

| | **Auto Parts** | | | **Electronics** |
Stem	**Leaf**		**Stem**	**Leaf**
0	7 8 8 9 9 9 9		0	8
1	0 0 0 1 2 2 3 3 3 4 4 5 5		1	0 1 2 2 2 4 4 4 5 6 9 9
			2	0 2 4 7 7
	Key: Leaf units are ones		3	9
			4	
			5	
			6	
			7	
			8	5

Key: Leaf units are ones

b.   From the stem and leaf displays, the P/E ratios are always less than 16 for the auto parts
     industry, while for electronics, most of the numbers are less than 27 with two numbers
     even larger. Therefore, the P/E ratios of firms in the electronics industry tend to be
     larger than the P/E ratios for the auto parts industry.

2.21   a.   The relative frequency bar chart for
            1970 is:

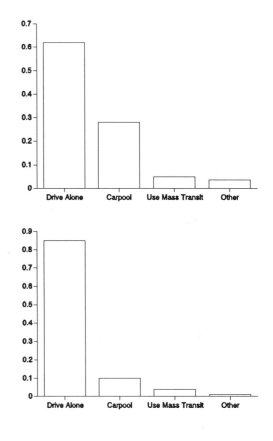

The relative frequency bar chart for
1990 is:

b. The combined bar charts are as follows:

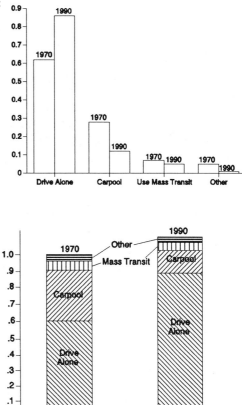

c. The stacked bar charts are as follows:

d. First, there are many more commuters in 1990 than in 1970 (10,000 versus 6000). In 1990, 85% of the commuters drove alone, while in 1970, only 62% drove alone. The percent of commuters using carpools or mass transit has fallen sharply in 1990 compared to 1970 (14% from 34%).

2.23 The sample mean is:

$$\bar{x} = \frac{\sum_{i=1}^{n} x_i}{n} = \frac{18 + 10 + 15 + 13 + 17 + 15 + 12 + 15 + 18 + 16 + 11}{11}$$

$$= \frac{160}{11} = 1.545$$

The median is the middle number when the data are arranged in order. The data arranged in order are: 10, 11, 12, 13, 15, 15, 15, 16, 17, 18, 18. The middle number is the 6th number, which is 15.

Assume the data are a sample. The mode is the observation that occurs most frequently. For this sample, the mode is 15, which occurs 3 times.

2.25   The median is the middle number once the data have been arranged in order. If n is even, there is not a single middle number. Thus, to compute the median, we take the average of the middle two numbers. If n is odd, there is a single middle number. The median is this middle number.

A data set with 5 measurements arranged in order is 1, 3, 5, 6, 8. The median is the middle number, which is 5.

A data set with 6 measurements arranged in order is 1, 3, 5, 5, 6, 8. The median is the average of the middle two numbers which is

$$\frac{5 + 5}{2} = \frac{10}{2} = 5$$

2.27   Quite often when dealing with opinions, guesses, and estimates, there tend to be some very high or very low answers. Since the mean is more sensitive to extreme values, the median would represent a more accurate description of the responses.

2.29   a.   We can identify the sample observations using the notation:

$$x_1 = 110, x_2 = 90, \dots , x_{19} = 162$$

Then, the sample mean is

$$\overline{x} = \frac{\sum\limits_{i=1}^{n} x_i}{n} = \frac{110 + 90 + \cdots + 162}{19} = \frac{3420}{19} = 180$$

Since $n = 19$ is odd, the median will be the middle observation (x_{10}) when the data are arranged from smallest to largest.

The data arranged from smallest to largest are:

90	162
110	176
110	181
114	187
125	200
129	200
158	230
159	274
162	290
	363

The median is 162.

b.   Since the median is less than the mean, the data set is skewed to the right.

c.   No, the median will not always be an actual value in the data set. When n is odd, it will always be an actual value; but when n is even, we average the two middle values in the data set. If these two values are not the same value, the median will not be an actual value of the data set.

2.31 a. The sample mean is

$$\bar{x} = \frac{\sum_{i=1}^{n} x_i}{n} = \frac{92.5 + 63.3 + \cdots + 69.7}{20} = \frac{1400.2}{20} = 70.01$$

Since $n = 20$ is even, the median is the average of the middle two numbers when the data are arranged from smallest to largest.

The data arranged from smallest to largest are:

63.3	64.9	69.0	74.9
63.4	65.5	69.7	74.9
63.9	65.7	70.6	76.8
64.6	67.6	70.8	76.8
64.6	68.5	72.4	92.5

The median is $\dfrac{x_{(10)} + x_{(11)}}{2} = \dfrac{68.5 + 69.0}{2} = \dfrac{137.5}{2} = 68.75$

The mode is the measurement that occurs with the greatest frequency. In this data set, 64.6 and 76.8 are both modes. (They both occur twice.)

b. The highest price in the data set is 92.5. By eliminating this price from the data set, the sample mean $= \bar{x} = \dfrac{\sum x}{n} = \dfrac{1307.7}{19} = 68.83$

Since $n = 19$ is odd, the median is the middle number $(x_{(10)})$ when the data are arranged from smallest to largest.

The median $= x_{(10)} = 68.5$

The modes are 64.6 and 76.8.

By dropping the highest value in the data set, the mean and median decreased while the mode did not change.

c. By eliminating the highest two prices (92.5 and 76.8) and the lowest two prices (63.3 and 63.4), the 80% trimmed mean is

$$\bar{x} = \frac{\sum_{i=1}^{n} x_i}{n} = \frac{1104.2}{16} = 69.0125$$

2.33 The mean value per coupon is the total value of the coupons $\left(\sum_{i=1}^{n} x \right)$ divided by the number of coupons redeemed (n).

$$\bar{x} = \frac{\sum_{i=1}^{n} x_i}{n} = \frac{2.24 \text{ billion}}{6.49 \text{ billion}} = \$0.35$$

2.35   a.   The range, variance, and standard deviation measure the variability of the data set. They give a measure of the spread of the data. The larger the value, the more spread out the data. The range gives you an idea how wide the data set is, the standard deviation tells you approximately how far the values are from the mean, and the variance gives the sum of the squared distances from the mean divided by the number of observations minus one.

b.   An advantage of using the range is that it is so easy to compute. A disadvantage is that it only uses the smallest and largest values of the data set. No information is used from the other observations of the data set.

An advantage of the variance is that it does take all of the values of the data set into account. It gives the sum of the squared distances from the mean divided by the number of observations minus one. A disadvantage is that it is expressed in square units. This makes it hard to interpret. It is also harder to calculate than the range.

An advantage of the standard deviation is that it also takes all of the observations into account. It is also expressed in the original units of the problem so that it is easier to interpret than the variance. It gives a measure of the distance the observations are from the mean. A disadvantage is that it is hard to calculate.

2.37   a. $\displaystyle \overline{x} = \frac{\sum_{i=1}^{n} x_i}{n} = \frac{50}{25} = 2$

$$s^2 = \frac{\sum_{i=1}^{n} x_i^2 - \frac{\left(\sum_{i=1}^{n} x_i\right)^2}{n}}{n-1} = \frac{1000 - \frac{50^2}{25}}{25-1} = \frac{900}{24} = 37.5$$

$s = \sqrt{37.5} = 6.124$

b. $\displaystyle \overline{x} = \frac{\sum_{i=1}^{n} x_i}{n} = \frac{100}{80} = 1.25$

$$s^2 = \frac{\sum_{i=1}^{n} x_i^2 - \frac{\left(\sum_{i=1}^{n} x_i\right)^2}{n}}{n-1} = \frac{270 - \frac{100^2}{80}}{80-1} = \frac{145}{79} = 1.835$$

$s = \sqrt{1.835} = 1.355$

2.39   a. $\sum_{i=1}^{n} x_i = 10 + 1 + 0 + 0 + 20 = 31$

$\sum_{i=1}^{n} x_i^2 = 10^2 + 1^2 + 0^2 + 0^2 + 20^2 = 501$

$\bar{x} = \dfrac{\sum_{i=1}^{n} x_i}{n} = \dfrac{31}{5} = 6.2$

$s^2 = \dfrac{\sum_{i=1}^{n} x_i^2 - \dfrac{\left[\sum_{i=1}^{n} x_i\right]^2}{n}}{n - 1} = \dfrac{501 - \dfrac{31^2}{5}}{5 - 1} = \dfrac{308.8}{4} = 77.2$

$s = \sqrt{77.2} = 8.786$

b. $\sum_{i=1}^{n} x_i = 5 + 9 + (-1) + 100 = 113$

$\sum_{i=1}^{n} x_i^2 = 5^2 + 9^2 + (-1)^2 + 100^2 = 10{,}107$

$\bar{x} \dfrac{\sum_{i=1}^{n} x_i}{n} = \dfrac{113}{4} = 28.25$

$s^2 = \dfrac{\sum_{i=1}^{n} x_i^2 - \dfrac{\left[\sum_{i=1}^{n} x_i\right]^2}{n}}{n - 1} = \dfrac{10107 - \dfrac{113^2}{4}}{4 - 1} = \dfrac{6914.75}{3} = 2304.92$

$s = \sqrt{2304.92} = 48.010$

2.41   This is one possibility for the two data sets.

Data Set 1: 1, 1, 2, 2, 3, 3, 4, 4, 5, 5

Data Set 2: 1, 1, 1, 1, 1, 5, 5, 5, 5, 5

$\bar{x}_1 = \dfrac{\sum_{i=1}^{n} x_i}{n} = \dfrac{1 + 1 + 2 + 2 + 3 + 3 + 4 + 4 + 5 + 5}{10} = \dfrac{30}{10} = 3$

$\bar{x}_2 = \dfrac{\sum_{i=1}^{n} x_i}{n} = \dfrac{1 + 1 + 1 + 1 + 1 + 5 + 5 + 5 + 5 + 5}{10} = \dfrac{30}{10} = 3$

Therefore, the two data sets have the same mean. The variances for the two data sets are:

$$s_1^2 = \frac{\sum\limits_{i=1}^{n} x_i^2 - \dfrac{\left[\sum\limits_{i=1}^{n} x_i\right]^2}{n}}{n-1} = \frac{110 - \dfrac{30^2}{10}}{9} = \frac{20}{9} = 2.2222$$

$$s_2^2 = \frac{\sum\limits_{i=1}^{n} x_i^2 - \dfrac{\left[\sum\limits_{i=1}^{n} x_i\right]^2}{n}}{n-1} = \frac{130 - \dfrac{30^2}{10}}{9} = \frac{40}{9} = 4.4444$$

The dot diagrams for the two data sets are shown below.

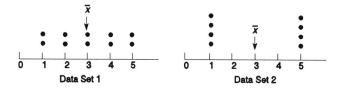

2.43   The variance of a data set can never be negative. The variance of a sample is the sum of the **squared** deviations from the mean divided by $n-1$. The square of any number, positive or negative, is always positive. Thus, the variance will be positive.

The variance is usually greater than the standard deviation. However, it is possible for the variance to be smaller than the standard deviation. If the variance is between 0 and 1, the variance will be smaller than the standard deviation. For example, suppose the data set is .8, .7, .9, .5, and .3. The sample mean is

$$\bar{x} = \frac{\sum\limits_{i=1}^{n} x_i}{n} = \frac{.8 + .7 + .9 + .5 + .3}{5} = \frac{3.2}{5} = .64$$

The sample variance is

$$s^2 = \frac{\sum\limits_{i=1}^{n} x_i^2 - \dfrac{\left[\sum\limits_{i=1}^{n} x_i\right]^2}{n}}{n-1} = \frac{2.28 - \dfrac{3.2^2}{5}}{5-1} = \frac{2.28 - 2.048}{4} = \frac{.325}{4} = 0.58$$

The standard deviation is

$$s = \sqrt{.058} = .241$$

2.45   a.   A worker's overall time to complete the operation under study is determined by adding the subtask-time averages.

Worker A

The average for subtask 1 is: $\bar{x} = \dfrac{\sum_{i=1}^{n} x_i}{n} = \dfrac{211}{7} = 30.14$

The average for subtask 2 is: $\bar{x} = \dfrac{\sum_{i=1}^{n} x_i}{n} = \dfrac{21}{7} = 3$

Worker A's overall time is $30.14 + 3 = 33.14$.

Worker B

The average for subtask 1 is: $\bar{x} = \dfrac{\sum_{i=1}^{n} x_i}{n} = \dfrac{213}{7} = 30.43$

The average for subtask 2 is: $\bar{x} = \dfrac{\sum_{i=1}^{n} x_i}{n} = \dfrac{29}{7} = 4.14$

Worker B's overall time is $30.43 + 4.14 = 34.57$.

b.   **Worker A**

$$s = \sqrt{\dfrac{\sum_{i=1}^{n} x_i^2 - \dfrac{\left(\sum_{i=1}^{n} x_i\right)^2}{n}}{n-1}} = \sqrt{\dfrac{6455 - \dfrac{211^2}{7}}{7-1}} = \sqrt{15.8095} = 3.98$$

Worker B

$$s = \sqrt{\dfrac{\sum_{i=1}^{n} x_i^2 - \dfrac{\left(\sum_{i=1}^{n} x_i\right)^2}{n}}{n-1}} = \sqrt{\dfrac{6487 - \dfrac{213^2}{7}}{7-1}} = \sqrt{.9524} = .98$$

c.   The standard deviations represent the amount of variability in the time it takes the worker to complete subtask 1.

d.   **Worker A**

$$s = \sqrt{\dfrac{\sum_{i=1}^{n} x_i^2 - \dfrac{\left(\sum_{i=1}^{n} x_i\right)^2}{n}}{n-1}} = \sqrt{\dfrac{67 - \dfrac{21^2}{7}}{7-1}} = \sqrt{.6667} = .82$$

Worker B

$$s = \sqrt{\frac{\sum_{i=1}^{n} x_i^2 - \frac{\left(\sum_{i=1}^{n} x_i\right)^2}{n}}{n - 1}} = \sqrt{\frac{147 - \frac{29^2}{7}}{7 - 1}} = \sqrt{4.4762} = 2.12$$

e.   I would choose workers similar to worker B to perform subtask 1. Worker B has a slightly higher average time on subtask 1 (A: $\bar{x} = 30.14$, B: $\bar{x} = 30.43$). But, Worker B has a smaller variability in the time it takes to complete subtask 1 (part **b**). He or she is more consistent in the time needed to complete the task.

I would choose workers similar to Worker A to perform subtask 2. Worker A has a smaller average time on subtask 2 (A: $\bar{x} = 3$, B: $\bar{x} = 4.14$). Worker A also has a smaller variability in the time needed to complete subtask 2 (part **d**).

2.47   Chebyshev's theorem can be applied to any data set. The Empirical Rule applies only to data sets that are mound-shaped—that are approximately symmetric, with a clustering of measurements about the midpoint of the distribution and that tail off as one moves away from the center of the distribution.

2.49   a.   Using Chebyshev's theorem, at least $\left[1 - \frac{1}{1^2}\right]100\% = 0(100\%) = 0\%$ observations are within 1 standard deviation of the mean. Thus, nothing can be said about the number of observations in the interval $\bar{x} - s$ to $\bar{x} + s$.

b.   Using Chebyshev's theorem, at least $\left[1 - \frac{1}{2^2}\right]100\% = (3/4)(100\%) = 75\%$ of the observations are within 2 standard deviations of the mean. Thus, at least 75% of the observations are in the interval $\bar{x} - 2s$ to $\bar{x} + 2s$.

c.   Using Chebyshev's theorem, at least $\left[1 - \frac{1}{3^2}\right]100\% = (8/9)(100\%) = 88.9\%$ of the observations are within 3 standard deviations of the mean. Thus, at least 88.9% of the observations are in the interval $\bar{x} - 3s$ to $\bar{x} + 3s$.

2.51   Using Chebyshev's theorem, at least 8/9 of the measurements will fall within 3 standard deviations of the mean. Thus, the range of the data would be around 6 standard deviations. Using the Empirical Rule, approximately 95% of the observations are within 2 standard deviations of the mean. Thus, the range of the data would be around 4 standard deviations. We would expect the standard deviation to be somewhere between Range/6 and Range/4.

For our data, the range $= 760 - 135 = 625$.

The Range/6 $= 625/6 = 104.17$ and Range/4 $= 625/4 = 156.25$.

Therefore, we would estimate that the standard deviation of the data set is between 104.17 and 156.25.

It would therefore not be feasible to have a standard deviation of 25. If the standard deviation were 25, the data set would span $625/25 = 25$ standard deviations. This would be extremely unlikely.

2.53 a. $$\bar{x} = \frac{\sum\limits_{i=1}^{n} x_1}{n} = \frac{3.72}{25} = .1488$$

$$s^2 = \frac{\sum\limits_{i=1}^{n} x_i^2 - \frac{\left[\sum\limits_{i=1}^{n} x_i\right]^2}{n}}{n-1} = \frac{1.2088 - \frac{3.72^2}{25}}{25-1} = \frac{.655264}{24} = .0273$$

$$s = \sqrt{.0273} = .1652$$

b. According to Chebyshev's theorem, at least $(1 - 1/k^2) \times 100\%$ of the measurements will fall within $\bar{x} \pm ks$ when k is any number greater than 1.

For $\bar{x} \pm .75s$, $k = .75$. Since k is less than 1, it is possible that none of the measurements will fall within $\bar{x} \pm .75s$. Chebyshev's theorem doesn't apply in this case.

For $\bar{x} \pm 2.5s$, $k = 2.5$. We would expect at least $(1 - 1/2.5^2) \times 100 = 84\%$ of the measurements to fall within this interval.

For $\bar{x} \pm 4s$, $k = 4$. We would expect at least $(1 - 1/4^2) \times 100 = 93.75\%$ of the measurements to fall within this interval.

c. $\bar{x} \pm .75s \Rightarrow .1448 \pm .75(.1652) \Rightarrow .1488 \pm .1239 \Rightarrow (.0249, .2727)$

In this data set, $17/25 \times 100 = 68\%$ of the measurements fall within $\bar{x} \pm .75s$.

$\bar{x} \pm 2.5s = .1488 \pm 2.5(.1652) \Rightarrow .1488 \pm .413 \Rightarrow (-.2642, .5618)$

In this data set, $24/25 \times 100 = 96\%$ of the measurements fall within $\bar{x} \pm 2.5s$.

$\bar{x} \pm 4s \Rightarrow .1488 \pm 4(.1652) \Rightarrow .1488 \pm .6608 \Rightarrow (-.512, .8096)$

In this data set, all of the measurements (100%) fall within $\bar{x} \pm 4s$.

These results seem to be comparable with the results of part b.

d. The data represent the ratio of the target firm's sales to the bidder firm's sales. Since all the data values are less than one, 0% of the mergers involved a target firm with larger sales than the bidder firm.

2.55 Since we do not know if the distribution of the number of cases of sparkling water lost per week due to breakage last year is mound-shaped, we need to apply Chebyshev's theorem. We know $\mu = 30.4$ and $\sigma = 3.8$. Therefore,

$$\mu \pm 2\sigma \Rightarrow 30.4 \pm 2(3.8) \Rightarrow 30.4 \pm 7.6 \Rightarrow (22.8, 38.0)$$

. . . .

According to Chebyshev's theorem, for at least 3/4 or .75 of the weeks last year, the number of cases lost falls within the interval. Thus, for at most .25 of the weeks, the number of cases lost falls outside of $\mu \pm 2\sigma$. Therefore, for at most $.25 \times 52 = 13$ weeks last year, the number of cases lost due to breakage was more than 38 cases.

2.57   Since $\bar{x} = 385$ and $s = 15$,

$$\bar{x} \pm 3s \Rightarrow 385 \pm 3(15) \Rightarrow 385 \pm 45 \Rightarrow (340, 430)$$

According to Chebyshev's theorem, at least $1 - \dfrac{1}{3^2} = \dfrac{8}{9} =$ or .889 of the days the number of vehicles on the road falls within this interval.

2.59 $\bar{x} = 125$, $s = 15$

a. $\bar{x} \pm 3s \Rightarrow 125 \pm 3(15) \Rightarrow 125 \pm 45 \Rightarrow (80, 170)$

Since nothing is known about the distribution of utility bills, we need to apply Chebyshev's theorem. According to Chebyshev's theorem, the fraction of all three-bedroom homes with gas or electric energy that have bills within this interval is at least

$$1 - \frac{1}{3^2} = \frac{8}{9}$$

b.   If it is reasonable to assume the distribution of utility bills is mound-shaped, we can apply the Empirical Rule.

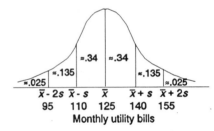

As illustrated in the figure at right, approximately $.135 + .025 = .16$ of three-bedroom homes would have monthly bills less than $110.

c.   Yes, these three values do suggest that solar energy units might result in lower utility bills. This is evident since if the solar energy units do not decrease the utility bills, only approximately .025 of the utility bills would be under $95. But all three of the sampled bills from houses with solar energy units were under $95.

2.61   Since we do not know if the distribution of the heights of the trees is mound-shaped, we need to apply Chebyshev's theorem. We know $\mu = 30$ and $\sigma = 3$. Therefore,

$$\mu \pm 3\sigma \Rightarrow 30 \pm 3(3) \Rightarrow 30 \pm 9 \Rightarrow (21, 39)$$

According to Chebyshev's theorem, at least $1 - \dfrac{1}{3^2} = \dfrac{8}{9} =$ or .89 of the tree heights on this piece of land fall within this interval. However, the buyer will only purchase the land if at least $\dfrac{1000}{5000}$ or .20 of the tree heights are at least 40 feet tall. Therefore, the buyer should not buy the piece of land.

2.63 a. Since we know $\bar{x}$ and s, the sample z-score is

$$z = \frac{x - \bar{x}}{s} = \frac{31 - 24}{7} = \frac{7}{7} = 1$$

b. Since we know $\bar{x}$ and s, the sample z-score is

$$z = \frac{x - \bar{x}}{s} = \frac{95 - 101}{4} = \frac{-6}{4} = -1.5$$

c. Since we know μ and σ, the population z-score is

$$z = \frac{x - \mu}{\sigma} = \frac{5 - 2}{1.7} = \frac{3}{1.7} = 1.765$$

d. Since we know μ and σ, the population z-score is

$$z = \frac{x - \mu}{\sigma} = \frac{14 - 7}{5} = \frac{-3}{5} = -.6$$

2.65 a. From the problem, $\mu = 2.7$ and $\sigma = .5$

$$z = \frac{x - \mu}{\sigma} \Rightarrow z\sigma = x - \mu \Rightarrow x = \mu + z\sigma$$

For $z = 2.0$, $x = 2.7 + 2.0(.5) = 3.7$

For $z = -1.0$, $x = 2.7 - 1.0(.5) = 2.2$

For $z = .5$, $x = 2.7 + .5(.5) = 2.95$

For $z = -2.5$, $x = 2.7 - 2.5(.5) = 1.45$

b. For $z = -1.6$, $x = 2.7 - 1.6(.5) = 1.9$

c. If we assume the distribution of GPAs is approximately mound-shaped, we can use the Empirical Rule.

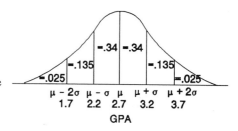

From the Empirical Rule, we know that $\approx .025$ or $\approx 2.5\%$ of the students will have GPAs above 3.7 (with $z = 2$). Thus, the GPA corresponding to summa cum laude (top 2.5%) will be greater than 3.7 ($z > 2$).

We know that $\approx .16$ or 16% of the students will have GPAs above 3.2 ($z = 1$). Thus, the limit on GPAs for cum laude (top 16%) will be greater than 3.2 ($z > 1$).

We must assume the distribution is mound shaped.

2.67   a.   To calculate the U.S. merchandise trade balance for each of the ten countries, take the exports minus imports.

U.S. Merchandise Trade Balance

Country	(in billions)
Brazil	-3.825
Egypt	1.745
France	-2.787
Italy	-5.510
Japan	-56.326
Mexico	-5.689
Panama	0.387
Soviet Union	1.055
Sweden	-2.864
Turkey	0.622

b.   To find the z-scores, we must first calculate the sample mean and standard deviation.

$$\bar{x} = \frac{\sum\limits_{i=1}^{n} x_i}{n} = \frac{-73.152}{10} = -7.3142$$

$$s^2 = \frac{\sum\limits_{i=1}^{n} x_i^2 - \frac{\left[\sum\limits_{i=1}^{n} x_i\right]^2}{n}}{n-1} = \frac{3270.68965 - \frac{(-73.152)^2}{10}}{10 - 1} = \frac{2735.56814}{9}$$

$$= 303.952$$

$$s = \sqrt{303.952} = 17.4342$$

Japan: $z = \dfrac{x - \bar{x}}{s} = \dfrac{-56.326 - (-7.3152)}{17.4342} = -2.81$

The relative position of the U.S. trade balance with Japan is 2.81 standard deviations below the mean. This indicates that this measurement is small compared to the other U.S. trade balances.

Soviet Union: $z = \dfrac{x - \bar{x}}{s} = \dfrac{1.055 - (-7.3152)}{17.4342} = .48$

The relative position of the U.S. trade balance with the Soviet Union is .48 standard deviation above the mean. This indicates that this measurement is larger than the average of the U.S. trade balances.

2.69 In 1989, Control Data Corporation ranked 153 out of 500. Therefore, $(500 - 153)/500 \times 100\% = 69.4\%$ of the corporations were ranked below Control Data Corporation. Control Data Corporation was at the 69.4th percentile. In 1984, Control Data Corporation ranked 71 out of 500. Thus, $(500 - 71)/500 \times 100\% = 85.8\%$ of the corporations were ranked below Control Data Corporation. Control Data Corporation was at the 85.8th percentile.

2.71 a. From the Empirical Rule, we know that approximately 68% of the measurements will fall within $\bar{x} \pm s$, 95% within $\bar{x} \pm 2s$, and almost all within $\bar{x} \pm 3s$.

Since $\bar{x} = 35$ and $s = \sqrt{9} = 3$,

$$\bar{x} \pm s \Rightarrow 35 \pm 3 \Rightarrow (32, 38)$$

Therefore, approximately 68% of the measurements should fall between 32 and 38.

$$\bar{x} \pm 3s \Rightarrow 35 \pm 3(3) \Rightarrow 35 \pm 9 \Rightarrow (26, 34)$$

Therefore, almost all (100%) of the measurements should fall between 26 and 44.

b. $z = \dfrac{x - \bar{x}}{s}$ $-1.33 = \dfrac{x - 35}{3}$
$$-3.99 = x - 35$$
$$31.01 = x \text{ due to rounding}$$

The store sold 31 VCRs.

c. We will calculate the z-score for $x = 41$ for both stores.

Large department store:
$$z = \frac{x - \bar{x}}{s} - \frac{41 - 35}{3} = \frac{6}{3} = 2$$

Rival department store:
$$z = \frac{x - \bar{x}}{s} = \frac{41 - 35}{2} = \frac{6}{2} = 3$$

The store with the lowest z-score is more likely to sell more than 41 VCRs. Thus, the first store is more likely to sell 41 or more VCRs.

2.73 a. When the income x is \$190,
$$z = \frac{x - \mu}{\sigma} = \frac{190 - 170}{10} = 2$$

Using Chebyshev's theorem, at least 3/4 of the incomes fall within 2 standard deviations of the mean. Therefore, at most $1/4 \times 100\% = 25\%$ of the employees should expect an income over \$190 per week.

When $x = 160$,
$$z = \frac{x - \mu}{\sigma} = \frac{160 - 170}{10} = -1$$

Using Chebyshev's theorem, it is possible that none of the incomes will fall within 1 standard deviation of the mean. Therefore, we cannot determine the percentage of employees that should expect an income under $160 per week.

When $x = 200$,
$$z = \frac{x - \mu}{\sigma} = \frac{200 - 170}{10} = 3$$

At least 8/9 of the incomes fall within 3 standard deviations of the mean. Therefore, at most, $1/9 \times 100\% = 11.1\%$ of the employees should expect an income over $200 per week.

b. When your income x is $185,
$$z = \frac{x - \mu}{\sigma} = \frac{185 - 170}{10} = 1.5$$

My salary would be 1.5 standard deviations above the mean salary.

2.75 The 25th percentile, or lower quartile, is the measurement that has 25% of the measurements below it and 75% of the measurements above it. The 50th percentile, or median, is the measurement that has 50% of the measurements below it and 50% of the measurements above it. The 75th percentile, or upper quartile, is the measurement that has 75% of the measurements below it and 25% of the measurements above it.

2.77 a. Median is approximately 39.

b. Q_L is approximately 31.5 (Lower Quartile)
Q_U is approximately 45 (Upper Quartile)

c. IQR $= Q_U - Q_L \approx 45 - 31.5 \approx 13.5$

d. The data set is skewed to the left since the left whisker is longer.

e. 50% of the measurements are to the right of the median and 75% are to the left of the upper quartile.

2.79 Using Minitab, the box plot for the sample is given below.

```
                -----------------
       --------I    +          I----------------      *    *
                -----------------
-------+---------+---------+---------+---------+--------C1
   1.20      1.40      1.60      1.80      2.00
```

2.81 a. From the problem, $\bar{x} = 26.2$ and $s = 4.56$.

The highest salary is 40 (thousand).

The z-score is $z = \dfrac{x - \bar{x}}{s} = \dfrac{40 - 26.2}{4.56} = 3.03$

Therefore, the highest salary is 3.03 standard deviations above the mean.

The lowest salary is 14.0 (thousand).

The z-score is $z = \dfrac{x - \bar{x}}{s} = \dfrac{14.0 - 26.2}{4.56} = -2.68$

Therefore, the lowest salary is 2.68 standard deviations below the mean.

The mean salary offer is 26.2 (thousand).

The z-score is $z = \dfrac{x - \bar{x}}{s} = \dfrac{26.2 - 26.2}{4.56} = 0$

The z-score for the mean salary offer is 0 standard deviations from the mean.

Yes, the highest salary offer is unusually high. If this is a mound-shaped distribution, almost all the salaries should have z-scores between -3 and 3. A z-score of 3.03 would be very unusual.

b. Using Minitab, the box plot is:

```
                    -------
      0       * * *--------I   + I---     *   * * 0    0
                    -------
      ----+---------+---------+---------+---------+---------+--
      15.0      20.0      25.0      30.0      35.0      40.0
```

The salary offers 14,000, 37,200, and 40,000 are potentially faulty observations. These values are outliers since they are all outside the outer fences.

2.83 a.
```
             --------------
      ---------I    +    I------------     * **          *
             --------------
      +---------+---------+---------+---------+---------+------
      0       12       24       36       48       60
```

The median of the data set is approximately 18; the IQR is approximately $25.2 - 10.8 = 14.4$; and the range is $63.6 - 0 = 3.6$. The whiskers are about the same length, but there are 4 observations in the upper part of the box plot beyond the inner fence. This would cause the frequency distribution of the data set to be skewed to the right.

b. The 4 customers whose down times are indicated with an asterisk have unusually lengthy down times. These are customers 238, 268, 269, and 264.

c. To find the z-scores for these customers, we must first compute $\bar{x}$ and s.

$$\bar{x} = \frac{\sum\limits_{i=1}^{n} x_i}{n} = \frac{815}{40} = 20.375$$

$$s^2 = \frac{\sum\limits_{i=1}^{n} x_i^2 - \frac{\left[\sum\limits_{i=1}^{n} x_i\right]^2}{n}}{n-1} = \frac{24129 - \frac{(815)^2}{40}}{39} = \frac{7523.375}{39} = 192.9071$$

$$s = \sqrt{192.9071} = 13.8891$$

Customer 238: $z = \dfrac{x - \bar{x}}{s} = \dfrac{47 - 20.375}{13.8891} = 1.92$

The down time for customer 238 is 1.92 standard deviations above the mean.

Customer 268: $z = \dfrac{x - \bar{x}}{s} = \dfrac{49 - 20.375}{13.8891} = 2.06$

The down time for customer 268 is 2.06 standard deviations above the mean.

Customer 269: $z = \dfrac{x - \bar{x}}{s} = \dfrac{50 - 20.375}{13.8891} = 2.13$

The down time for customer 269 is 2.13 standard deviations above the mean.

Customer 264: $z = \dfrac{x - \bar{x}}{s} = \dfrac{64 - 20.375}{13.8891} = 3.14$

The down time for customer 264 is 3.14 standard deviations above the mean.

2.85   a.   Ratio; time is measured on a numerical scale where the zero point has meaning. A time of 20 minutes is four times as long as a time of 5 minutes.

b.   Nominal; style of music is measured with nonranked categories such as rock, classical, big band, etc.

c.   Interval; arrival time is measured on a numerical scale where the zero point has no meaning. An arrival time of 5:00 P.M. is not twice as long as an arrival time of 2:30 P.M.

d.   Ordinal; a rating is measured on a nonnumerical scale that can be ranked. Good is better than fair, and fair is better than poor.

e.   (a)   Quantitative; time is measured on a numerical scale.
     (b)   Qualitative; style is measured on a nonnumerical scale.
     (c)   Quantitative; arrival time is measured on a numerical scale.
     (d)   Qualitative; rating is measured on a nonnumerical scale.

2.87   a. $\displaystyle\sum_{i=1}^{n} x_i^2 = 11^2 + 1^2 + 2^2 + 8^2 + 7^2 = 239$

$\displaystyle\sum_{i=1}^{n} x_i = 11 + 1 + 2 + 8 + 7 = 29$

$\displaystyle\left[\sum_{i=1}^{n} x_i\right]^2 = 29^2 = 841$

b. $\displaystyle\sum_{i=1}^{n} x_i^2 = 15^2 + 15^2 + 2^2 + 6^2 + 12^2 = 634$

$\displaystyle\sum_{i=1}^{n} x_i = 15 + 15 + 2 + 6 + 12 = 50$

$\displaystyle\left[\sum_{i=1}^{n} x_i\right]^2 = 50^2 = 2500$

c. $\displaystyle\sum_{i=1}^{n} x_i^2 = (-1)^2 + 2^2 + 0^2 + (-4)^2 + (-8)^2 + 13^2 = 254$

$\displaystyle\sum_{i=1}^{n} x_i = -1 + 2 + 0 + (-4) + (-8) + 13 = 2$

$\displaystyle\left[\sum_{i=1}^{n} x_i\right]^2 = 2^2 = 4$

d. $\displaystyle\sum_{i=1}^{n} x_i^2 = 100^2 + 0^2 + 0^2 + 2^2 = 10{,}004$

$\displaystyle\sum_{i=1}^{n} x_i = 100 + 0 + 0 + 2 = 102$

$\displaystyle\left[\sum_{i=1}^{n} x_i\right]^2 = 102^2 = 10{,}404$

2.89   a.   One reason the plot may be interpreted differently is that no scale is given on the vertical axis. Also, since the plot almost reaches the horizontal axis at 3 years, it is obvious that the bottom of the plot has been cut off. Another important factor omitted is who responded to the survey.

b.   A scale should be added to the vertical axis. Also, that scale should start at 0.

2.91

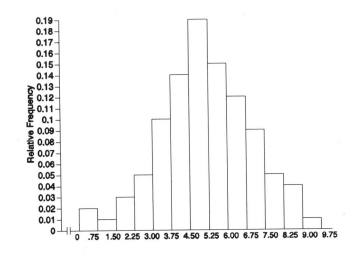

2.93 a. The "normal" Minnesota temperature is the mean temperature from 1961 through 1990. It is:

$$\bar{x} = \frac{\sum x}{n} = \frac{349.5}{30} = 11.65$$

b. $$s = \sqrt{\frac{\sum x^2 - \frac{(\sum x)^2}{n}}{n - 1}} = \sqrt{\frac{5351.87 - \frac{349.5^2}{30}}{30 - 1}} = \sqrt{\frac{1280.195}{29}} = \sqrt{44.144655}$$

$$= 6.6441$$

The standard deviation gives an indication of how the data are dispersed around the mean. For most distributions, most of the observations are within 3 standard deviations of the mean.

c. According to Chebyshev's theorem, at least $\left[1 - \frac{1}{k^2}\right] 100\% = \left[1 - \frac{1}{2^2}\right] 100\% =$

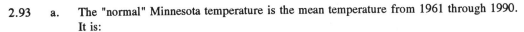

$\left[\frac{3}{4}\right] 100\% = 75\%$ of the data are within 2 standard deviations of the mean. The interval from 2 standard deviations below the mean to 2 standard deviations above the mean is:

$$\bar{x} \pm 2s \Rightarrow 11.65 \pm 2(6.6441) \Rightarrow 11.65 \pm 13.2882 \Rightarrow (-1.6382, 24.9382).$$

All but one of the data points are in the above interval. Thus, $29/30 = (.967)100\% = 96.7\%$ of the data points are within 2 standard deviations of the mean.

2.95 a. Because the median (8 ppb) is less than the mean (10 ppb), the distribution is skewed to the right.

b. Using the Empirical Rule:

Approximately 68% of the observations will fall in the interval
$$\bar{x} \pm s \Rightarrow 10 \pm 3 \Rightarrow (7, 13)$$

Approximately 95% of the observations will fall in the interval
$$\bar{x} \pm 2s \Rightarrow 10 \pm 2(3) \Rightarrow 10 \pm 6 \Rightarrow (4, 16)$$

Approximately all of the observations will fall in the interval
$$\bar{x} \pm 3s \Rightarrow 10 \pm 3(3) \Rightarrow 10 \pm 9 \Rightarrow (1, 19)$$

Using Chebyshev's theorem:

At least $1 - \dfrac{1}{2^2} = \dfrac{3}{4}$ of the observations will fall in the interval
$$\bar{x} \pm 2s \Rightarrow 10 \pm 2(3) \Rightarrow 10 \pm 6 \Rightarrow (4, 16)$$

At least $1 - \dfrac{1}{3^2} = \dfrac{8}{9}$ of the observations will fall in the interval
$$\bar{x} \pm 3s \Rightarrow 10 \pm 3(3) \Rightarrow 10 \pm 9 \Rightarrow (1, 19)$$

c. Using the Empirical Rule, only approximately 2.5% of the homes will have levels below the background level of 4. Therefore, approximately .975(50) = 48.75 or 49 homes will have levels above the background level of 4 ppb.

Using Chebyshev's theorem, at least 3/4 or 3/4 of 50 = 37.5 or 38 homes will have levels above the background level of 4 ppb.

d. The z-score is $z = \dfrac{x - \bar{x}}{s} = \dfrac{20 - 10}{3} = 3.33$. It is not likely that this observation came from the same distribution as the other 50. We know that almost all of the observations are within 3 standard deviations of the mean and this observation is 3.33 standard deviations above the mean.

2.97 a. The population of interest is the set of all gasoline stations in the United States.

b. The sample of interest is the set of 200 stations selected.

c. The variable of interest is the price of regular unleaded gasoline at each station.

d. μ = the mean price of regular unleaded gasoline at all stations in the United States.

σ = the standard deviation of the price of regular unleaded gasoline at all stations in the United States.

$\bar{x}$ = the mean price of regular unleaded gasoline at the 200 selected stations.

s = the standard deviation of the price of regular unleaded gasoline at the 200 selected stations.

e. The mean price of a gallon of regular unleaded gasoline at the 200 selected stations is $1.39, and the standard deviation is $0.12. The distribution of the prices is probably approximately mound-shaped. About half of the stations probably have prices above $1.39 and about half have prices below $1.39.

f. For $x = \$1.09$, $z = \dfrac{x - \bar{x}}{s} = \dfrac{1.09 - 1.39}{.12} = \dfrac{-.30}{.12} = -2.5$

A price of $1.09 per gallon is 2.5 standard deviations below the mean. Not many stations will have prices below this value.

2.99 a. Frequency bar chart.

b. It presents the number of napkins (out of 1000) that fall into each of 4 categories.

c. Of the 1000 napkins printed, 700 were successful. Another way of saying this is $700/1000 \times 100\% = 70\%$ of the imprints were successful.

2.101 The standard deviation is expressed in the units of the problem while the variance is in square units. Therefore, it is easier to interpret the standard deviation than the variance.

2.103 A z-score locates a measurement in a data set by expressing the measurement in terms of the number of standard deviations it is above or below the mean.

2.105 The mode is often not an acceptable measure of central tendency. If the data set has few observations, no value may occur more than one time. If the data set is of moderate size, several different values may occur the same number of times, giving little information about the center of the distribution.

2.107 a. First, we must compute the Total processing times by adding the processing times of the three departments. The total processing times are as follows:

Request	Total Processing Time	Request	Total Processing Time	Request	Total Processing Time
1	13.3	17	19.4*	33	23.4*
2	5.7	18	4.7	34	14.2
3	7.6	19	9.4	35	14.3
4	20.0*	20	30.2	36	24.0*
5	6.1	21	14.9	37	6.1
6	1.8	22	10.7	38	7.4
7	13.5	23	36.2*	39	17.7*
8	13.0	24	6.5	40	15.4
9	15.6	25	10.4	41	16.4
10	10.9	26	3.3	42	9.5
11	8.7	27	8.0	43	8.1
12	14.9	28	6.9	44	18.2*
13	3.4	29	17.2*	45	15.3
14	13.6	30	10.2	46	13.9
15	14.6	31	16.0	47	19.9*
16	14.4	32	11.5	48	15.4
				49	14.3*
				50	19.0

The stem-and-leaf displays with the appropriate leaves circled are as follows:

Stem-and-leaf of Mkt
Leaf Unit = 0.10

```
    6     0   112446
    7     1   3
   14     2   ⓪024699
   16     3   2⑤
   22     4   ⓪⓪①577
  (10)    5   0344556889
   18     6   000②2247⑨⑨
    8     7   003⑧
    4     8   ⓪7
    2     9
    2    10   0
    1    11   0
```

Stem-and-leaf of Engr
Leaf Unit = 0.10

```
    7     0   4466699
   14     1   333378⑧
   19     2   ①22④6
   23     3   1568
   (5)    4   24688
   22     5   233
   19     6   ⓪12③9
   14     7   ②②379
    9     8
    9     9   66
    7    10   0
    6    11   3
    5    12   02③
    2    13   ⓪
    1    14   ④
```

Stem-and-leaf of Accnt
Leaf Unit = 0.10

```
   19     0   11111111111②2333444
   (8)    0   55556⑧88
   23     1   00
   21     1   7⑨
   19     2   00②3
   15     2
   15     3   23
   13     3   78
   11     4
   11     4
   11     5
   11     5   8
   10     6   2
    9     6
    9     7   ⓪
    8     7
    8     8   4
   HI   ⑨⑨, ⑩⑤, ⑬⑤, 144,
        ⑱②, 220, ⑨⑩⑩
```

Stem-and-leaf of Total
Leaf Unit = 1.00

```
    1     0   1
    3     0   33
    5     0   45
   11     0   666677
   17     0   888999
   21     1   0000
   (5)    1   33333
   24     1   4④44445555
   14     1   66⑦⑦
   10     1   ⑧9⑨⑨
    6     2   ⓪
    5     2   ③
    4     2   ④4
   HI   30, ㉚
```

Of the 50 requests, 10 were lost. For each of the three departments, the processing times for the lost requests are scattered throughout the distributions. The processing times for the departments do not appear to be related to whether the request was lost or not. However, the total processing times for the lost requests appear to be clustered towards the high side of the distribution. It appears that if the total processing time could be kept under 17 days, 76% of the data could be maintained, while reducing the number of lost requests to 1.

b.   For the Marketing department, if the maximum processing time was set at 6.5 days, 78% of the requests would be processed, while reducing the number of lost requests by 4.
For the Engineering department, if the maximum processing time was set at 7.0 days, 72% of the requests would be processed, while reducing the number of lost requests by 5. For the Accounting department, if the maximum processing time was set at 8.5 days, 86% of the requests would be processed, while reducing the number of lost requests by 5.

2.109   Pareto Analysis involves the categorization of items and the determination of which categories contain the most observations. These are the "vital few" categories. Pareto analysis is used in industry today as a problem identification tool. Managers and workers use it to identify the most important problems or causes of problems that plague them. Knowledge of the "vital few" problems permits management to prioritize and focus their problem-solving efforts.

2.111 $\bar{x} = 75, s^2 = 36, s = 6$

a. $\bar{x} \pm s \Rightarrow 75 \pm 6 \Rightarrow (69, 81)$

Since we do not know the shape of the distribution, using Chebyshev's theorem we can say that at least $1 - \dfrac{1}{1^2} = 0$ will fall within this interval.

b. $\bar{x} \pm 2s \Rightarrow 75 \pm 2(6) \Rightarrow 75 \pm 12 \Rightarrow (63, 87)$

By Chebyshev's theorem, at least $1 - \dfrac{1}{2^2} = .75$ of the measurements should fall within this interval. So at most, $.25 \times 100\% = 25\%$ of the trainees would be expected to fail.

2.113 The time series plot is:

From the plot, it is evident that the claim that "your hood will be open for less than 12 minutes when we serve your car" is not justified. Of the 25 service times recorded, only 7 were less than 12 minutes.

2.115 a. $\quad s \approx \dfrac{\text{Range}}{4} \approx \dfrac{30000}{4} \approx \$7500$

b. It is more likely that division B's sales next month will be over \$120,000 since it is fewer standard deviations from the mean of \$110,000.

$$(z = \frac{120{,}000 - 110{,}000}{7500} = 1.33 \text{ is smaller in magnitude than}$$

$$z = \frac{90{,}000 - 220{,}000}{7500} = -2.67)$$

c. Yes, it is possible for division B's sales next month to be over \$160,000. But it very unlikely since

$$z = \frac{x - \bar{x}}{s} = \frac{160{,}000 - 110{,}000}{7500} = 6.67$$

This is a very large z-score.

2.117 No information about the variability of a data set can be obtained from this formula, because $\sum (x_i - \bar{x}) = 0$ for all data sets. The data set from Exercise 2.87a is 11, 1, 2, 8, 7.

$$\bar{x} = \frac{\sum\limits_{i=1}^{n} x_i}{n} = \frac{29}{5} = 5.8$$

$$\sum_{i=1}^{n} (x_i - \bar{x}) = (11 - 5.8) + (1 + 5.8) + (2 - 5.8) + (8 - 5.8) + (7 - 5.8)$$

$$= 5.2 + (-4.8) + (-3.8) + 2.2 + 1.2$$

$$= 0$$

Methods for Describing Sets of Data

So, $V = \dfrac{\sum\limits_{i=1}^{n}\left(x_i - \bar{x}\right)}{n} = \dfrac{0}{5} = 0$

V will always be 0 regardless of the data set.

2.119   a.

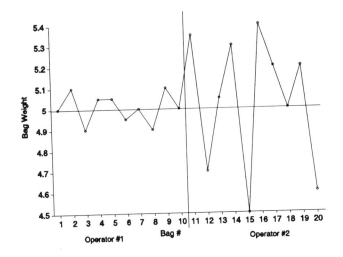

b.   From the plot, it is evident that there is more variability in the fill weights of operator #2 than in those of operator #1. The fill weights of operator #2 vary from 4.50 to 5.40 while the fill weights of operator #1 vary from 4.90 to 5.10. However, each operator had only 3 fill weights less than 5 pounds.

c.   From the plot, 30% or 6 of the 20 bags were underfilled, while 50% or 10 of the 20 bags were overfilled. Of the bags underfilled, only 3 were underfilled by more than .1 pound. All three of these bags were filled by operator #2. Thus, this could account for the customer complaints.

d.   Again, it appears that operator #1 is much more consistent filling the bags than is operator #2. Even though operator #2 tends to underfill bags by more than operator #1, operator #2 also tends to overfill the bags by more than operator #1.

e.   It might be better to randomly select bags from each operator. It might be that the operators need to get "warmed up." That is, the variability of the weights of the bags might be greater at the beginning of a shift than in the middle or end of a shift. The bags should be randomly selected from the entire shift.

2.121   a.   Using Chebyshev's theorem with $k = 2$, we would expect at least $1 - \dfrac{1}{2^2} = .75$ of the

measurements to lie within 2 standard deviations of the mean.

Chapter 2

b. Using Chebyshev's theorem with $k = 3$, we would expect at least $1 - \dfrac{1}{3^2} = .89$ of the measurements to lie within 3 standard deviations of the mean.

c. When $k = 1.5$, we would expect at least $1 - \dfrac{1}{1.5^2} = .56$ of the measurements to lie within 1.5 standard deviations of the mean.

d. When $k = 2.75$, we would expect at least $1 - \dfrac{1}{2.75^2} = .87$ of the measurements to fall within 2.75 standard deviations of the mean. Therefore, at most $1 - .87 = .13$ of the measurements would lie more than 2.75 standard deviations from the mean.

2.123 a. We will characterize the magnitude using the sample mean and the variability using the sample variance of the 12 market concentration ratios for each of the 6 countries.

United States:

$$\bar{x} = \frac{\sum\limits_{i=1}^{n} x_i}{n} = \frac{493}{12} = 41.08\%$$

$$s^2 = \frac{\sum\limits_{i=1}^{n} x_i^2 - \dfrac{\left(\sum\limits_{i=1}^{n} x_i\right)^2}{n}}{n-1} = \frac{23885 - \dfrac{493^2}{12}}{12 - 1} = 330.08$$

Canada:

$$\bar{x} = \frac{\sum\limits_{i=1}^{n} x_i}{n} = \frac{850}{12} = 70.83\%$$

$$s^2 = \frac{\sum\limits_{i=1}^{n} x_i^2 - \dfrac{\left(\sum\limits_{i=1}^{n} x_i\right)^2}{n}}{n-1} = \frac{66030 - \dfrac{850^2}{12}}{12 - 1} = 529.24$$

United Kingdom:

$$\bar{x} = \frac{\sum\limits_{i=1}^{n} x_i}{n} = \frac{725}{12} = 60.42\%$$

$$s^2 = \frac{\sum\limits_{i=1}^{n} x_i^2 - \dfrac{\left(\sum\limits_{i=1}^{n} x_i\right)^2}{n}}{n-1} = \frac{50779 - \dfrac{725^2}{12}}{12 - 1} = 634.27$$

Sweden:

$$\overline{x} = \frac{\sum\limits_{i=1}^{n} x_i}{n} = \frac{1001}{12} = 83.42\%$$

$$s^2 = \frac{\sum\limits_{i=1}^{n} x_i^2 - \dfrac{\left[\sum\limits_{i=1}^{n} x_i\right]^2}{n}}{n-1} = \frac{89123 - \dfrac{1001^2}{12}}{12-1} = 511.17$$

France:

$$\overline{x} = \frac{\sum\limits_{i=1}^{n} x_i}{n} = \frac{796}{12} = 66.33\%$$

$$s^2 = \frac{\sum\limits_{i=1}^{n} x_i^2 - \dfrac{\left[\sum\limits_{i=1}^{n} x_i\right]^2}{n}}{n-1} = \frac{64372 - \dfrac{796^2}{12}}{12-1} = 1051.88$$

West Germany:

$$\overline{x} = \frac{\sum\limits_{i=1}^{n} x_i}{n} = \frac{673}{12} = 56.08\%$$

$$s^2 = \frac{\sum\limits_{i=1}^{n} x_i^2 - \dfrac{\left[\sum\limits_{i=1}^{n} x_i\right]^2}{n}}{n-1} = \frac{47723 - \dfrac{673^2}{12}}{12-1} = 907.17$$

b. We will characterize the magnitude using the sample mean and the variability using the sample variance of the 6 market concentration ratios for each of the 12 industries.

Brewing:

$$\overline{x} = \frac{\sum\limits_{i=1}^{n} x_i}{n} = \frac{325}{6} = 54.17$$

$$s^2 = \frac{\sum\limits_{i=1}^{n} x_i^2 - \dfrac{\left[\sum\limits_{i=1}^{n} x_i\right]^2}{n}}{n-1} = \frac{20809 - \dfrac{325^2}{6}}{6-1} = 640.97$$

Cigarettes:

$$\bar{x} = \frac{\sum_{i=1}^{n} x_i}{n} = \frac{546}{6} = 91.00$$

$$s^2 = \frac{\sum_{i=1}^{n} x_i^2 - \frac{\left[\sum_{i=1}^{n} x_i\right]^2}{n}}{n-1} = \frac{50396 - \frac{546^2}{6}}{6-1} = 142.00$$

Fabric weaving:

$$\bar{x} = \frac{\sum_{i=1}^{n} x_i}{n} = \frac{214}{6} = 35.67$$

$$s^2 = \frac{\sum_{i=1}^{n} x_i^2 - \frac{\left[\sum_{i=1}^{n} x_i\right]^2}{n}}{n-1} = \frac{9458 - \frac{214^2}{6}}{6-1} = 365.07$$

Paints:

$$\bar{x} = \frac{\sum_{i=1}^{n} x_i}{n} = \frac{244}{6} = 40.67$$

$$s^2 = \frac{\sum_{i=1}^{n} x_i^2 - \frac{\left[\sum_{i=1}^{n} x_i\right]^2}{n}}{n-1} = \frac{13560 - \frac{244^2}{6}}{6-1} = 727.47$$

Petroleum refining:

$$\bar{x} = \frac{\sum_{i=1}^{n} x_i}{n} = \frac{375}{6} = 62.5$$

$$s^2 = \frac{\sum_{i=1}^{n} x_i^2 - \frac{\left[\sum_{i=1}^{n} x_i\right]^2}{n}}{n-1} = \frac{26771 - \frac{375^2}{6}}{6-1} = 666.70$$

Shoes (except rubber):

$$\bar{x} = \frac{\sum\limits_{i=1}^{n} x_i}{n} = \frac{122}{6} = 20.33$$

$$s^2 = \frac{\sum\limits_{i=1}^{n} x_i^2 - \frac{\left[\sum\limits_{i=1}^{n} x_i\right]^2}{n}}{n-1} = \frac{2840 - \frac{122^2}{6}}{6-1} = 71.87$$

Glass Bottles:

$$\bar{x} = \frac{\sum\limits_{i=1}^{n} x_i}{n} = \frac{515}{6} = 85.83$$

$$s^2 = \frac{\sum\limits_{i=1}^{n} x_i^2 - \frac{\left[\sum\limits_{i=1}^{n} x_i\right]^2}{n}}{n-1} = \frac{45259 - \frac{515^2}{6}}{6-1} = 210.97$$

Cement:

$$\bar{x} = \frac{\sum\limits_{i=1}^{n} x_i}{n} = \frac{406}{6} = 67.67$$

$$s^2 = \frac{\sum\limits_{i=1}^{n} x_i^2 - \frac{\left[\sum\limits_{i=1}^{n} x_i\right]^2}{n}}{n-1} = \frac{31498 - \frac{406^2}{6}}{6-1} = 805.07$$

Ordinary steel:

$$\bar{x} = \frac{\sum\limits_{i=1}^{n} x_i}{n} = \frac{364}{6} = 60.67$$

$$s^2 = \frac{\sum\limits_{i=1}^{n} x_i^2 - \frac{\left[\sum\limits_{i=1}^{n} x_i\right]^2}{n}}{n-1} = \frac{23846 - \frac{364^2}{6}}{6-1} = 352.67$$

Antifriction bearings:

$$\bar{x} = \frac{\sum\limits_{i=1}^{n} x_i}{n} = \frac{484}{6} = 80.67$$

$$s^2 = \frac{\sum\limits_{i=1}^{n} x_i^2 - \frac{\left[\sum\limits_{i=1}^{n} x_i\right]^2}{n}}{n-1} = \frac{40994 - \frac{484^2}{6}}{6-1} = 390.27$$

Refrigerators:

$$\bar{x} = \frac{\sum\limits_{i=1}^{n} x_i}{n} = \frac{465}{6} = 77.50$$

$$s^2 = \frac{\sum\limits_{i=1}^{n} x_i^2 - \frac{\left[\sum\limits_{i=1}^{n} x_i\right]^2}{n}}{n-1} = \frac{37051 - \frac{465^2}{6}}{6-1} = 202.70$$

Storage batteries:

$$\bar{x} = \frac{\sum\limits_{i=1}^{n} x_i}{n} = \frac{478}{6} = 79.67$$

$$s^2 = \frac{\sum\limits_{i=1}^{n} x_i^2 - \frac{\left[\sum\limits_{i=1}^{n} x_i\right]^2}{n}}{n-1} = \frac{39430 - \frac{478^2}{6}}{6-1} = 269.87$$

c.   On the average, the most competition within its industries is in the United States since the average market concentration ratio is smallest in the U.S.

On the average, the least competition within its industries is in Sweden since the average market concentration ratio is largest in Sweden.

d.   The three most competitive industries are shoes, fabric weaving, and paints since they have the smallest averages; and the three least competitive industries are cigarettes, glass bottles, and antifriction bearings since they have the largest average market concentration ratios.

2.125 a. Using Minitab, the stem-and-leaf display is:

```
Stem-and-leaf of C1
Leaf Unit = 0.10     N = 46
   4    0   34④④
 (25)   0   5⑤⑤⑤⑤⑤⑤⑤56666⑥⑥6⑦⑦⑦7⑦8888⑧9
  16    1   000011222③④34
   4    1   7⑦
   2    2
   2    2
   2    3
   2    3   9
   1    4
   1    4   7
```

b. The leaves that represent those brands that carry the American Dental Association seal are circled above.

c. It appears that the cost of the brands approved by the ADA tend to have the lower costs. Thirteen of the twenty brands approved by the ADA, or $(13/20) \times 100\% = 65\%$ are less than the median cost.

CHAPTER THREE

. . . .
Probability

3.1   A simple event is an outcome of an experiment that cannot be decomposed into a simpler outcome. An event is a collection of one or more simple events.

3.3   The sample space for the experiment is

$$S = \{E_1, E_2, E_3, E_4, E_5\}$$

Therefore,

$$P(S) = P(E_1) + P(E_2) + P(E_3) + P(E_4) + P(E_5) = 1$$

a. $P(E_3) = 1 - [P(E_1) + P(E_2) + P(E_4) + P(E_5)]$
         $= 1 - [.1 + .2 + .3 + .1]$
         $= 1 - .7$
         $= .3$

b. $P(E_1) + P(E_2) + P(E_3) + P(E_4) + P(E_5) = 1$
                $P(E_3) + .1 + P(E_3) + .2 + .2 = 1$
                            $2P(E_3) + .5 = 1$
                                $2P(E_3) = .5$
                                    $P(E_3) = .25$

c. $P(E_1) + P(E_2) + P(E_3) + P(E_4) + P(E_5) = 1$
                    $.1 + .1 + P(E_3) + .1 + .1 = 1$
                            $P(E_3) + .4 = 1$
                                $P(E_3) = .6$

3.5   a.   The simple events are:

E_1 = loan is under \$2,000
E_2 = loan is \$2,000–\$4,999
E_3 = loan is \$5,000–\$7,999
E_4 = loan is \$8,000 or over

b. $P(E_4) = \dfrac{92}{500} = .184$

c. $P(\text{loan is less than \$5,000}) = P(E_1) + P(E_2) = \dfrac{35}{500} + \dfrac{73}{500} = \dfrac{108}{500} = .216$

3.7   There are $\begin{bmatrix} 6 \\ 3 \end{bmatrix} = \dfrac{6!}{3!3!} = \dfrac{6 \cdot 5 \cdot 4 \cdot 3 \cdot 2 \cdot 1}{3 \cdot 2 \cdot 1 \cdot 3 \cdot 2 \cdot 1} = 20$ possible ways to select 3 cars

from 6. Only one of these combinations includes all three lemons, so the probability that dealer A receives all three lemons is 1/20.

3.11   a.   The simple events are the types of credit cards and are: Visa and Mastercard, Diners Club, Carte Blanche, and Choice.

b.   There were $6.0 + 2.2 + .3 + 1.0 = 9.5$ million credit cards issued by Citicorp.

P(Visa and Mastercard issued) $= 6/9.5 = 60/95 = .632$
P(Diners Club issued) $= 2.2/9.5 = 22/95 = .232$
P(Carte Blanche issued) $= .3/9.5 = 3/95 = .032$
P(Choice issued) $= 1.0/9.5 = 10/95 = .105$

c. P(customer uses one of Citicorp's own credit cards)
        $= P$(Diners Club) $+ P$(Carte Blanche) $+ P$(Choice)
        $= \dfrac{22}{95} + \dfrac{3}{95} + \dfrac{10}{95}$
        $= \dfrac{35}{95} = .368$

3.13   a.   The four classifications are:

(1)   Raise a broad mix of crops
(2)   Raise livestock
(3)   Use chemicals sparingly
(4)   Use techniques for regenerating the soil

Let us define the following events:

A_1:   Raise a broad mix of crops
A_2:   Do not raise a broad mix of crops
B_1:   Raise livestock
B_2:   Do not raise livestock
C_1:   Use chemical sparingly
C_2:   Do not use chemical sparingly
D_1:   Use techniques for regenerating the soil
D_2:   Do not use techniques for regenerating the soil

Each farmer is classified as using or not using each of the 4 techniques. Thus, the simple events are:

$A_1B_1C_1D_1,\ A_1B_1C_1D_2,\ A_1B_1C_2D_1,\ A_1B_2C_1D_1,\ A_2B_1C_1D_1,\ A_1B_1C_2D_2,$
$A_1B_2C_1D_2,\ A_2B_1C_1D_2,\ A_1B_2C_2D_1,\ A_2B_1C_2D_1,\ A_2B_2C_1D_1,\ A_1B_2C_2D_2,$
$A_2B_1C_2D_2,\ A_2B_2C_1D_2,\ A_2B_2C_2D_1,\ A_2B_2C_2D_2$

b. Since there are 16 classification sets or 16 simple events, the probability of any one simple event is 1/16. The probability that a farmer will be classified as unlikely on all four criteria is

$$P(A_2B_2C_2D_2) = 1/16$$

c. The probability that a farmer will be classified as likely on at least three of the criteria is

$$P(A_1B_1C_1D_1) + P(A_1B_1C_1D_2) + P(A_1B_1C_2D_1) + P(A_1B_2C_1D_1) + P(A_2B_1C_1D_1)$$
$$= 1/16 + 1/16 + 1/16 + 1/16 + 1/16 = 5/16$$

3.15   a. The odds in favor of a Snow Chief win are $\frac{1}{3}$ to $1 - \frac{1}{3} = \frac{2}{3}$ or 1 to 2.

b. If the odds in favor of Snow Chief are 1 to 1, then the probability that Snow Chief wins is $\frac{1}{1 + 1} = \frac{1}{2}$.

c. If the odds against Snow Chief are 3 to 2, then the odds in favor of Snow Chief are 2 to 3. Therefore, the probability that Snow Chief wins is $\frac{2}{2 + 3} = \frac{2}{5}$.

3.17 Two events which are mutually exclusive have no simple events in common. A Venn diagram of two mutually exclusive events is:

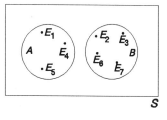

3.19 $A = \{(H, H, H), (H, H, T), (H, T, H), (H, T, T), (T, H, H), (T, H, T), (T, T, H)\}$
$B = \{(H, H, T), (H, T, H), (T, H, H)\}$
$C = \{(T, T, H), (T, H, T), (H, T, T)\}$

There are 8 possible outcomes when tossing three coins, so the probability of each simple event is 1/8.

a. $P(A \cup B) = P(H, H, H) + P(H, H, T) + P(H, T, H) + P(H, T, T) + P(T, H, H)$
                $+ P(T, H, T) + P(T, T, H)$
$$= \frac{1}{8} + \frac{1}{8} + \frac{1}{8} + \frac{1}{8} + \frac{1}{8} + \frac{1}{8} + \frac{1}{8} = \frac{7}{8}$$

b. $P(A \cap B) = P(H, H, T) + P(H, T, H) + P(T, H, H)$
$$= \frac{1}{8} + \frac{1}{8} + \frac{1}{8} = \frac{3}{8}$$

c. $P(A \cap C) = P(T, T, H) + P(T, H, T) + P(H, T, T)$

$$= \frac{1}{8} + \frac{1}{8} + \frac{1}{8} = \frac{3}{8}$$

d. $P(B \cap C) = 0$ (Events B and C have no elements in common)

e. Yes. The event $B \cap C$ has no simple events in it.

f. $P(B \cup C) = P(H, H, T) + P(H, T, H) + P(T, H, H) + P(H, T, T) + P(T, H, T)$
$\qquad + P(T, T, H)$

$$= \frac{1}{8} + \frac{1}{8} + \frac{1}{8} + \frac{1}{8} + \frac{1}{8} + \frac{1}{8} = \frac{6}{8} = \frac{3}{4}$$

3.21 a. $A \cap F$

b. $B \cup C$

c. A^c

3.23 a. $B \cap C$

b. A^c

c. $C \cup B$

d. $A \cap C^c$

3.25 a. A simple event is an event that cannot be decomposed into two or more other events. In this example, there are nine simple events. Let 1 be Warehouse 1, 2 be Warehouse 2, 3 be Warehouse 3, and let R be Regular, S be Stiff, and E be Extra stiff. The simple events are:

       (1, R) (1, S) (1, E)
       (2, R) (2, S) (2, E)
       (3, R) (3, S) (3, E)

b. The *sample space* of an experiment is the collection of all its simple events.

c. $P(C) = P(3, R) + P(3, S) + P(3, E)$
$\qquad = \quad .11 \quad + \quad .07 \quad + \quad .06$
$\qquad = \quad .24$

d. $P(F) = P(1, E) + P(2, E) + P(3, E)$
$\qquad = \quad 0 \quad + \quad .04 \quad + \quad .06$
$\qquad = .10$

e. $P(A) = P(1, R) + P(1, S) + P(1, E)$
$\qquad = \quad .41 \quad + \quad .06 \quad + \quad 0$
$\qquad = .47$

f. $P(D) = P(1, R) + P(2, R) + P(3, R)$
$$= .41 + .10 + .11$$
$$= .62$$

g. $P(E) = P(1, S) + P(2, S) + P(3, S)$
$$= .06 + .15 + .07$$
$$= .28$$

3.27 a. The 20 simple events may be listed as follows:

E_1: (Under 2000, 12) E_{11}: (5000–7999, 12)
E_2: (Under 2000, 24) E_{12}: (5000–7999, 24)
E_3: (Under 2000, 36) E_{13}: (5000–7999, 36)
E_4: (Under 2000, 42) E_{14}: (5000–7999, 42)
E_5: (Under 2000, 48) E_{15}: (5000–7999, 48)
E_6: (2000–4999, 12) E_{16}: (8000 or more, 12)
E_7: (2000–4999, 24) E_{17}: (8000 or more, 24)
E_8: (2000–4999, 36) E_{18}: (8000 or more, 36)
E_9: (2000–4999, 42) E_{19}: (8000 or more, 42)
E_{10}: (2000–4999, 48) E_{20}: (8000 or more, 48)

where the first entry within the parentheses indicates the amount of the loan (in dollars), and the second entry indicates the length of the loan (in months).

b. P(Selected loan will be for $8000 or more)
$$= P(E_{16}) + P(E_{17}) + P(E_{18}) + P(E_{19}) + P(E_{20})$$
$$= 0 + 0 + \frac{4}{500} + \frac{37}{500} + \frac{51}{500} = \frac{92}{500} = .184$$

Yes, this is the same answer obtained in part b of Exercise 3.5.

c. $P(E_{18}) = \dfrac{4}{500} = .008$

d. P(Selected loan is for three or four years)
$$= P(\text{Selected loan is for 36 or 48 months})$$
$$= P(E_3) + P(E_5) + P(E_8) + P(E_{10}) + P(E_{13}) + P(E_{15}) + P(E_{18}) + P(E_{20})$$
$$= 0 + 0 + \frac{20}{500} + 0 + \frac{89}{500} + \frac{114}{500} + \frac{4}{500} + \frac{51}{500} = \frac{278}{500} = .556$$

e. P(Selected loan is a 42-month loan for $2000 or more)
$$= P(E_9) + P(E_{14}) + P(E_{19})$$
$$= \frac{31}{500} + \frac{95}{500} + \frac{37}{500} = \frac{163}{500} = .326$$

3.29 a. The sale was paid by credit card, or the merchandise purchased was women's wear.

$$P(A \cup B) = .06 + .41 + .09 + .22 + .03 = .81$$

b. The merchandise purchased was women's wear or men's wear.

$$P(B \cup C) = .06 + .41 + .09 + .09 = .65$$

c. The merchandise purchased was women's wear, and it was paid by credit card.

$$P(B \cap A) = .41$$

d. The merchandise purchased was sportswear, and it was paid by credit card.

$$P(D \cap A) = .22$$

e. The pairs of events that are mutually exclusive have no simple events in common.

$A \cap B$ contains the event 'Sale was paid by credit card and merchandise purchased was women's wear.' Therefore, A and B are not mutually exclusive.

$A \cap C$ contains the event 'Sale was paid by credit card and merchandise purchased was men's wear.' Therefore, A and C are not mutually exclusive.

$A \cap D$ contains the event 'Sale was paid by credit card and merchandise purchased was sportswear.' Therefore, A and D are not mutually exclusive.

$B \cap C$ contains no simple events. Therefore, B and C are mutually exclusive.

$B \cap D$ contains no simple events. Therefore, B and D are mutually exclusive.

$C \cap D$ contains no simple events. Therefore, C and D are mutually exclusive.

3.31 The following events are defined:

A: {male worker}
B: {female worker}
C: {service worker}
D: {managerial/professional worker}
E: {operator/fabricator worker}
F: {technical/sales/administrative worker}

a. $P(A \cap C) = .052$

b.
$$\begin{aligned} P(D) &= P(A \cap D) + P(B \cap D) \\ &= .142 + .117 \\ &= .259 \end{aligned}$$

c.
$$\begin{aligned} P[(B \cap D) \cup (B \cap E)] &= .117 + .040 \\ &= .157 \end{aligned}$$

d.
$$\begin{aligned} P(F^c) &= 1 - P(F) = 1 - [P(A \cap F) + P(B \cap F)] \\ &= 1 - (.108 + .200) \\ &= 1 - .308 \\ &= .692 \end{aligned}$$

3.33   a. $P(A) = P(E_1) + P(E_2) + P(E_3)$
             $= .2 + .3 + .3$
             $= .8$

             $P(B) = P(E_2) + P(E_3) + P(E_5)$
             $= .3 + .3 + .1$
             $= .7$

             $P(A \cap B) = P(E_2) + P(E_3)$
             $= .3 + .3$
             $= .6$

       b. $P(E_1 \mid A) = \dfrac{P(E_1 \cap A)}{P(A)} = \dfrac{P(E_1)}{P(A)} = \dfrac{.2}{.8} = .25$

             $P(E_2 \mid A) = \dfrac{P(E_2 \cap A)}{P(A)} = \dfrac{P(E_2)}{P(A)} = \dfrac{.3}{.8} = .375$

             $P(E_3 \mid A) = \dfrac{P(E_3 \cap A)}{P(A)} = \dfrac{P(E_3)}{P(A)} = \dfrac{.3}{.8} = .375$

             The original simple event probabilities are in the proportion .2 to .3 to .3 or 2 to 3 to 3.

             The conditional probabilities for these simple events are in the proportion .25 to .375 to .375 or 2 to 3 to 3.

       c.   (1) $P(B \mid A) = P(E_2 \mid A) + P(E_3 \mid A)$
                   $= .375 + .375$ (from part b)
                   $= .75$

             (2) $P(B \mid A) = \dfrac{P(A \cap B)}{P(A)} = \dfrac{.6}{.8} = .75$ (from part a)

             The two methods do yield the same result.

       d.   If A and B are independent events, $P(B \mid A) = P(B)$.

             From part c, $P(B \mid A) = .75$. From part a, $P(B) = .7$.

             Since $.75 \neq .7$, A and B are not independent events.

3.35   a.   If A and B are independent, then $P(A \mid B) = P(A)$.

             $P(A \mid B) = \dfrac{P(A \cap B)}{P(B)} = \dfrac{P(E_3)}{P(E_2 \cup E_3 \cup E_4)} = \dfrac{.15}{.31 + .15 + .22} = \dfrac{.15}{.68} = .2206$

             $P(A) = P(E_1) + P(E_3) = .22 + .15 = .37$
             $.2206 \neq .37$; therefore, A and B are not independent.

b.	If A and C are independent, then $P(A \mid C) = P(A)$.

$$P(A \mid C) = \frac{P(A \cap C)}{P(C)} = \frac{P(E_1)}{P(E_1) + P(E_5)} = \frac{.22}{.22 + .10} = .6875$$

$P(A) = P(E_1) + P(E_3) = .22 + .15 = .37$

$.37 \neq .6875$; therefore, A and C are not independent.

c.	If B and C are independent, then $P(B \mid C) = P(B)$.

$$P(B \mid C) = \frac{P(B \cap C)}{P(C)} = \frac{0}{.32} = 0$$

$P(B) = P(E_2) + P(E_3) + P(E_4) = .31 + .15 + .22 = .68$

$0 \neq .68$; therefore, B and C are not independent.

3.37	a.	Dependent events are **not** always mutually exclusive. Let $P(A) = .3$, $P(B) = .2$ and $P(A \cap B) = .1$. $P(A \mid B) = \dfrac{P(A \cap B)}{P(B)} = \dfrac{.1}{.2} = .5$

Since $P(A \mid B) \neq P(A)(.5 \neq .3)$, A and B are dependent. However, since $P(A \cap B) = .1 \neq 0$, A and B are not mutually exclusive.

b.	Mutually exclusive events are **always** dependent. Let A and B be mutually exclusive events with $P(A) = .3$ and $P(B) = .2$. Since A and B are mutually exclusive, $P(A \cap B) = 0$. $P(A \mid B) = \dfrac{P(A \cap B)}{P(B)} = \dfrac{0}{.2} = 0$

Since $P(A) \neq P(A \mid B)(.3 \neq 0)$, A and B are dependent.

c.	Independent events are **not** always mutually exclusive. Let A and B be independent events with $P(A) = .3$ and $P(B) = .2$. Since A and B are independent, $P(A \cap B) = P(A)P(B) = .3(.2) = .06$. Since $.06 \neq 0$, A and B are not mutually exclusive.

d.	Mutually exclusive events are **not** always independent. Let A and B be mutually exclusive events with $P(A) = .3$ and $P(B) = .2$. Since A and B are mutually exclusive, $P(A \cap B) = 0$. $P(A \mid B) = \dfrac{P(A \cap B)}{P(B)} = \dfrac{0}{.2} = 0$

Since $P(A) \neq P(A \mid B)(.3 \neq 0)$, A and B are not independent.

3.39	The 36 possible outcomes obtained when tossing two dice are listed below:

$(1, 1)\ (1, 2)\ (1, 3)\ (1, 4)\ (1, 5)\ (1, 6)$
$(2, 1)\ (2, 2)\ (2, 3)\ (2, 4)\ (2, 5)\ (2, 6)$
$(3, 1)\ (3, 2)\ (3, 3)\ (3, 4)\ (3, 5)\ (3, 6)$
$(4, 1)\ (4, 2)\ (4, 3)\ (4, 4)\ (4, 5)\ (4, 6)$
$(5, 1)\ (5, 2)\ (5, 3)\ (5, 4)\ (5, 5)\ (5, 6)$
$(6, 1)\ (6, 2)\ (6, 3)\ (6, 4)\ (6, 5)\ (6, 6)$

$A = \{(1, 2), (1, 4), (1, 6), (2, 1), (2, 3), (2, 5), (3, 2), (3, 4), (3, 6), (4, 1), (4, 3), (4, 5),$
$\quad (5, 2), (5, 4), (5, 6), (6, 1), (6, 3), (6, 5)\}$

$B = \{(2, 6), (3, 5), (3, 6), (4, 4), (4, 5), (4, 6), (5, 3), (5, 4), (5, 5), (5, 6), (6, 2), (6, 3),$
$\quad (6, 4), (6, 5)\}$

$A \cap B = \{(3, 6), (4, 5), (5, 4), (5, 6), (6, 3), (6, 5)\}$

If A and B are independent, then $P(A \mid B) = P(A)$.

$$P(A \mid B) = \frac{P(A \cap B)}{P(B)} = \frac{\frac{6}{36}}{\frac{14}{36}} = \frac{6}{14} = \frac{3}{7}$$

$$P(A) = \frac{18}{36} = \frac{1}{2}$$

$\frac{3}{7} \neq \frac{1}{2}$; therefore, events A and B are not independent

3.41   Define the following events:

       A: {Take tough action early}
       B: {Take tough action late}
       C: {Never take tough action}
       D: {Wisconsin}
       E: {Illinois}
       F: {Arkansas}
       G: {Louisiana}

a. $P(D \cup G) = P(D) + P(G)$ (since D and G are mutually exclusive)

$$= \frac{0}{151} + \frac{37}{151} + \frac{9}{151} + \frac{1}{151} + \frac{21}{151} + \frac{15}{151} = \frac{83}{151} = .550$$

b. $P((D \cup G)^c) = 1 - P(D \cup G)$

$$= 1 - \frac{83}{151} = \frac{68}{151} = .450$$

c. $P(C) = \frac{9}{151} + \frac{11}{151} + \frac{6}{151} + \frac{15}{151} = \frac{41}{151} = .272$

d. $P(F \cap C) = \frac{6}{151} = .040$

e. $P(C \mid F) = \frac{P(F \cap C)}{P(F)} = \frac{\frac{6}{151}}{\frac{33}{151}} = \frac{6}{33} = .182$

f. $P((F \cup G) \mid A) = \dfrac{P((F \cup G) \cap A)}{P(A)} = \dfrac{\frac{5}{151} + \frac{1}{151}}{\frac{7}{151}} = \dfrac{6}{7} = .857$

g. $P(C \mid F) = \dfrac{P(F \cap C)}{P(F)} = \dfrac{\frac{6}{151}}{\frac{33}{151}} = \dfrac{6}{33} = .182$

3.43 Define the following events.

A: {A defective case gets by inspector 1}
B: {A defective case gets by inspector 2}

We want to know

$A \cap B$: {A defective case gets by inspector 1 and inspector 2}.

We know that the two inspectors check the cases independently. Therefore,

$P(A \cap B) = P(A)P(B) = (.05)(.10) = .005$

3.45 Define the following events:

A: {Product A is profitable}
B: {Product B is profitable}

We know that

P(Individual product profitable) $= .18$
and
P(Two products profitable) $= .05$

a. $P(A) = .18$

b. $P(B^c) = 1 - P(B) = 1 - .18 = .82$

c. $P(A \cup B) = P(A) + P(B) - P(A \cap B)$
$= .18 + .18 - .05$
$= .31$

d. $P((A \cup B)^c) = 1 - P(A \cup B)$
$= 1 - .31$
$= .69$

e. $P(A \cup B) - P(A \cap B)$
$= .31 - .05$
$= .26$

3.47   Define the following events:

   A: {Selected firm implemented TQM}
   B: {Selected firm's sales increased}

From the information given, $P(A) = 30/100 = .3$, $P(B) = 60/100 = .6$, and $P(A \mid B) = 20/60 = 1/3$.

a. $P(A) = 30/100 = .3$
      $P(B) = 60/100 = .6$

b.   If A and B are independent, $P(A \mid B) = P(A)$. However, $P(A \mid B) = 1/3 \neq P(A) = .3$. Thus, A and B are not independent.

c.   Now, $P(A \mid B) = 18/60 = .3$. Since $P(A \mid B) = .3 = P(A) = .3$, A and B are independent.

3.49   Starting in row 5, column 2, of Table I of Appendix B and reading across, take the first 20 single digit numbers going left to right.

The 20 digits selected for the random sample are:

3, 9, 9, 7, 5, 8, 1, 8, 3, 7, 1, 6, 6, 5, 6, 0, 6, 1, 2, 1

3.51   a.   The possible pairs of accounts that could be obtained are:

         (0001, 0002) (0001, 0003) (0001, 0004) (0001, 0005)
              (0002, 0003) (0002, 0004) (0002, 0005)
                   (0003, 0004) (0003, 0005)
                        (0004, 0005)

        b.   There are 10 possible pairs of accounts that could be obtained. In a random sample, all 10 pairs of accounts have an equal chance of being selected. The probability of selecting any one of the 10 pairs is 1/10. Therefore, the probability of selecting accounts 0001 and 0004 is 1/10.

        c.   Since only 2 accounts have a balance of $1,000 (0001 and 0004), the probability of selecting 2 accounts that each have a balance of $1,000 is 1/10.

              Since there are only 3 accounts that do not have a balance of $1,000 (0002, 0003, and 0005), there are 3 possible pairs of accounts in which each has a balance other than $1,000 (0002, 0003), (0002, 0005), and (0003, 0005)). Therefore, the probability of selecting a pair of accounts in which each has a balance other than $1,000 is 3/10.

3.53   a.   Give each stock in the NYSE-Composite Transactions table of the Wall Street Journal a number (1 to m). Using Table I of Appendix B, pick a starting point and read down using the same number of digits as in m until you have n different numbers between 1 and m, inclusive.

3.55   (1)   The probabilities of all simple events must lie between 0 and 1, inclusive.
        (2)   The probabilities of all the simple events in the sample space must sum to 1.

3.57 $P(A \cap B) = .4, P(A \mid B) = .8$

Since the $P(A \mid B) = \dfrac{P(A \cap B)}{P(B)}$, substitute the given probabilities into the formula and solve for $P(B)$.

$$.8 = \frac{.4}{P(B)} \Rightarrow P(B) = \frac{.4}{.8} = .5$$

3.59   Define the following events:

         A: {The watch is accurate}
         N: {The watch is not accurate}

Assuming the manufacturer's claim is correct,

         $P(N) = .05$ and $P(A) = 1 - P(N) = 1 - .05 = .95$

The sample space for the purchase of 4 of the manufacturer's watches is listed below.

(A, A, A, A)	(N, A, A, A)	(A, N, N, A)	(N, A, N, N)
(A, A, A, N)	(A, A, N, N)	(N, A, N, A)	(N, N, A, N)
(A, A, N, A)	(A, N, A, N)	(N, N, A, A)	(N, N, N, A)
(A, N, A, A)	(N, A, A, N)	(A, N, N, N)	(N, N, N, N)

a.   All 4 watches being accurate as claimed is the simple event (A, A, A, A).

Assuming the watches purchased operate independently and the manufacturer's claim is correct,

$$P(A, A, A, A) = P(A)P(A)P(A)P(A) = (.95)^4 = .8145$$

b.   The simple events in the sample space that consist of exactly two watches failing to meet the claim are listed below.

(A, A, N, N)	(N, A, A, N)
(A, N, A, N)	(N, A, N, A)
(A, N, N, A)	(N, N, A, A)

The probability that exactly two of the four watches fail to meet the claim is the sum of the probabilities of these six simple events.

Assuming the watches purchased operate independently and the manufacturer's claim is correct,

$$P(A, A, N, N) = P(A)P(A)P(N)P(N) = (.95)(.95)(.05)(.05) = .00225625$$

All six of the simple events will have the same probability. Therefore, the probability that exactly two of the four watches fail to meet the claim when the manufacturer's claim is correct is

$$6(0.00225625) = .0135$$

c. The simple events in the sample space that consist of three of the four watches failing to meet the claim are listed below.

$$(A, N, N, N) \quad (N, N, A, N)$$
$$(N, A, N, N) \quad (N, N, N, A)$$

The probability that three of the four watches fail to meet the claim is the sum of the probabilities of the four simple events.

Assuming the watches purchased operate independently and the manufacturer's claim is correct,

$$P(A, N, N, N) = P(A)P(N)P(N)P(N) = (.95)(.05)(.05)(.05) = .00011875$$

All four of the simple events will have the same probability. Therefore, the probability that three of the four watches fail to meet the claim when the manufacturer's claim is correct is

$$4(.00011875) = .000475$$

If this event occurred, we would tend to doubt the validity of the manufacturer's claim since its probability of occurring is so small.

d. All 4 watches tested failing to meet the claim is the simple event (N, N, N, N).

Assuming the watches purchased operate independently and the manufacturer's claim is correct,

$$P(N, N, N, N) = P(N)P(N)P(N)P(N) = (.05)^4 = .00000625$$

Since the probability of observing this event is so small if the claim is true, we have strong evidence against the validity of the claim. However, we do not have conclusive proof that the claim is false. There is still a chance the event can occur (with probability .00000625) although it is extremely small.

3.61 Define the following events:

T: {Technical staff}
N: {Nontechnical staff}
U: {Under 20 years with company}
O: {Over 20 years with company}
R_1: {Retire at age 65}
R_2: {Retire at age 68}

The probabilities for each simple event are given in table form.

	U		O	
	T	N	T	N
R_1	$\dfrac{31}{200}$	$\dfrac{5}{200}$	$\dfrac{45}{200}$	$\dfrac{12}{200}$
R_2	$\dfrac{59}{200}$	$\dfrac{25}{200}$	$\dfrac{15}{200}$	$\dfrac{8}{200}$

Each simple event consists of 3 characteristics: type of staff (T or N), years with the company, (U or O), and age plan to retire (R_1 or R_2).

a. $P(T) = P(T \cap U \cap R_1) + P(T \cap U \cap R_2) + P(T \cap O \cap R_1) + P(T \cap O \cap R_2)$

$$= \frac{31}{200} + \frac{59}{200} + \frac{45}{200} + \frac{15}{200} = \frac{150}{200} = .75$$

b. $P(O) = P(O \cap T \cap R_1) + P(O \cap T \cap R_2) + P(O \cap N \cap R_1) + P(O \cap N \cap R_2)$

$$= \frac{45}{200} + \frac{15}{200} + \frac{12}{200} + \frac{8}{200} = \frac{80}{200} = .4$$

$$P(R_2 \cap O) = P(R_2 \cap O \cap T) + P(R_2 \cap O \cap N) = \frac{15}{200} + \frac{8}{200} = \frac{23}{200} = .115$$

Thus, $P(R_2 \mid O) = \dfrac{P(R_2 \cap O)}{P(O)} = \dfrac{.115}{.4} = .2875$

c. $P(T) = .75$ from a.

$$P(U \cap T) = P(U \cap T \cap R_1) + P(U \cap T \cap R_2) = \frac{31}{200} + \frac{59}{200} = \frac{90}{200} = .45$$

Thus, $P(U \mid T) = \dfrac{P(U \cap T)}{P(T)} = \dfrac{.45}{.75} = .6$

d. $P(O \cap N \cap R_1) = \dfrac{12}{200} = .06$

3.63 Define the following events:

A: {Acupoll predicts the success of a particular product}
B: {Product is successful}

From the problem, we know

$$P(A \mid B) = .89 \text{ and } P(B) = .90$$

Thus, $P(A \cap B) = P(A \mid B)P(B) = .89(.90) = .801$

3.65   Define the following events:

> G: {regularly use the golf course}
> T: {regularly use the tennis courts}

Given: $P(G) = .7$ and $P(T) = .5$

The event 'uses neither facility' can be written as $G^c \cap T^c$ or $(G \cup T)^c$. We are given $P(G^c \cap T^c) = P[(G \cup T)^c] = .05$. The complement of the event 'uses neither facility' is the event 'uses at least one of the two facilities' which can be written as $G \cup T$.

$$P(G \cup T) = 1 - P[(G \cup T)^c] = 1 - .05 = .95$$

From the additive rule, $P(G \cup T) = P(G) + P(T) - P(G \cap T)$
$$\Rightarrow .95 = .7 + .5 - P(G \cap T)$$
$$\Rightarrow P(G \cap T) = .25$$

a.   The Venn Diagram is:

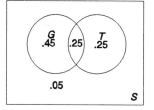

b. $P(G \cup T) = .95$ from above.

c. $P(G \cap T) = .25$ from above.

d. $P(G \mid T) = \dfrac{P(G \cap T)}{P(T)} = \dfrac{.25}{.5} = .5$

3.67   The statement would be valid if 1/50th of all U.S. citizens reside in New Hampshire. However, this is not true. Therefore, the statement is invalid.

3.69   Define the following events:

> O_1: Component #1 operates properly
> O_2: Component #2 operates properly
> O_3: Component #3 operates properly

$P(O_1) = 1 - P(O_1^c) = 1 - .12 = .88$
$P(O_2) = 1 - P(O_2^c) = 1 - .09 = .91$
$P(O_3) = 1 - P(O_3^c) = 1 - .11 = .89$

a. P(system operates properly)

$$= P(O_1 \cap O_2 \cap O_3)$$
$$= P(O_1)P(O_2)P(O_3) \text{ (since the three components operate independently)}$$
$$= (.88)(.91)(.89)$$
$$= .7127$$

b. P(system fails) = 1 − P(system operates properly)

$$= 1 - .7127 \text{ (see part } \mathbf{a}\text{)}$$
$$= .2873$$

3.71 Define the following events:

A: {product A is accepted by the public}
B: {product B is accepted by the public}

Given: $P(A \cap B^c) = .3$
$P(A^c \cap B) = .4$
$P(A \cap B) = .2$

The simple events of the sample space are:

A, B
A, B^c
A^c, B
A^c, B^c

Since the sum of the probabilities of these 4 events is 1,

$$P(A^c \cap B^c) = 1 - P(A \cap B) - P(A \cap B^c) - P(A^c \cap B)$$
$$= 1 - .2 - .3 - .4 = .1$$

This probability is not equal to .01. Therefore, we would disagree with the manager.

3.73 Let M_1, M_2, M_3, and M_4 represent the four minority applicants, and N_1 and N_2 represent the two nonminority applicants. The sample space for choosing two of the six applicants is:

(M_1, M_2)	(M_2, M_3)	(M_3, N_1)
(M_1, M_3)	(M_2, M_4)	(M_3, N_2)
(M_1, M_4)	(M_2, N_1)	(M_4, N_1)
(M_1, N_1)	(M_2, N_2)	(M_4, N_2)
(M_1, N_2)	(M_3, M_4)	(N_1, N_2)

Since the choice is random, all 15 of the simple events listed are equally likely. Therefore, they all have probability 1/15 of occurring.

a. $P(A) = P(N_1, N_2) = \dfrac{1}{15}$

b. $P(B) = P(M_1, M_2) + P(M_1, M_3) + P(M_1, M_4) + P(M_2, M_3) + P(M_2, M_4)$
$+ P(M_3, M_4)$

$$= \frac{1}{15} + \frac{1}{15} + \frac{1}{15} + \frac{1}{15} + \frac{1}{15} + \frac{1}{15} = \frac{6}{15}$$

c. $P(C) = 1 - P(A) = 1 - \dfrac{1}{15} = \dfrac{14}{15}$

d. $P(B \mid C) = \dfrac{P(B \cap C)}{P(C)} = \dfrac{P(B)}{P(C)} = \dfrac{\frac{6}{15}}{\frac{14}{15}} = \dfrac{6}{14}$

e. $P(D \mid C) = \dfrac{P(D \cap C)}{P(C)}$

$$= \frac{P(M_1, M_2) + P(M_1, M_3) + P(M_1, M_4) + P(M_1, N_1) + P(M_1, N_2)}{P(C)}$$

$$= \frac{\frac{5}{15}}{\frac{14}{15}} = \frac{5}{14}$$

3.75 Define the following events:

A: {salesperson sells computer on first visit}
B: {salesperson sells computer on second visit}

Given: $P(A) = .4$ and $P(B \mid A^c) = .65$

$P(A^c) = 1 - P(A) = 1 - .4 = .6$

$P(\text{Sale}) = P(A) + P(B \cap A^c) = P(A) + P(B \mid A^c)P(A^c)$
$= .4 + .65(.6) = .4 + .390 = .79$

3.77 From the Venn Diagram, you can see that

$P(A) = P(A \cap B^c) + P(A \cap B)$

Thus, $P(A \cap B^c) = P(A) - P(A \cap B)$

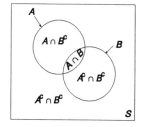

3.79 This statement is false. The outcomes of the first 20 tosses are independent of any future tosses. Anytime a fair coin is tossed, the probability of a head is 1/2 and the probability of a tail is 1/2.

CHAPTER FOUR

···

Discrete Random Variables

4.1   A random variable is a rule that assigns one and only one value to each simple event of an experiment.

4.3   a.   The amount of water flowing through the Hoover Dam in a year can take on an infinite number of different values. Thus, this is a continuous random variable.

       b.   The number of people who fly American Airlines in a day can take on a countable number of values. Thus, this is a discrete random variable.

       c.   The length of time it takes to assemble one Ford Thunderbird can take on an infinite number of different values. Thus, this is a continuous variable.

       d.   The number of patients that are admitted per week in a particular hospital can take on a countable number of values. Thus, this is a discrete random variable.

4.5   The number of occupied units in an apartment complex at any time is a discrete random variable, as is the number of shares of stock traded on the New York Stock Exchange on a particular day. Two examples of continuous random variables are the length of time to complete a building project and the weight of a truckload of oranges.

4.7   An economist might be interested in the percentage of the work force that is unemployed, or the current inflation rate, both of which are continuous random variables.

4.9   The manager of a clothing store might be concerned with the number of employees on duty at a specific time of day, or the number of articles of a particular type of clothing that are on hand.

4.11   a.   The eight simple events and the corresponding values of x are shown in the following table.

Simple Event	x
(H, H, H)	3
(H, H, T)	2
(H, T, H)	2
(T, H, H)	2
(T, T, H)	1
(T, H, T)	1
(H, T, T)	1
(T, T, T)	0

b. Since the 8 simple events are equally likely,

$$p(0) = P(x = 0) = P(T, T, T) = \frac{1}{8}$$

$$p(1) = P(x = 1) = P(T, T, H) + P(T, H, T) + P(H, T, T)$$
$$= \frac{1}{8} + \frac{1}{8} + \frac{1}{8} = \frac{3}{8}$$

$$p(2) = P(x = 2) = P(H, H, T) + P(H, T, H) + P(T, H, H)$$
$$= \frac{1}{8} + \frac{1}{8} + \frac{1}{8} = \frac{3}{8}$$

$$p(3) = P(x = 3) = P(H, H, H) = \frac{1}{8}$$

The probability distribution of x may now be summarized in tabular form:

x	$p(x)$
0	1/8
1	3/8
2	3/8
3	1/8

c. The probability distribution of x may also be presented in graphical form:

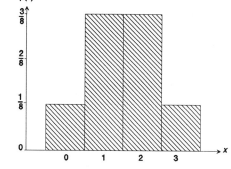

4.13 a. This is *not* a valid distribution because $\sum p(x) = .9 \neq 1$.

b. This is a *valid* distribution because $0 \leq p(x) \leq 1$ for all values of x and $\sum p(x) = 1$.

c. This is *not* a valid distribution because $p(4) = -.3 < 0$.

d. The sum of the probabilities over all possible values of the random variable is $1.1 > 1$, so this is *not* a valid probability distribution.

4.15   a.   The probability distribution for x in graphical form:

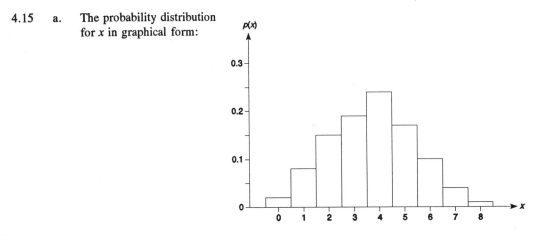

b. $P(x > 3) = p(4) + p(5) + p(6) + p(7) + p(8)$
$$= .24 + .17 + .10 + .04 + .01 = .56$$

$P(x > 4) = p(5) + p(6) + p(7) + p(8)$
$$= .17 + .10 + .04 + .01 = .32$$

4.17   Define the following events:

M: {Number matches} $P(M) = 1/10 = .1$
N: {Number does not match} $P(N) = 9/10 = .9$

Also, let x = number of matches in 3 tries. The sample space along with the values of x is:

	x
MMM	3
MMN	2
MNM	2
NMM	2
MNN	1
NMN	1
NNM	1
NNN	0

a. $P(x = 0) = P(NNN) = P(N)P(N)P(N) = .9^3 = .729$ (by independence)
$P(x = 3) = P(MMM) = P(M)P(M)P(M) = .1^3 = .001$

b. $P(x = 1) = P(MNN) + P(NMN) + P(NNM) = 3(.1)(.9)^2 = .243$
$P(x = 2) = P(MMN) + P(MNM) + P(NMM) = 3(.1)^2(.9) = .027$

c.   Yes. If it is not required that the order of the numbers drawn match the order in which you selected the numbers, then your pick, (1, 3, 7), would match (1, 3, 7), (1, 7, 3), (3, 1, 7), (3, 7, 1), (7, 1, 3), and (7, 3, 1). Thus, the probability that all your numbers match would be 6/1000 = .006 instead of .001.

4.19   a. $E(x) = \displaystyle\sum_{\text{All } x} xp(x) = 0(.05) + 1(.25) + 2(.30) + 5(.20) + 10(.20)$

$$= 0 + .25 + .60 + 1.00 + 2.00 = 3.85$$

b.

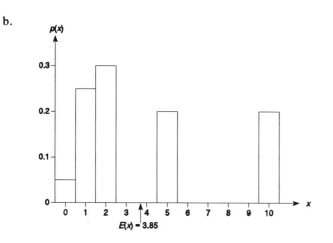

c.   We expect the value of x to be 3.85. This is the average value of x over a large number of trials.

4.21   a. $\mu = E(x) = \displaystyle\sum_{\text{All } x} xp(x) = 30(.05) + 40(.20) + 50(.10) + 60(.25) + 70(.15) + 80(.25)$

$$= 1.5 + 8 + 5 + 15 + 10.5 + 20 = 60$$

$\sigma^2 = E[(x - \mu)^2] = \displaystyle\sum_{\text{All } x} (x - \mu)^2 p(x)$

$= (30 - 60)^2(.05) + (40 - 60)^2(.20) + (50 - 60)^2(.10) + (60 - 60)^2(.25)$
$\quad + (70 - 60)^2(.15) + (80 - 60)^2(.25)$

$= 45 + 80 + 10 + 0 + 15 + 100 = 250$

$\sigma = \sqrt{\sigma^2} = \sqrt{250} = 15.811$

b.

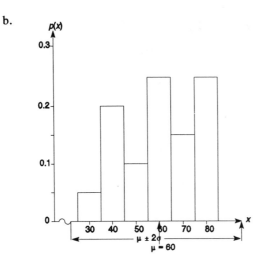

c. $\mu \pm 2\sigma \Rightarrow 60 \pm 2(15.811) \Rightarrow 60 \pm 31.622 \Rightarrow (28.378, 91.622)$

$$P(28.378 < x < 91.622) = p(30) + p(40) + p(50) + p(60) + p(70) + p(80)$$
$$= .05 + .20 + .10 + .25 + .15 + .25$$
$$= 1$$

4.23   a. $E(x) = \sum_{\text{All } x} xp(x)$

Firm A: $E(x) = 0(.01) + 500(.01) + 1000(.01) + 1500(.02) + 2000(.35) + 2500(.30)$
$$+ 3000(.25) + 3500(.02) + 4000(.01) + 4500(.01) + 5000(.01)$$
$$= 0 + 5 + 10 + 30 + 700 + 750 + 750 + 70 + 40 + 45 + 50$$
$$= 2450$$

Firm B: $E(x) = 0(.00) + 200(.01) + 700(.02) + 1200(.02) + 1700(.15) + 2200(.30)$
$$+ 2700(.30) + 3200(.15) + 3700(.02) + 4200(.02) + 4700(.01)$$
$$= 0 + 2 + 14 + 24 + 255 + 660 + 810 + 480 + 74 + 84 + 47$$
$$= 2450$$

b. $\sigma = \sqrt{\sigma^2} \qquad \sigma^2 = \sum_{\text{All } x} (x - \mu)^2 p(x)$

Firm A: $\sigma^2 = (0 - 2450)^2(.01) + (500 - 2450)^2(.01) + \cdots + (5000 - 2450)^2(.01)$
$$= 60,025 + 38,025 + 21,025 + 18,050 + 70,875 + 750 + 75,625$$
$$+ 22,050 + 24,025 + 42,025 + 65,025$$
$$= 437,500$$
$$\sigma = 661.44$$

Firm B: $\sigma^2 = (0 - 2450)^2(.00) + (200 - 2450)^2(.01) + \cdots + (4700 - 2450)^2(.01)$
$$= 0 + 50{,}625 + 61{,}250 + 31{,}250 + 84{,}375 + 18{,}750 + 84{,}375$$
$$+ 31{,}250 + 61{,}250 + 50{,}625$$
$$= 492{,}500$$
$$\sigma = 701.78$$

Firm B faces greater risk of physical damage because it has a higher variance and standard deviation.

4.25   a.   Let x = change in price of ABC at the close of business. First we need to find the probability distribution for x.

Since we are assuming that when stock ABC's price changes it is by $2, the probability distribution of x is:

x	-2	0	2
$p(x)$	.2	.2	.6

The expected change in ABC's price is

$$E(x) = \sum_{\text{All } x} xp(x) = -2(.2) + 0(.2) + 2(.6)$$
$$= -.4 + 0 + 1.2$$
$$= \$.80$$

b.   We are assuming that ABC's price changes by exactly $2 when it changes. Therefore, the change in ABC's price cannot equal its expected value of $.80. The change must be $-\$2$, $\$0$, or $\$2$.

4.27   a.   The probability distribution for the cost of treating the patient is:

Disease	Hepatitis	Cirrhosis	Gallstones	Pan. Cancer
Cost	$700	$1,110	$3,320	$16,450
p(Cost)	.4	.1	.45	.05

b. $E(\text{Cost}) = \sum \text{Cost } p(\text{Cost}) = 700(.4) + 1110(.1) + 3320(.45) + 16450(.05)$
$$= 280 + 111 + 1494 + 822.5 = \$2{,}707.50$$

This is the average cost per patient for many patients with the same symptoms.

c. $P(\text{Hepatitis} \mid \text{Hepatitis or Cirrhosis}) = .4/.5 = .8$
$P(\text{Cirrhosis} \mid \text{Hepatitis or Cirrhosis}) = .1/.5 = .2$

The probability distribution for the cost of treating the patient is:

Disease	Hepatitis	Cirrhosis
Cost	$700	$1,110
p(Cost)	.8	.2

d. $E(\text{Cost}) = \sum \text{Cost } p(\text{Cost}) = 700(.8) + 1110(.2)$
$$= 560 + 222 = \$782$$

This is the average cost per patient for many patients with the same symptoms.

4.29 a. Let x = total cost to the firm for the interviewing strategy. We will hire the first qualified candidate interviewed but will interview a maximum of 3. The candidates will be randomly selected from a group of 4 in which only 1 is qualified. The value of x is $1000 times the number of candidates interviewed.

Define the following events:

Q_i: {ith candidate is qualified}
N_i: {ith candidate is not qualified}

$P(x = 1000) = P(Q_1) = 1/4$
$P(x = 2000) = P(N_1 \cap Q_2) = P(Q_2 \mid N_1)P(N_1) = (1/3)(3/4) = 1/4$
$P(x = 3000) = P(N_1 \cap N_2 \cap Q_3) + P(N_1 \cap N_2 \cap N_3)$
$\qquad = P(Q_3 \mid N_1 \cap N_2)P(N_1 \cap N_2) + P(N_3 \mid N_1 \cap N_2)P(N_1 \cap N_2)$
$\qquad = P(Q_3 \mid N_1 \cap N_2)P(N_2 \mid N_1)P(N_1) + P(N_3 \mid N_1 \cap N_2)P(N_2 \mid N_1)P(N_1)$
$\qquad = 1/2(2/3)(3/4) + 1/2(2/3)(3/4) = 1/4 + 1/4 = 1/2$

The probability distribution of x is

x	$1,000	$2,000	$3,000
$p(x)$	.25	.25	.50

b. $P(\text{nobody hired}) = P(\text{qualified candidate is not chosen})$
$\qquad = P(N_1 \cap N_2 \cap N_3)$
$\qquad = P(N_3 \mid N_1 \cap N_2)P(N_1 \cap N_2)$
$\qquad = P(N_3 \mid N_1 \cap N_2)P(N_2 \mid N_1)P(N_1)$
$\qquad = \dfrac{1}{2}\left[\dfrac{2}{3}\right]\left[\dfrac{3}{4}\right] = \dfrac{1}{4} = .25$

c. $\mu = E(x) = \displaystyle\sum_{\text{All } x} xp(x) = 1,000(.25) + 2,000(.25) + 3,000(.50)$
$$= 250 + 500 + 1500$$
$$= \$2,250$$

d.   The expected cost of the interviewing strategy, $E(x)$, is the same as the mean of the probability distribution.

Therefore, $\mu = E(x) = \$2,250$.

4.31   a.   The random variable x is discrete since it can only assume a countable number of values (0, 1, 2, 3, 4, 5, 6).

b.   This is a binomial probability distribution with $n = 6$ and $p = .4$.

c.   In order to graph the probability distribution for x, we need to know the probabilities for every possible value of x. Using Table II of Appendix B with $n = 6$ and $p = .4$:

$$P(x = 0) = p(0) = .047$$
$$P(x = 1) = P(x \le 1) - P(x = 0) = .233 - .047 = .186$$
$$P(x = 2) = P(x \le 2) - P(x \le 1) = .544 - .233 = .311$$
$$P(x = 3) = P(x \le 3) - P(x \le 2) = .821 - .544 = .277$$
$$P(x = 4) = P(x \le 4) - P(x \le 3) = .959 - .821 = .138$$
$$P(x = 5) = P(x \le 5) - P(x \le 4) = .996 - .959 = .037$$
$$P(x = 6) = P(x \le 6) - P(x \le 5) = 1 - .996 = .004$$

The probability distribution of x in graphical form is:

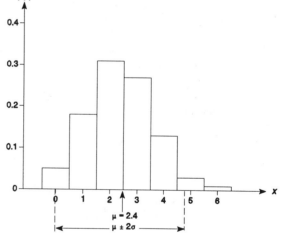

d. $\mu = np = 6(.4) = 2.4$

$\sigma = \sqrt{npq} = \sqrt{6(.4)(.6)} = \sqrt{1.44} = 1.2$

e. $\mu \pm 2\sigma \Rightarrow 2.4 \pm 2(1.2) \Rightarrow (0, 4.8)$

4.33   a.   Using Table II, Appendix B, with $n = 8$ and $p = .5$, $P(x = 5) = P(x \le 5) - P(x \le 4)$ $= .855 - .637 = .218$

b.   Using Table II, Appendix B, with $n = 7$ and $p = 1 - q = 1 - .3 = .7$, $P(x = 2) =$ $P(x \le 2) - P(x \le 1) = .029 - .004 = .025$

c. $P(x = 4) = \begin{pmatrix} 4 \\ 4 \end{pmatrix} \dfrac{4!}{4!(4-4)!} \cdot .6^4 \cdot .4^{(4-4)} = \dfrac{4 \cdot 3 \cdot 2 \cdot 1}{4 \cdot 3 \cdot 2 \cdot 1 \cdot 1} \cdot .6^4 \cdot .4^0$

$\qquad\qquad = 1(.1296)(1) = .1296$

4.35 From Exercise 4.34,

x	0	1	2	3	4	5	6
$p(x)$	.000007	.000131	.001178	.006478	.024291	.065587	.131173

x	7	8	9	10	11	12	13
$p(x)$	.196760	.221355	.184462	.110677	.045277	.011319	.001306

a. $P(x \le 5) = p(0) + p(1) + p(2) + p(3) + p(4) + p(5)$
$\qquad\qquad = .000007 + .000131 + .001178 + .006478 + .024291 + .065587$
$\qquad\qquad = .097672$

b. $P(x \ge 3) = 1 - P(x \le 2) = 1 - [p(0) + p(1) + p(2)]$
$\qquad\qquad = 1 - (.000007 + .000131 + .001178) = 1 - .001316 = .998684$

c. $P(x < 7) = p(0) + p(1) + p(2) + p(3) + p(4) + p(5) + p(6)$
$\qquad\qquad = .000007 + .000131 + .001178 + .006478 + .024291 + .065587$
$\qquad\qquad\quad + .131173$
$\qquad\qquad = .228845$

4.37 x is a binomial random variable with $n = 6$.

a. If the probability distribution of x is symmetric, $p(0) = p(6)$, $p(1) = p(5)$, and $p(2) = p(4)$.

Since $p(x) = \begin{pmatrix} n \\ x \end{pmatrix} p^x q^{n-x} \qquad x = 0, 1, \dots, n,$

when $n = 6$

$$\begin{pmatrix} 6 \\ 0 \end{pmatrix} p^0 q^6 = \begin{pmatrix} 6 \\ 6 \end{pmatrix} p^6 q^0$$

$$\Rightarrow \dfrac{6!}{0!6!} p^0 q^6 = \dfrac{6!}{6!0!} p^6 q^0$$
$$\Rightarrow q^6 = p^6$$
$$\Rightarrow p = q$$

Since $p + q = 1$, $p = .5$

Therefore, the probability distribution of x is symmetric when $p = .5$

b. If the probability distribution of x is skewed to the right, then the mean is greater than the median. Therefore, there are more small values in the distribution (0, 1) than large values (5, 6). Therefore, p must be smaller than .5. Let $p = .2$ and the probability distribution of x will be skewed to the right.

c. If the probability distribution of x is skewed to the left, then the mean is smaller than the median. Therefore, there are more large values in the distribution $(5, 6)$ than small values $(0, 1)$. Therefore, p must be larger than .5. Let $p = .8$ and the probability distribution of x will be skewed to the left.

d. In part **a**, x is a binomial random variable with $n = 6$ and $p = .5$.

$$p(x) = \binom{6}{x} .5^x .5^{6-x} \quad x = 0, 1, 2, 3, 4, 5, 6$$

$$p(0) = \binom{6}{0} .5^0 .5^6 = \frac{6!}{0!6!} .5^6 = 1(.5)^6 = .0156$$

$$p(1) = \binom{6}{1} .5^1 .5^5 = \frac{6!}{1!5!} .5^6 = 6(.5)^6 = .0938$$

$$p(2) = \binom{6}{2} .5^2 .5^4 = \frac{6!}{2!4!} .5^6 = 15(.5)^6 = .2344$$

$$p(3) = \binom{6}{3} .5^3 .5^3 = \frac{6!}{3!3!} .5^6 = 20(.5)^6 = .3125$$

$p(4) = p(2) = .2344$ (since the distribution is symmetric)
$p(5) = p(1) = .0938$
$p(6) = p(0) = .0156$

The probability distribution of x in tabular form is

x	0	1	2	3	4	5	6
$p(x)$	.0156	.0938	.2344	.3125	.2344	.0938	.0156

$\mu = np = 6(.5) = 3$

The graph of the probability distribution of x when $n = 6$ and $p = .5$ follows:

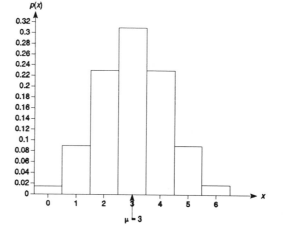

In part **b**, x is a binomial random variable with $n = 6$ and $p = .2$.

$$p(x) = \begin{bmatrix} 6 \\ x \end{bmatrix} .2^x .8^{6-x} \quad x = 0, 1, 2, 3, 4, 5, 6$$

$$p(0) = \begin{bmatrix} 6 \\ 0 \end{bmatrix} .2^0 .8^6 = 1(1).8^6 = .2621$$

$$p(1) = \begin{bmatrix} 6 \\ 1 \end{bmatrix} .2^1 .8^5 = 6(.2)(.8)^5 = .3932$$

$$p(2) = \begin{bmatrix} 6 \\ 2 \end{bmatrix} .2^2 .8^4 = 15(.2)^2(.8)^4 = .2458$$

$$p(3) = \begin{bmatrix} 6 \\ 3 \end{bmatrix} .2^3 .8^3 = 20(.2)^3(.8)^3 = .0819$$

$$p(4) = \begin{bmatrix} 6 \\ 4 \end{bmatrix} .2^4 .8^2 = 15(.2)^4(.8)^2 = .0154$$

$$p(5) = \begin{bmatrix} 6 \\ 5 \end{bmatrix} .2^5 .8^1 = 6(.2)^5(.8)^1 = .0015$$

$$p(6) = \begin{bmatrix} 6 \\ 6 \end{bmatrix} .2^6 .8^0 = 1(.2)^6 1 = .0001$$

The probability distribution of x in tabular form is

x	0	1	2	3	4	5	6
$p(x)$	.2621	.3932	.2458	.0819	.0154	.0015	.0001

$\mu = np = 6(.2) = 1.2$

The graph of the probability distribution of x when $n = 6$ and $p = .2$ follows:

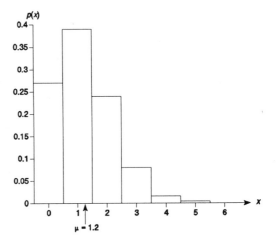

In part c, x is a binomial random variable with $n = 6$ and $p = .8$.

$$p(x) = \begin{bmatrix} 6 \\ x \end{bmatrix} .8^x.2^{6-x} \quad x = 0, 1, 2, 3, 4$$

$$p(0) = \begin{bmatrix} 6 \\ 0 \end{bmatrix} .8^0.2^6 = 1(1).2^6 = .0001$$

$$p(1) = \begin{bmatrix} 6 \\ 1 \end{bmatrix} .8^1.2^5 = 6(.8)(.2)^5 = .0015$$

$$p(2) = \begin{bmatrix} 6 \\ 2 \end{bmatrix} .8^2.2^4 = 15(.8)^2(.2)^4 = .0154$$

$$p(3) = \begin{bmatrix} 6 \\ 3 \end{bmatrix} .8^3.2^3 = 20(.8)^3(.2)^3 = .0819$$

$$p(4) = \begin{bmatrix} 6 \\ 4 \end{bmatrix} .8^4.2^2 = 15(.8)^4(.2)^2 = .2458$$

$$p(5) = \begin{bmatrix} 6 \\ 5 \end{bmatrix} .8^5.2^1 = 6(.8)^5(.2) = .3932$$

$$p(6) = \begin{bmatrix} 6 \\ 6 \end{bmatrix} .8^6.2^0 = 1(.8)^6(1) = .2621$$

The probability distribution of x in tabular form is

x	0	1	2	3	4	5	6
$p(x)$	.0001	.0015	.0154	.0819	.2458	.3932	.2621

Note: The distribution of x when $n = 6$ and $p = .2$ is the reverse of the distribution of x when $n = 6$ and $p = .8$.

$\mu = np = 6(.8) = 4.8$

The graph of the probability distribution of x when $n = 6$ and $p = .8$ follows:

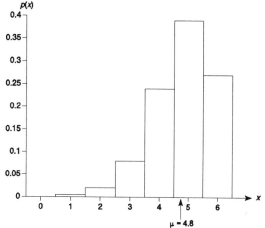

e.   In general, when $p = .5$, a binomial distribution will be symmetric regardless of the value of n. When p is less than $.5$, the binomial distribution will be skewed to the right; and when p is greater than $.5$, it will be skewed to the left. (Refer to parts **a**, **b**, and **c**).

4.39   Define the following events:

A: {Taxpayer is audited}
B: {Taxpayer has income less than \$25,000}
C: {Taxpayer has income of \$25,000 or more}
D: {Taxpayer has income of \$100,000 or more}

a.   From the information given in the problem,

$$P(A \mid B) = 6/1000 = .006$$
$$P(A \mid C) = 14/1000 = .014$$
$$P(A \mid D) = 46/1000 = .046$$

b.   Let x = number of taxpayers with incomes under \$25,000 who are audited. Then x is a binomial random variable with $n = 5$ and $p = .006$.

$$P(x = 1) = \binom{5}{1} \frac{5!}{1!(5-1)!} .006^1 .994^{(5-1)} = \frac{5 \cdot 4 \cdot 3 \cdot 2 \cdot 1}{1 \cdot 4 \cdot 3 \cdot 2 \cdot 1} .006^1 .994^4$$

$$= 5(.006)(.976215137) = .0293$$

$$P(x > 1) = 1 - [P(x = 0) + P(x = 1)]$$

$$= 1 - \left[\binom{5}{0} \frac{5!}{0!(5-0)!} .006^0 .994^{(5-0)} + .0293 \right]$$

$$= 1 - \left[\frac{5 \cdot 4 \cdot 3 \cdot 2 \cdot 1}{1 \cdot 5 \cdot 4 \cdot 3 \cdot 2 \cdot 1} .006^0 .994^5 + .0293 \right]$$

$$= 1 - [1(1)(.9704) + .0293] = 1 - .9997 = .0003$$

c.   Let x = number of taxpayers with incomes \$25,000 or more who are audited. Then x is a binomial random variable with $n = 5$ and $p = .014$.

$$P(x = 1) = \binom{5}{1} \frac{5!}{1!(5-1)!} .014^1 .986^{(5-1)} = \frac{5 \cdot 4 \cdot 3 \cdot 2 \cdot 1}{1 \cdot 4 \cdot 3 \cdot 2 \cdot 1} .014^1 .986^4$$

$$= 5(.014)(.945165062) = .0662$$

$$P(x > 1) = 1 - [P(x = 0) + P(x = 1)]$$

$$= 1 - \left[\binom{5}{0} \frac{5!}{0!(5-0)!} .014^0 .986^{(5-0)} + .0662 \right]$$

$$= 1 - \left[\frac{5 \cdot 4 \cdot 3 \cdot 2 \cdot 1}{1 \cdot 5 \cdot 4 \cdot 3 \cdot 2 \cdot 1} .014^0 .986^5 + .0662 \right]$$

$$= 1 - [1(1)(.9319) + .0662] = 1 - .9981 = .0019$$

d. Let x = number of taxpayers with incomes under \$25,000 who are audited. Then x is a binomial random variable with $n = 2$ and $p = .006$.

Let y = number of taxpayers with incomes \$100,000 or more who are audited. Then y is a binomial random variable with $n = 2$ and $p = .046$.

$$P(x = 0) = \binom{2}{0} \frac{2!}{0!(2-0)!} .006^0 .994^{(2-0)} = \frac{2 \cdot 1}{1 \cdot 2 \cdot 1} .006^0 .994^2$$
$$= 1(1)(.988036) = .988036$$

$$P(y = 0) = \binom{2}{0} \frac{2!}{0!(2-0)!} .046^0 .954^{(2-0)} = \frac{2 \cdot 1}{1 \cdot 2 \cdot 1} .046^0 .954^2$$
$$= 1(1)(.910116) = .910116$$

$$P(x = 0)P(y = 0) = .988036(.910116) = .8992$$

e. We must assume that the variables defined as x and y are binomial random variables. We must assume that the trials are identical, the probability of success is the same from trial to trial, and that the trials are independent.

4.41 a. We must assume that the probability that a specific type of ball meets the requirements is always the same from trial to trial and the trials are independent. To use the binomial probability distribution, we need to know the probability that a specific type of golf ball meets the requirements.

b. For a binomial distribution,

$$\mu = np$$
$$\sigma = \sqrt{npq}$$

In this example, n = two dozen = $2 \cdot 12 = 24$.

$p = .10$ (Success here means the golf ball *does not* meet standards.)
$q = .90$

$\mu = np = 24(.10) = 2.4$
$\sigma = \sqrt{npq} = \sqrt{24(.10)(.90)} = 1.47$

c. In this situation,

p = Probability of success
  = Probability golf ball *does* meet standards
  = .90
$q = 1 - .90 = .10$
$n = 24$
$E(x) = \mu = np = 24(.90) = 21.60$
$\sigma = \sqrt{npq} = \sqrt{24(.10)(.90)} = 1.47$ (Note that this is the same as in part **b**.)

. . . .
Discrete Random Variables

4.43 The random variable x = number of defective fuses is a binomial random variable with $n = 25$. We will accept a lot if $x < 3$ or $x \leq 2$.

Using Table II, Appendix B:

a. $P(\text{accepting a lot}) = P(x \leq 2) = 1$ when $p = 0$

b. $P(\text{accepting a lot}) = P(x \leq 2) = .998$ when $p = .01$

c. $P(\text{accepting a lot}) = P(x \leq 2) = .537$ when $p = .10$

d. $P(\text{accepting a lot}) = P(x \leq 2) = .009$ when $p = .30$

e. $P(\text{accepting a lot}) = P(x \leq 2) \approx 0$ when $p = .50$

f. $P(\text{accepting a lot}) = P(x \leq 2) \approx 0$ when $p = .80$

g. $P(\text{accepting a lot}) = P(x \leq 2) \approx 0$ when $p = .95$

h. $P(\text{accepting a lot}) = P(x \leq 2) \approx 0$ when $p = 1$

A graph of the operating characteristic curve for this sampling plan is shown at right.

4.45 a. If $n = 800$ and $p = .65$, $\mu = np = 800(.65) = 520$

$$\sigma = \sqrt{npq} = \sqrt{800(.65)(.35)} = 13.4907$$

b. Converting x to a z-score, we get

$$z = \frac{x - \mu}{\sigma} = \frac{472 - 520}{13.4907} = -3.56$$

If the newspaper's claim is true, it would be very unlikely to observe $x \leq 472$. For most distributions, most of the observations are within 3 standard deviations of the mean.

c. No. If the newspaper's claim is correct, we have seen a very unlikely event. Therefore, we would conclude that the newspaper's claim is probably incorrect.

4.47   a.   When $\lambda = 5$, $P(x \leq 3) = .265$

b.   When $\lambda = 3$, $P(x \leq 3) = .647$

c.   When $\lambda = 1$, $P(x \leq 3) = .981$

d.   It increases. As the expected value λ gets smaller, the $P(x \leq 3)$ would increase.

4.49   a.   To graph the Poisson probability distribution with $\lambda = 5$, we need to calculate $p(x)$ for $x = 0$ to 15. Using Table III, Appendix B,

$p(0) = .007$
$p(1) = P(x \leq 1) - P(x \leq 0) = .040 - .007 = .033$
$p(2) = P(x \leq 2) - P(x \leq 1) = .125 - .040 = .085$
$p(3) = P(x \leq 3) - P(x \leq 2) = .265 - .125 = .140$
$p(4) = P(x \leq 4) - P(x \leq 3) = .440 - .265 = .175$
$p(5) = P(x \leq 5) - P(x \leq 4) = .616 - .440 = .176$
$p(6) = P(x \leq 6) - P(x \leq 5) = .762 - .616 = .146$
$p(7) = P(x \leq 7) - P(x \leq 6) = .867 - .762 = .105$
$p(8) = P(x \leq 8) - P(x \leq 7) = .932 - .867 = .065$
$p(9) = P(x \leq 9) - P(x \leq 8) = .968 - .932 = .036$
$p(10) = P(x \leq 10) - P(x \leq 9) = .986 - .968 = .018$
$p(11) = P(x \leq 11) - P(x \leq 10) = .995 - .986 = .009$
$p(12) = P(x \leq 12) - P(x \leq 11) = .998 - .995 = .003$
$p(13) = P(x \leq 13) - P(x \leq 12) = .999 - .998 = .001$
$p(14) = P(x \leq 14) - P(x \leq 13) = 1.000 - .999 = .001$
$p(15) = P(x \leq 15) - P(x \leq 14) = 1.000 - 1.000 = .000$

The graph is shown at right:

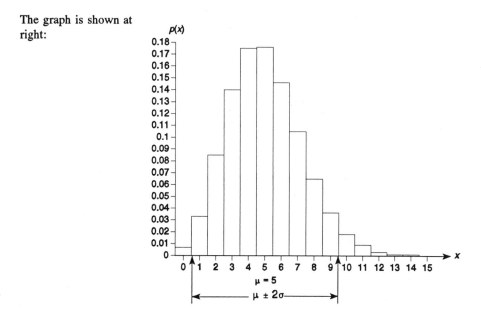

b. $\mu = \lambda = 5$

$\sigma = \sqrt{\lambda} = \sqrt{5} = 2.2361$

$\mu \pm 2\sigma \Rightarrow 5 \pm 2(2.2361) \Rightarrow 5 \pm 4.4722 \Rightarrow (.5278, 9.4722)$

c. $P(.5278 < x < 9.4722)$

$= P(1 \le x \le 9) = P(x \le 9) - P(x = 0)$

$= .968 - .007 = .961$

4.51 a. $\sigma = \sqrt{\sigma^2} = \sqrt{\lambda} = \sqrt{4} = 2$

b. $P(x > 10) = 1 - P(x \le 10)$

$= 1 - .977$ (Table III, Appendix B)

$= .003$

No. The probability that a sample of air from the plant exceeds the EPA limit is only .003. Since this value is very small, it is not very likely that this will occur.

c. The experiment consists of counting the number of parts per million of vinyl chloride in air samples. We must assume the probability of a part of vinyl chloride appearing in a million parts of air is the same for each million parts of air. We must also assume the number of parts of vinyl chloride in one million parts of air is independent of the number in any other one million parts of air.

4.53 $\mu = \lambda = 3.4$, using Table III, Appendix B:

$P(x = 2) = P(x \le 2) - P(x \le 1)$

$= .340 - .147 = .193$

$P(x \ge 3) = 1 - P(x \le 2)$

$= 1 - .340 = .660$

We need to assume that the probability that an accident occurs in a particular month is the same for all months. The number of accidents that occur in a particular month must all be independent of the number occurring in other months.

4.55 $\mu = \lambda = 3$, using Table III, Appendix B:

$P(x = 3) = P(x \le 3) - P(x \le 2) = .647 - .423 = .224$

$P(x = 0) = .050$

The probability that no bulbs fail in one hour is .050. If we let y = number of one hour intervals out of 8 that have no bulbs fail, then y is a binomial random variable with $n = 8$ and $p = .05$. Then, the probability that no bulbs fail in an 8 hour shift is

$$P(y = 8) = \begin{pmatrix} 8 \\ 8 \end{pmatrix} .05^8 .95^{(8-8)} = \frac{8!}{8!(8-8)!} .05^8 .95^0$$

$$= \frac{8 \cdot 7 \cdot 6 \cdot 5 \cdot 4 \cdot 3 \cdot 2 \cdot 1}{8 \cdot 7 \cdot 6 \cdot 5 \cdot 4 \cdot 3 \cdot 2 \cdot 1 \cdot 1} .05^8 .95^0 = .05^8$$

We must assume that the 8 one-hour intervals are independent and identical, and that the probability that no bulbs fail is the same for each one-hour interval.

4.57 Let x = the number of major medical claims the health insurance company must pay.

The random variable x is a binomial random variable with $n = 1000$, $p = .001$, and $q = 1 - .001 = .999$.

$$np = 1000(.001) = 1 \text{ which is } \leq 7$$

Therefore, the Poisson probability distribution provides a good approximation to the binomial probability distribution.

$$
\begin{aligned}
P(x \geq 1) &= 1 - P(x = 0) \\
&= 1 - .368 \qquad \text{Table III, Appendix B, with } \lambda = 1 \\
&= .632
\end{aligned}
$$

4.59 $p(x) = \begin{bmatrix} n \\ x \end{bmatrix} p^x q^{n-x} \quad x = 0, 1, 2, \ldots, n$

a. $P(x = 3) = p(3) = \begin{bmatrix} 7 \\ 3 \end{bmatrix} .5^3 .5^4 = \dfrac{7!}{3!4!} .5^3 .5^4 = 35(.125)(.0625) = .2734$

b. $P(x = 3) = p(3) = \begin{bmatrix} 4 \\ 3 \end{bmatrix} .8^3 .2^1 = \dfrac{4!}{3!1!} .8^3 .2^1 = 4(.512)(.2) = .4096$

c. $P(x = 1) = p(1) = \begin{bmatrix} 15 \\ 1 \end{bmatrix} .1^1 .9^{14} = \dfrac{15!}{1!14!} .1^1 .9^{14} = 15(.1)(.228768) = .3432$

4.61 Using Table III, Appendix B:

a. When $\lambda = 14$, $p(11) = P(x = 11) = P(x \leq 11) - P(x \leq 10) = .260 - .176 = .084$

b. When $\lambda = 11$, $p(14) = P(x = 14) = P(x \leq 14) - P(x \leq 13) = .854 - .781 = .073$

c. When $\lambda = 13$, $p(7) = P(x = 7) = P(x \leq 7) - P(x \leq 6) = .054 - .026 = .028$

4.63 a. Discrete — The number of damaged inventory items is countable.

b. Continuous — The average monthly sales can take on any value within an acceptable limit.

c. Continuous — The number of square feet can take on any positive value.

d. Continuous — The length of time we must wait can take on any positive value.

4.65 a. $E(x) = \sum x p(x)$

$$
\begin{aligned}
&= 37,500(.14) + 112,500(.3) + 225,000(.2) + 400,000(.17) + 750,000(.1) \\
&\quad + 3,000,000(.09) \\
&= 5250 + 33,750 + 45,000 + 68,000 + 75,000 + 270,000 \\
&= 497,000
\end{aligned}
$$

. . . .
Discrete Random Variables

b. $E[(x - \mu)^2] = \sum (x - \mu)^2 p(x)$

$$= (37,500 - 497,000)^2(.14) + (112,500 - 497,000)^2(.3)$$
$$+ (225,000 - 497,000)^2(.2) + (400,000 - 497,000)^2(.17)$$
$$+ (750,000 - 497,000)^2(.1) + (3,000,000 - 497,000)^2(.09)$$
$$= 29,559,635,000 + 44,352,075,000 + 14,796,800,000 + 1,599,530,000$$
$$+ 6,400,900,000 + 563,850,810,000$$
$$= 660,559,750,000$$
$$= 6.6055975 \times 10^{11}$$

c. The average sales volume for the fabricare firms is $497,000. This is what we expect the sales volume of a randomly selected fabricare firm to be.

The variance in the sales volume of the fabricare firms is 6.6056×10^{11}. This is a measure of the spread of the sales volumes of the firms.

4.67 a. For company A,

$$E(x) = \sum_{\text{All } x} xp(x) = 2(.05) + 3(.15) + 4(.20) + 5(.35) + 6(.25)$$
$$= .10 + .45 + .80 + 1.75 + 1.50$$
$$= 4.60$$

For company B,

$$E(x) = \sum_{\text{All } x} xp(x) = 2(.15) + 3(.30) + 4(.30) + 5(.20) + 6(.05)$$
$$= .30 + .90 + 1.20 + 1.00 + .30$$
$$= 3.70$$

b. The expected profit equals the expected value of x times the profit for each job.

For company A,

$$4.6(\$10,000) = \$46,000$$

For company B,

$$3.7(\$15,000) = \$55,500$$

c. For company A,

$$\sigma^2 = \sum_{\text{All } x} (x - \mu)^2 p(x) = (2 - 4.6)^2.05 + (3 - 4.6)^2.15 + (4 - 4.6)^2.20$$
$$+ (5 - 4.6)^2.35 + (6 - 4.6)^2.25$$
$$= .338 + .384 + .072 + .056 + .49$$
$$= 1.34$$

$$\sigma = \sqrt{\sigma^2} = \sqrt{1.34} = 1.16$$

For company B,

$$\sigma^2 = \sum_{\text{All } x} (x - \mu)^2 p(x) = (2 - 3.7)^2.15 + (3 - 3.7)^2.30 + (4 - 3.7)^2.30$$
$$+ (5 - 3.7)^2.20 + (6 - 3.7)^2.05$$
$$= .4335 + .147 + .027 + .338 + .2645$$
$$= 1.21$$

$$\sigma = \sqrt{\sigma^2} = \sqrt{1.21} = 1.10$$

d. For company A, the graph of $p(x)$ is given at right.

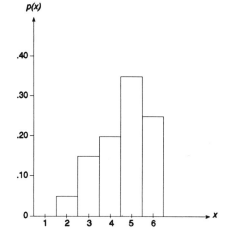

For company A,

$$\mu \pm 2\sigma \Rightarrow 4.6 \pm 2(1.16) \Rightarrow 4.6 \pm 2.32 \Rightarrow (2.28, 6.92)$$

$$P(2.28 < x < 6.92) = p(3) + p(4) + p(5) + p(6)$$
$$= .15 + .20 + .35 + .25$$
$$= .95$$

For company B, the graph of $p(x)$ is given at right.

For company B,

$$\mu \pm 2\sigma \Rightarrow 3.70 \pm 2(1.10) \Rightarrow 3.67 \pm 2.2 \Rightarrow (1.5, 5.9)$$

$$P(1.5 < x < 5.9) = p(2) + p(3) + p(4) + p(5)$$
$$= .15 + .30 + .30 + .20$$
$$= .95$$

4.69   Let x = the number of defective units in the sample. The random variable x is a binomial random variable with $n = 10$ and $p = .11$.

$$p(x) = \begin{bmatrix} 10 \\ x \end{bmatrix} .11^x .89^{10-x} \quad x = 0, 1, 2, \dots , 10$$

P(correct decision)

$= P$(reject the lot)   (since more than 10% of the units are defective)

$= P(x \geq 2)$

$= 1 - P(x \leq 1)$

$= 1 - [p(0) + p(1)]$

$$= 1 - \left[\begin{bmatrix} 10 \\ 0 \end{bmatrix} .11^0 .89^{10} + \begin{bmatrix} 10 \\ 1 \end{bmatrix} .11^1 .89^9 \right]$$

$= 1 - [.312 + .385]$

$= 1 - .697$

$= .303$

4.71   Let x = the number of typewriters the outlet sells tomorrow. The random variable x is a Poisson random variable with $\mu = \lambda = 2.4$. Using Table III, Appendix B:

P(outlet runs out of typewriters tomorrow)

$= P(x > 5)$

$= 1 - P(x \leq 5)$

$= 1 - .964$

$= .036$

4.73   a.   Let x = the number of sales made today and y = the number of sales made tomorrow. The random variable x is a binomial random variable with $n = 5$ and $p = .7$, and y is a binomial random variable with $n = 20$ and $p = .7$

$P(x = 4 \cap y > 10)$

$= P(x = 4)P(y > 10)$ (Assuming the number of sales is independent from day to day.)

$= [P(x \leq 4) - P(x \leq 3)][1 - P(y \leq 10)]$

$= (.832 - .472)(1 - .048)$ (Table II, Appendix B)

$= .36(.952)$

$= .34272$

b.   Let x = the number of sales made in two days. The random variable x is a binomial random variable with $n = 9$ and $p = .7$.

$$P(x = 2) = P(x \leq 2) - P(x \leq 1)$$
$$= .004 - .000$$
$$= .004 \quad \text{(Table II, Appendix B)}$$

4.75   The random variable x is a binomial random variable with $n = 25$, $p = .20$, and $q = 1 - p = .80$. (Assuming that whether a person refuses to take part in the poll is independent of any other person refusing.)

a. $\mu = np = 25(.20) = 5$
$\sigma^2 = npq = 25(.20)(.80) = 4$

b. $P(x \leq 5) = .617$ \quad Table II, Appendix B

c. $P(x > 10) = 1 - P(x \leq 10)$
$= 1 - .994$
$= .006$

CHAPTER FIVE

· ·

Continuous Random Variables

5.1   a. $f(x) = \dfrac{1}{d - c}$ $(c \leq x \leq d)$

$$\frac{1}{d - c} = \frac{1}{45 - 20} = \frac{1}{25} = .04$$

So, $f(x) = .04$ $(20 \leq x \leq 45)$.

       b. $\mu = \dfrac{c + d}{2} = \dfrac{20 + 45}{2} = \dfrac{65}{2} = 32.5$

$$\sigma = \frac{d - c}{\sqrt{12}} = \frac{45 - 20}{\sqrt{12}} = 7.2169$$

$$\sigma^2 = (7.2169)^2 = 52.0833$$

       c.

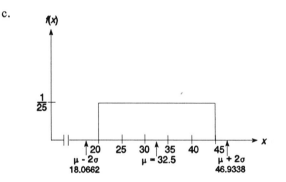

$$\mu \pm 2\sigma \Rightarrow 32.5 \pm 2(7.2169) \Rightarrow (18.0662, 46.9338)$$

$$P(18.0662 < x < 49.9338) = P(20 < x < 45)$$
$$= (45 - 20).04 = 1$$

5.3   a. $f(x) = \dfrac{1}{d - c}$ $(c \leq x \leq d)$

$$\frac{1}{d - c} = \frac{1}{7 - 3} = \frac{1}{4}$$

$$f(x) = \frac{1}{4} \qquad (3 \leq x \leq 7)$$

b. $\mu = \dfrac{c + d}{2} = \dfrac{3 + 7}{2} = \dfrac{10}{2} = 5$

$\sigma = \dfrac{d - c}{\sqrt{12}} = \dfrac{7 - 3}{\sqrt{12}} = \dfrac{4}{\sqrt{12}} = 1.155$

$\sigma^2 = (1.155)^2 = 1.333$

c.

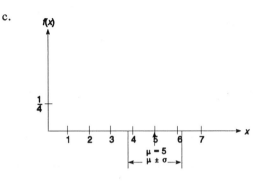

$\mu \pm \sigma \Rightarrow 5 \pm 1.155 \Rightarrow (3.845, 6.155)$

$P(a < x < b) = P(3.845 < x < 6.155) = \dfrac{b - a}{d - c} = \dfrac{6.155 - 3.845}{7 - 3} = \dfrac{2.31}{4}$
$= .5775$

5.5 $f(x) = \dfrac{1}{d - c} = \dfrac{1}{200 - 100} = \dfrac{1}{100} = .01 \quad (100 \le x \le 200)$

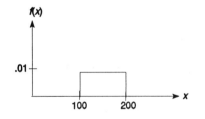

$\mu = \dfrac{c + d}{2} = \dfrac{100 + 200}{2} = \dfrac{300}{2} = 150$

$\sigma = \dfrac{d - c}{\sqrt{12}} = \dfrac{200 - 100}{\sqrt{12}} = \dfrac{100}{\sqrt{12}} = 28.8675$

a. $\mu \pm 2\sigma \Rightarrow 150 \pm 2(28.8675) \Rightarrow 150 \pm 57.735 \Rightarrow (92.265, 207.735)$

$P(x < 92.265) + P(x > 207.735) = P(x < 100) + P(x > 200)$
$= \quad 0 \quad + \quad 0$
$= 0$

b. $\mu \pm 3\sigma \Rightarrow 150 \pm 3(28.8675) \Rightarrow 150 \pm 86.6025 \Rightarrow (63.3975, 236.6025)$

$P(63.3975 < x < 236.6025) = P(100 < x < 200) = (200 - 100)(.01) = 1$

c. From a, $\mu \pm 2\sigma \Rightarrow (92.265, 207.735)$.

$P(92.265 < x < 207.735) = P(100 < x < 200) = (200 - 100)(.01) = 1$

5.7 To construct a relative frequency histogram for the data, we can use 7 measurement classes.

$$\text{Interval width} = \frac{\text{Largest number} - \text{smallest number}}{\text{Number of classes}}$$

$$= \frac{98.0716 - .7434}{7} = 13.9$$

We will use an interval width of 14 and a starting value of .74335.

The measurement classes, frequencies, and relative frequencies are given in the table below.

Class	Measurement Class	Class Frequency	Class Relative Frequency
1	.74335 − 14.74335	6	6/40 = .15
2	14.74335 − 28.74335	4	.10
3	28.74335 − 42.74335	6	.15
4	42.74335 − 56.74335	6	.15
5	56.74335 − 70.74335	5	.125
6	70.74335 − 84.74335	4	.10
7	84.74335 − 98.74335	9	.225
		40	1.000

The histogram looks like the data could be from a uniform distribution. The last class (84.74335 − 98.74335) has a few more observations in it than we would expect. However, we cannot expect a perfect graph from a sample of only 40 observations.

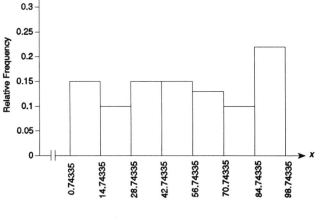

5.9 Let x = the closing price of a particular stock. The random variable x is a uniform random variable.

$$f(x) = \frac{1}{d-c} \quad (c \le x \le d)$$

$$\frac{1}{d-c} = \frac{1}{40-30} = \frac{1}{10} = .1$$

So, $f(x) = .1 \ (30 \le x \le 40)$

a.

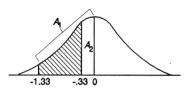

b. $P(x > 35) = (40 - 35)(.1) = .5$

$P(x > 38) = (40 - 38)(.1) = .2$

$P(34 \le x \le 36) = (36 - 34)(.1) = .2$

$P(x > 50) = 0$ (Since $30 \le x \le 40$)

5.11 Let x = distance the groove is from the end of the spindle. Then x has a uniform distribution with $c = 0$ and $d = 18$.

$$f(x) = \frac{1}{d-c} \quad (c \le x \le d)$$

$$f(x) = \frac{1}{18-0} = \frac{1}{18} \quad (0 \le x \le 18)$$

The probability that the spindle can be salvaged is

$$P(x \le 4) + P(x \ge 14) = (4 - 0)\frac{1}{18} + (18 - 14)\frac{1}{18} = \frac{4}{18} + \frac{4}{18} = \frac{8}{18} = .4444$$

5.13 It is helpful to draw sketches to assist in finding the required areas. The areas can be found using Table IV, Appendix B.

a. The area between $z = -1.33$ and $z = -.33$ is equal to $A_1 - A_2$,
where A_1 = area between -1.33 and $0 = .4082$
A_2 = area between $-.33$ and 0
$= .1293$

Thus, the required area is
$.4082 - .1293 = .2789$

b. The area between $z = -2$ and $z = 2$ is equal to

$$A_1 + A_2 = .4772 + .4772$$
$$= .9544$$

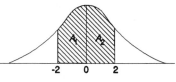

c. The area between $z = -2.33$ and $z = -1.33$ is equal to

$$A_1 - A_2 = .4901 - .4082$$
$$= .0819$$

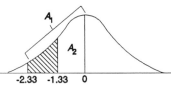

d. The area between $z = -3$ and $z = 3$ is equal to

$$A_1 + A_2 = .4987 + .4987$$
$$= .9974$$

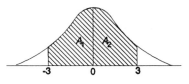

e. The area between $z = -2.25$ and $z = 0$ is equal to

$$A_1 = .4878$$

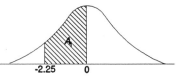

f. The area between $z = 0$ and $z = 2.25$ is equal to

$$A_1 = .4878$$

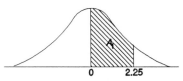

5.15 a. $P(z = 1) = 0$, since a single point does not have an area.

b. $P(z \leq 1) = P(z \leq 0) + P(0 < z \leq 1)$
$ = A_1 + A_2$
$ = .5 + .3413$ (Table IV, Appendix B)
$ = .8413$

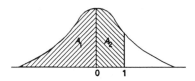

c. $P(z < 1) = P(z \leq 1) = .8413$ (Refer to part b.)

d. $P(z > 1) = 1 - P(z \leq 1) = 1 - .8413 = .1587$ (Refer to part b.)

5.17 Using Table IV in Appendix B:

a. $P(-1 \leq z < 1) = 2P(0 \leq z \leq 1)$
$\phantom{P(-1 \leq z < 1)} = 2(.3413) = .6826$

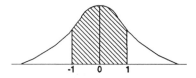

b. $P(-1.96 < z < 1.96) = 2P(0 < z < 1.96)$
$$= 2(.4750) = .95$$

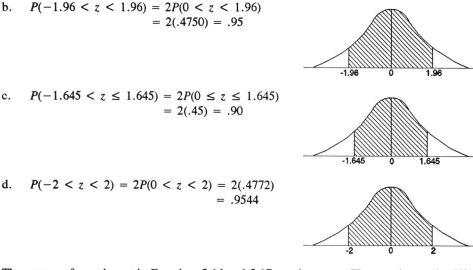

c. $P(-1.645 < z \leq 1.645) = 2P(0 \leq z \leq 1.645)$
$$= 2(.45) = .90$$

d. $P(-2 < z < 2) = 2P(0 < z < 2) = 2(.4772)$
$$= .9544$$

The answers for each part in Exercises 5.16 and 5.17 are the same. The questions only differ by the sign ("$<$" or "$\leq$"). These are equivalent events when using the normal distribution.

5.19   Using Table IV of Appendix B:

a. $P(z \leq z_0) = .2090$

$A = .5000 - .2090 = .2910$

Look up the area .2910 in the body of Table IV;
$z_0 = -.81$.

(z_0 is negative since the graph shows z_0 is on the left side of 0.)

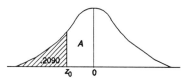

b. $P(z \leq z_0) = .7090$

$P(z \leq z_0) = P(z \leq 0) + P(0 \leq z \leq z_0)$
$$= .5 + P(0 \leq z \leq z) = .7090$$

Therefore, $P(0 \leq z \leq z_0) = .7090 - .5 = .2090$

Look up the area .2090 in the body of Table IV;
$z_0 \approx .55$.

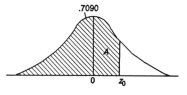

c. $P(-z_0 \leq z < z_0) = .8472$

$P(-z_0 \leq z < z_0) = 2P(0 \leq z \leq z_0)$
$2P(0 \leq z \leq z_0) = .8472$

Therefore, $P(0 \leq z \leq z_0) = .4236$.

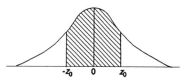

Look up the area .4236 in the body of Table IV; $z_0 = 1.43$.

d. $P(-z_0 \le z < z_0) = .1664$

$P(-z_0 \le z \le z_0) = 2P(0 \le z \le z_0)$
$2P(0 \le z \le z_0) = .1664$

Therefore, $P(0 \le z \le z_0) = .0832$.

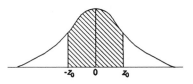

Look up the area .0832 in the body of Table IV; $z_0 = .21$.

e. $P(z_0 \le z \le 0) = .4798$

$P(z_0 \le z \le 0) = P(0 \le z \le -z_0)$

Look up the area .4798 in the body of Table IV;
$z_0 = -2.05$.

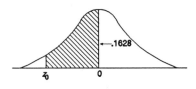

f. $P(-1 < z < z_0) = .5328$

$P(-1 < z < z_0)$
$\quad = P(-1 < z < 0) + P(0 < z < z_0)$
$\quad = .5328$

$P(0 < z < 1) + P(0 < z < z_0) = .5328$

Thus, $P(0 < z < z_0) = .5328 - .3413 = .1915$

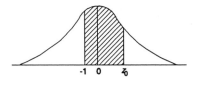

Look up the area .1915 in the body of Table IV; $z_0 = .50$.

5.21 From Exercise 5.20, x is described by a normal distribution with $\mu = 30$ and $\sigma = 4$. The number of standard deviations away from the mean of x is the z-score for the given x value.

a. If $x = 25$, $z = \dfrac{x - \mu}{\sigma} = \dfrac{25 - 30}{4} = \dfrac{-5}{4} = -1.25$

Therefore, x is 1.25 standard deviations below the mean

b. If $x = 37.5$, $z = \dfrac{x - \mu}{\sigma} = \dfrac{37.5 - 30}{4} = \dfrac{7.5}{4} = 1.875$

Therefore, x is 1.875 standard deviations above the mean.

c. If $x = 30$, $z = \dfrac{x - \mu}{\sigma} = \dfrac{30 - 30}{4} = 0$

Therefore, x is 0 standard deviations from the mean (x is the mean).

d. If $x = 36$, $z = \dfrac{x - \mu}{\sigma} = \dfrac{36 - 30}{4} = \dfrac{6}{4} = 1.5$

Therefore, x is 1.5 standard deviations above the mean.

5.23   Using Table IV of Appendix B:

a.   To find the probability that x assumes a value more than 2 standard deviations from μ:

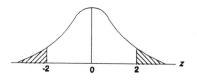

$P(x < \mu - 2\sigma) + P(x > \mu + 2\sigma)$
$\quad = P(z < -2) + P(z > 2)$
$\quad = 2P(z > 2)$
$\quad = 2(.5000 - .4772)$
$\quad = 2(.0228) = .0456$

To find the probability that x assumes a value more than 3 standard deviations from μ:

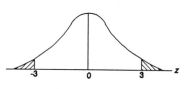

$P(x < \mu - 3\sigma) + P(x > \mu + 3\sigma)$
$\quad = P(z < -3) + P(z > 3)$
$\quad = 2P(z > 3)$
$\quad = 2(.5000 - .4987)$
$\quad = 2(.0013) = .0026$

b.   To find the probability that x assumes a value within 1 standard deviation of its mean:

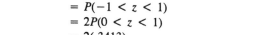

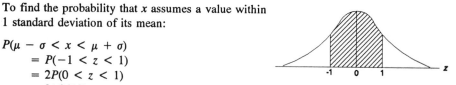

$P(\mu - \sigma < x < \mu + \sigma)$
$\quad = P(-1 < z < 1)$
$\quad = 2P(0 < z < 1)$
$\quad = 2(.3413)$
$\quad = .6826$

To find the probability that x assumes a value within 2 standard deviations of μ:

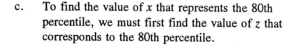

$P(\mu - 2\sigma < x < \mu + 2\sigma)$
$\quad = P(-2 < z < 2)$
$\quad = 2P(0 < z < 2)$
$\quad = 2(.4772)$
$\quad = .9544$

c.   To find the value of x that represents the 80th percentile, we must first find the value of z that corresponds to the 80th percentile.

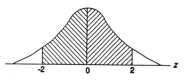

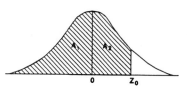

$P(z < z_0) = .80$. Thus, $A_1 + A_2 = .80$. Since $A_1 = .50$, $A_2 = .80 - .50 = .30$. Using the body of Table IV, $z_0 = .84$. To find x, we substitute the values into the z-score formula:

$$z = \frac{x - \mu}{\sigma}$$

$$.84 = \frac{x - 1000}{10} \Rightarrow x = .84(10) + 1000 = 1008.4$$

To find the value of x that represents the 10th percentile, we must first find the value of z that corresponds to the 10th percentile.

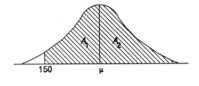

$P(z < z_0) = .10$. Thus, $A_1 = .50 - .10 = .40$. Using the body of Table IV, $z_0 = -1.28$. To find x, we substitute the values into the z-score formula:

$$z = \frac{x - \mu}{\sigma}$$

$$-1.28 = \frac{x - 1000}{10} \Rightarrow x = -1.28(10) + 1000 = 987.2$$

5.25   The random variable x has a normal distribution with $\sigma = 25$.

We know $P(x > 150) = .90$. So, $A_1 + A_2 = .90$. Since $A_2 = .50$, $A_1 = .90 - .50 = .40$. Look up the area .40 in the body of Table IV; (take the closest value) $z_0 = -1.28$.

To find μ, substitute all the values into the z-score formula:

$$z = \frac{x - \mu}{\sigma}$$

$$-1.28 = \frac{150 - \mu}{\sigma}$$

$$\mu = 150 + 25(1.28) = 182$$

5.27   Let x = the lifetimes of the participants in the plan. The random variable x is approximately normal with $\mu = 68$ and $\sigma = 3.5$.

a. $z = \frac{x - \mu}{\sigma} = \frac{70 - 68}{3.5} = .57$

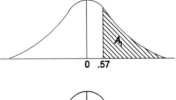

$P(x > 70) = P(z > .57)$
$= .5000 - .2157$
$= .2843$ (Table IV, Appendix B)

b. $z = \frac{x - \mu}{\sigma} \quad \frac{75 - 68}{3.5} = 2$

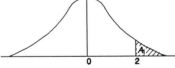

$P(x > 75) = P(z > 2)$
$= .5000 - .4772$
$= .0228$ (Table IV, Appendix B)

c. Only 15% of plan participants will receive payment beyond age 71.64.

We must find the age, x_0, such that $P(x > x_0)$
$= P(z > z_0) = .15$. $A_1 = .5 - .15 = .35$. From
the body of Table IV, $z_0 = 1.04$.

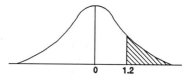

Using the z-score formula,

$$z = \frac{x - \mu}{\sigma}$$

$$1.04 = \frac{x - 68}{3.5} \Rightarrow x = 1.04(3.5) + 68 = 71.64$$

5.29 Let x = the wage rates. Then x is normally distributed with $\mu = 10.50$ and $\sigma = 1.25$.

a. $z = \frac{x - \mu}{\sigma} = \frac{12 - 10.5}{1.25} = 1.2$

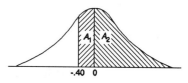

Thus, $P(x > 12)$
$= P(z > 1.2)$
$= .5000 - .3849$ (Table IV, Appendix B)
$= .1151$

b. $z = \frac{x - \mu}{\sigma} = \frac{10 - 10.5}{1.25} = -.40$

Thus, $P(x > 10)$
$= P(z > -.40)$
$= A_1 + A_2$
$= P(-.40 < z < 0) + P(z > 0)$
$= P(0 < z < .40) + .5000$ (Table IV, Appendix B)
$= .6554$

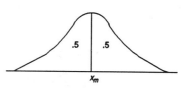

c. $P(x \geq x_m) = P(x \leq x_m) = .5$

Therefore, $x_m = \mu = 10.50$

(Recall from Section 2.4 that in a symmetric
distribution, the mean equals the median.)

5.31 Let x = monthly rate of return to stock ABC and y = monthly rate of return to stock XYZ.
The random variable x is normally distributed with $\mu = .05$ and $\sigma = .03$ and y is normally
distributed with $\mu = .07$ and $\sigma = .05$. You have $100 invested in each stock.

a. The average monthly rate of return for ABC stock is .05.

The average monthly rate of return for XYZ stock is .07.

Therefore, stock XYZ has the higher average monthly rate of return.

b. $E(x) = .05$ for each $1.

Since we have $100 invested in stock ABC, the monthly rate of return would be $100(.05) = $5.

Therefore, the expected value of the investment in stock ABC at the end of 1 month is $100 + 5 = $105.

$E(y) = .07$ for each $1.

Since we have $100 invested in stock XYZ, the monthly rate of return would be $100(.07) = $7.

Therefore, the expected value of the investment in stock XYZ at the end of 1 month is $100 + 7 = $107.

c.   We need to find the probability of incurring a loss for each stock and compare them.

P(incurring a loss on stock ABC)   P(incurring a loss on stock XYZ)
= P(monthly rate of return is   = P(monthly rate of return is
    negative on stock ABC)   negative on stock XYZ)
= $P(x < 0)$ = $P(y < 0)$

$$z = \frac{x - \mu}{\sigma} = \frac{0 - .05}{.03} = -1.67 \qquad z = \frac{y - \mu}{\sigma} = \frac{0 - .07}{.05} = -1.4$$

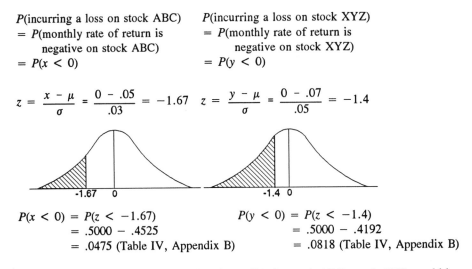

$$\begin{aligned} P(x < 0) &= P(z < -1.67) \\ &= .5000 - .4525 \\ &= .0475 \text{ (Table IV, Appendix B)} \end{aligned} \qquad \begin{aligned} P(y < 0) &= P(z < -1.4) \\ &= .5000 - .4192 \\ &= .0818 \text{ (Table IV, Appendix B)} \end{aligned}$$

Since the probability of incurring a loss is smaller for stock ABC, stock ABC would have a greater protection against occurring a loss next month.

5.33   Let x = the amount of dye discharged. The random variable x is normally distributed with $\sigma^2 = .16 \ (\sigma = .4)$.

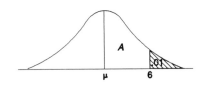

We want P(shade is unacceptable) $\leq .01$

$\Rightarrow P(x > 6) \leq .01$

Then $A = .50 - .01 = .49$. Look up the area .49 in the body of Table IV, Appendix B; (take the closest value) $z_0 = 2.33$.

To find μ, substitute into the z-score formula:

$$z = \frac{x - \mu}{\sigma}$$

$$2.33 = \frac{6 - \mu}{.4}$$

$$\mu = 6 - .4(2.33) = 5.068$$

5.35   a.   If $\lambda = 1$, $a = 1$, then $e^{-\lambda a} = e^{-1} = .367879$

        b.   If $\lambda = 1$, $a = 2.5$, then $e^{-\lambda a} = e^{-2.5} = .082085$

        c.   If $\lambda = 2.5$, $a = 3$, then $e^{-\lambda a} = e^{-7.5} = .000553$

        d.   If $\lambda = 5$, $a = .3$, then $e^{-\lambda a} = e^{-1.5} = .223130$

5.37   Using Table V in Appendix B:

        a. $P(x \leq 3) = 1 - P(x > 3) = 1 - e^{-2.5(3)} = 1 - e^{-7.5} = 1 - .000553 = .999447$

        b. $P(x \leq 4) = 1 - P(x > 4) = 1 - e^{-2.5(4)} = 1 - e^{-10} = 1 - .000045 = .999955$

        c. $P(x \leq 1.6) = 1 - P(x > 1.6) = 1 - e^{-2.5(1.6)} = 1 - e^{-4} = 1 - .018316 = .981684$

        d. $P(x \leq .4) = 1 - P(x > .4) = 1 - e^{-2.5(.4)} = 1 - e^{-1} = 1 - .367879 = .632121$

5.39 $f(x) = \lambda e^{-\lambda x} = e^{-x}$ $(x > 0)$

$$\mu = \frac{1}{\lambda} = \frac{1}{1} = 1, \ \sigma = \frac{1}{\lambda} = \frac{1}{1} = 1$$

        a. $\mu \pm 3\sigma \Rightarrow 1 \pm 3(1) \Rightarrow (-2, 4)$

              Since $\mu - 3\sigma$ lies below 0, find the probability that x is more than $\mu + 3\sigma = 4$.

              $P(x > 4) = e^{-1(4)} = e^{-4} = .018316$ (using Table V in Appendix B)

        b. $\mu \pm 2\sigma \Rightarrow 1 \pm 2(1) \Rightarrow (-1, 3)$

              Since $\mu - 2\sigma$ lies below 0, find the probability that x is between 0 and 3.

$$P(x < 3) = 1 - P(x \geq 3) = 1 - e^{-1(3)} = 1 - e^{-3} = 1 - .049787 = .950213$$
                                (using Table V in Appendix B)

        c. $\mu \pm .5\sigma \Rightarrow 1 \pm .5(1) \Rightarrow (.5, 1.5)$

$$P(.5 < x < 1.5) = P(x > .5) - P(x > 1.5)$$
$$= e^{-.5} - e^{-1.5}$$
$$= .606531 - .223130$$
$$= .383401 \quad \text{(using Table V in Appendix B)}$$

5.41   Let x = the shelf-life of bread. The mean of an exponential distribution is $\mu = 1/\lambda$. We know that $\mu = 2$; therefore, $\lambda = .5$.

We want to find:

$$P(x > 3) = e^{-.5(3)} = e^{-1.5} = .223130 \quad \text{(using Table V in Appendix B)}$$

5.43   Let x = the number of minutes to treat a patient in the emergency room. The mean of an exponential distribution is $\mu = 1/\lambda$. We know that $\mu = 58$; therefore, $\lambda = 1/58 = .0172$.

a. $P(x > 58) = e^{-.0172(58)} = e^{-1} = .367879$

$$P(x > 1.5(60)) = P(x > 90) = e^{-\lambda a} = e^{-.0172(90)}$$
$$= e^{-1.55} = .212248$$
$$\text{(using Table V in Appendix B)}$$

b. $P(x > 58)P(x > 58)P(x > 58) = .367879^3 = .049787$
$$\text{(using Table V in Appendix B)}$$

c.   The median time to treat a patient is less than 58 minutes since $P(x > 58) = .367879$ and the probability gets larger for smaller x values.

d. $e^{-.7}$ is closest to .5 on Table V in Appendix B.

Therefore, $.7 = \lambda a$
$.7 = .01724a$
$a = 40.6$

Thus, the median is approximately 40.6.

5.45   a. $R(x) = e^{-\lambda x} = R(x) = e^{-.5x}$

b. $P(x \geq 4) = e^{-.5(4)} = e^{-2} = .135335$ (Table V, Appendix B)

c. $\mu = \dfrac{1}{\lambda} = \dfrac{1}{.5} = 2$

$P(x > \mu) = P(x > 2) = e^{-.5(2)} = e^{-1} = .367879$ (Table V, Appendix B)

d.   For all exponential distributions, $\mu = \dfrac{1}{\lambda}$

$P(x > \mu) = P\left[x > \dfrac{1}{\lambda}\right] = e^{-\lambda(1/\lambda)} = e^{-1} = .367879$. Thus, regardless of the value of λ, the probability that x is larger than the mean is always .367879.

e. $P(x > 5) = e^{-.5(5)} = e^{-2.5} = .082085$ (Table V, Appendix B)

If 10,000 units are sold, approximately $10,000(.082085) = 820.85$ will perform satisfactorily for more than 5 years.

$$P(x \leq 1) = 1 - P(x > 1) = 1 - e^{-.5(1)} = 1 - e^{-.5} = 1 - .606531 = .39469$$

If 10,000 units are sold, approximately $10,000(.393469) = 3934.69$ will fail within 1 year.

f. $P(x < a) \leq .05$
$\Rightarrow 1 - P(x \geq a) \leq .05$
$\Rightarrow P(x \geq a) \geq .95$
$\Rightarrow e^{-.5a} \geq .95$

Using Table V, Appendix B, $e^{-.05}$ is closest to .95 (yet larger).

Thus, $.05 = .5a \Rightarrow a = .1$

The warranty should be for approximately .1 year or $.1(365) = 36.5$ or 37 days.

5.47 It is appropriate to approximate a binomial distribution with a normal distribution when n is large. As a rule of thumb: the interval $\mu \pm 3\sigma$ should lie in the range 0 to n in order for the normal approximation to be adequate.

5.49 a. In order to approximate the binomial distribution with the normal distribution, the interval $\mu \pm 3\sigma \Rightarrow np \pm 3\sqrt{npq}$ should lie in the range 0 to n.

When $n = 25$ and $p = .7$,

$$np \pm 3\sqrt{npq} \Rightarrow 25(.7) \pm 3\sqrt{25(.7)(1 - .7)}$$
$$\Rightarrow 17.5 \pm 3\sqrt{5.25}$$
$$\Rightarrow 17.5 \pm 6.87$$
$$\Rightarrow (10.63, 24.37)$$

Since the interval calculated does lie in the range 0 to 25, we can use the normal approximation.

b. $\mu = np = 25(.7) = 17.5$
$\sigma^2 = npq = 25(.7)(.3) = 5.25$

c. $P(x \geq 15) = 1 - P(x \leq 14)$
$= 1 - .098$
$= .902$ (Table II, Appendix B)

d. $z = \dfrac{(a - .5) - \mu}{\sigma} = \dfrac{14.5 - 17.5}{\sqrt{5.25}} = -1.31$

$P(x \geq 9) \approx P(z \geq -1.31)$
$= .5000 + .4049$
$= .9049$
(Using Table IV in Appendix B.)

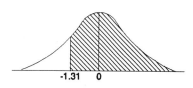

-1.31   0

5.51 x is a binomial random variable with $n = 1000$ and $p = .50$.

$$\mu \pm 3\sigma \Rightarrow np \pm 3\sqrt{npq} \Rightarrow 1000(.50) \pm 3\sqrt{1000(.5)(.5)}$$
$$\Rightarrow 500 \pm 3(15.8114)$$
$$\Rightarrow (452.5658, 547.4342)$$

Since the interval lies in the range 0 to 1000, we can use the normal approximation to approximate the probabilities.

a. $z = \dfrac{(a + .5) - \mu}{\sigma} = \dfrac{500.5 - 500}{15.8114} = .03$

(Using Table IV in Appendix B)

$P(x > 500) \approx P(z > .03)$
    $= .5000 - .012 = .488$

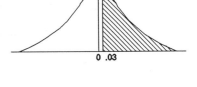

b. $z = \dfrac{(a + .5) - \mu}{\sigma} = \dfrac{489.5 - 500}{15.8114} = -.66$

$z = \dfrac{(a - .5) - \mu}{\sigma} = \dfrac{499.5 - 500}{15.8114} = -.03$

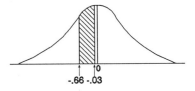

$P(490 \leq x < 500) \approx P(-.66 \leq z < -.03)$
    $= P(-.66 \leq z \leq 0) - P(-.03 \leq z \leq 0)$
    $= P(0 \leq z \leq .66) - P(0 \leq z \leq .03)$
    $= .2454 - .012 = .2334$

(Using Table IV in Appendix B)

c. $P(x > 1000) = 0$

Since $n = 1000$, the random variable x can only take on the values 0, 1, 2, ... , 1000.

5.53   a.   Let x = the number of workers on the job on a particular day out of 50 workers. The random variable x is a binomial random variable with $n = 50$ and $p = .80$ (if 20% are absent, 80% are on the job).

90% of 50 workers $= .9(50) = 45$

$\mu = np = 50(.8) = 40$
$\sigma^2 = npq = 50(.8)(.2) = 8$

$\sigma = \sqrt{\sigma^2} = \sqrt{8} = 2.8284$

$z = \dfrac{(a + .5) - \mu}{\sigma} = \dfrac{44.5 - 40}{2.8284} = 1.59$

$P(x \geq 45) \approx P(z \geq 1.59)$
    $= .5000 - .4441 = .0559$

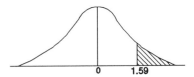

b.　$\mu \pm 3\sigma \Rightarrow np \pm 3\sqrt{npq} \Rightarrow 40 \pm 3(2.8284)$ (part a) $\Rightarrow 40 \pm 8.485 \Rightarrow (31.515, 48.485)$

Since the interval lies in the range 0 to 50, we can use the normal approximation to approximate the probability in part **a**.

c.　If the absentee rate is 2%, then 98% of the workers are on the job. Hence, x is a binomial random variable with $n = 50$ and $p = .98$.

$$\mu \pm 3\sigma \Rightarrow np \pm 3\sqrt{npq} \Rightarrow 50(.98) \pm 3\sqrt{50(.98)(1 - .98)}$$
$$\Rightarrow 49 \pm 2.9698$$
$$\Rightarrow (46.0302, 51.9698)$$

Since the interval does not lie in the range 0 to 50, we should not use the normal approximation to approximate the probability in part **a**.

5.55　a.　We must assume that whether one smoke detector is defective is independent of whether any other smoke detector is defective. Also the probability of a smoke detector being defective must remain constant for all smoke detectors. These assumptions seem to be satisfied since a random sample was taken.

b.　The random variable x is a binomial random variable with $n = 2000$ and $p = .40$.

$$\mu \pm 3\sigma \Rightarrow np \pm 3\sqrt{npq} \Rightarrow 2000(.40) \pm 3\sqrt{2000(.40)(1 - .40)}$$
$$\Rightarrow 800 \pm 3(21.9089) \Rightarrow 800 \pm 65.7267$$
$$\Rightarrow (734.2733, 865.7267)$$

Since the interval does lie in the range 0 to 2000, we can use the normal approximation to approximate the probability.

$$z = \frac{(a + .5) - \mu}{\sigma} = \frac{4.5 - 800}{21.9089} = -36.31$$
$$P(x \le 4) \approx P(z \le -36.31)$$
$$\approx .5 - .5 = 0$$

c.　No, it is not likely that 40% of their detectors are defective. If 40% really were defective, then the probability of four or fewer defectives is approximately zero. But there were only four defectives. Therefore, it is very unlikely that 40% are defective.

d.　Yes, it is possible that 40% of the detectors are defective. The probability of four or fewer defectives is approximately zero. It is possible but very, very unlikely.

5.57　a.　$f(x) = \dfrac{1}{d - c} = \dfrac{1}{90 - 10} = \dfrac{1}{80}$, $10 \le x \le 90$

b.　$\mu = \dfrac{c + d}{2} = \dfrac{10 + 90}{2} = 50$

$\sigma = \dfrac{d - c}{\sqrt{12}} = \dfrac{90 - 10}{\sqrt{12}} = 23.094011$

c. The interval $\mu \pm 2\sigma \Rightarrow 50 \pm 2(23.094) \Rightarrow 50 \pm 46.188 \Rightarrow (3.812, 96.188)$ is indicated on the graph.

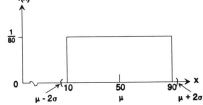

d. $P(x \leq 60) = \text{Base(height)} = (60 - 10)\dfrac{1}{80} = \dfrac{5}{8} = .625$

e. $P(x \geq 90) = 0$

f. $P(x \leq 80) = \text{Base(height)} = (80 - 10)\dfrac{1}{80} = \dfrac{7}{8} = .875$

g. $P(\mu - \sigma \leq x \leq \mu + \sigma) = P(50 - 23.094 \leq x \leq 50 + 23.094)$
$$= P(26.906 \leq x \leq 73.094)$$
$$= \text{Base(height)}$$
$$= (73.094 - 26.906)\left[\dfrac{1}{80}\right] = \dfrac{46.188}{80} = .577$$

h. $P(x > 75) = \text{Base(height)} = (90 - 75)\dfrac{1}{80} = \dfrac{15}{80} = .1875$

5.59 a. $P(x \geq .44) = .5000 - P(0 \leq z \leq .44)$
$$= .5000 - .1700 = .3300$$

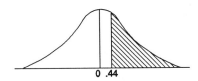

b. $P(x \leq -1.33) = .5 - P(-1.33 < z < 0)$
$$= .5 - .4082 = .0918$$

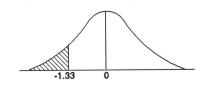

c. $P(-1.64 \leq z \leq 1.96)$
$$= P(-1.64 \leq z \leq 0) + P(0 \leq z \leq 1.96)$$
$$= .4495 + .4750$$
$$= .9245$$

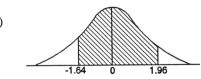

d. $P(1.64 \leq z \leq 1.96)$
$= P(0 \leq z \leq 1.96) - P(0 \leq z \leq 1.64)$
$= .4750 - .4495$
$= .0255$

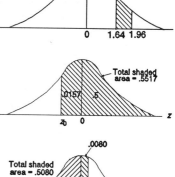

5.61 a. $P(z \geq z_0) = .5517$
$\Rightarrow P(z_0 \leq z \leq 0) = .5517 - .5 = .0517$

Looking up the area .0517 in Table IV, $z_0 = -.13$.

b. $P(z \leq z_0) = .5080$
$\Rightarrow P(0 \leq z \leq z_0) = .5080 - .5 = .0080$
Looking up the area .0080 in Table IV,
$\Rightarrow \qquad z_0 = .02$

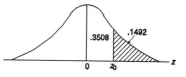

c. $P(z \geq z_0) = .1492$
$\Rightarrow P(0 \leq z \leq z_0) = .5 - .1492 = .3508$
Looking up the area .3508 in Table IV,
$\Rightarrow \qquad z_0 = 1.04$

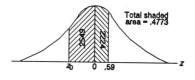

d. $P(z_0 \leq z \leq .59) = .4773$
$\Rightarrow P(z_0 \leq z \leq 0) + P(0 \leq z \leq .59) = .4773$

$P(0 \leq z \leq .59) = .2224$

Thus, $P(z_0 \leq z \leq 0) = .4773 - .2224 = .2549$
Looking up the area .2549 in Table IV, $z_0 = -.69$

5.63 x has an exponential distribution with $\lambda = .3$.

a. $P(x \leq 2) = 1 - P(x > 2) = 1 - e^{-.3(2)}$
$= 1 - e^{-.6} = 1 - .548812 = .451188$

b. $P(x > 3) = e^{-.3(3)} = e^{-.9} = .406570$

c. $P(x = 1) = 0$. Since x is continuous, there is no probability at a single point.

d. $P(x \leq 7) = 1 - P(x > 7) = 1 - e^{-.3(7)}$
$= 1 - e^{-2.1} = 1 - .122456 = .877544$

e. $P(4 \leq x \leq 12) = P(x \geq 4) - P(x > 12)$
$= e^{-.3(4)} - e^{-.3(12)}$
$= e^{-1.2} - e^{-3.6}$
$= .301194 - .027324$
$= .27387$

Continuous Random Variables

f. $P(x = 2.5) = 0$. Since x is continuous, there is no probability at a single point.

5.65   Let x be the noise level per jet takeoff in a neighborhood near the airport. The random variable x is approximately normally distributed with $\mu = 100$ and $\sigma = 6$.

a. $z = \dfrac{x - \mu}{\sigma} = \dfrac{108 - 100}{6} = 1.33$

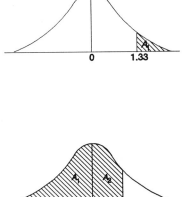

$$P(x > 108) = P(z > 1.33)$$
$$= .5000 - P(0 \leq z \leq 1.33)$$
$$= .5000 - .4082$$
$$= .0918$$

b. $P(x = 100) = 0$

c.   Given $P(x < 105) = .95$ and $\sigma = 6$,

$$P(x < 105) = P(x < \mu) + P(\mu < x < 105)$$
$$= .5 + .45 = A_1 + A_2$$
Looking up the area $A_2 = .45$ in Table IV, $z_0 = 1.645$.

Since $z = 1.645$, $x = 105$ and $\sigma = 6$,

$$z = \frac{x - \mu}{\sigma}$$
$$\Rightarrow 1.645 = \frac{105 - \mu}{6}$$
$$\Rightarrow 9.87 = 105 - \mu$$

Hence, $\mu = 95.13$

Since $\mu = 100$, the mean level of noise must be lowered $100 - 95.13 = 4.87$ decibels.

5.67   Let y be the profit on a metal part that is produced. Then y is $10, $-2, or $-1, depending where it falls with respect to the tolerance limits.

Let x be the tensile strength of a particular metal part. The random variable x is normally distributed with $\mu = 25$ and $\sigma = 2$.

$$z = \frac{x - \mu}{\sigma} = \frac{21 - 25}{2} = -2$$
$$z = \frac{x - \mu}{\sigma} = \frac{30 - 25}{2} = 2.5$$

$$P(y = 10) = P(x \text{ falls within the tolerance limits})$$
$$= P(21 < x < 30) = P(-2 < z < 2.5)$$
$$= P(-2 < z < 0) + P(0 < z < 2.5)$$
$$= P(0 < z < 2) + P(0 < z < 2.5)$$
$$= .4772 + .4938$$
$$= .9710$$

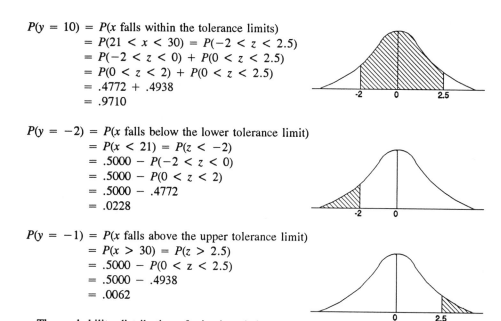

$$P(y = -2) = P(x \text{ falls below the lower tolerance limit})$$
$$= P(x < 21) = P(z < -2)$$
$$= .5000 - P(-2 < z < 0)$$
$$= .5000 - P(0 < z < 2)$$
$$= .5000 - .4772$$
$$= .0228$$

$$P(y = -1) = P(x \text{ falls above the upper tolerance limit})$$
$$= P(x > 30) = P(z > 2.5)$$
$$= .5000 - P(0 < z < 2.5)$$
$$= .5000 - .4938$$
$$= .0062$$

The probability distribution of y is given below:

y	10	-2	-1
$p(y)$	.9710	.0228	.0062

$$E(y) = \sum yp(y) = 10(.9710) + -2(.0228) + -1(.0062)$$
$$= 9.71 - .0456 - .0062$$
$$= \$9.6582$$

5.69 Let x be the time a worker is unemployed in weeks. The random variable x is an exponential random variable with $\lambda = .075$.

a. $\mu = 1/\lambda = 1/.075 = 13.33$ weeks

b. $P(x \geq 2) = e^{-2\lambda} = e^{-2(.075)} = e^{-.15} = .860708$
$P(x > 6) = e^{-6\lambda} = e^{-6(.075)} = e^{-.45} = .637628$

c. $P(x < 12) = 1 - P(x \geq 12) = 1 - e^{-12(.075)} = 1 - e^{-.9} = 1 - .40657 = .59343$

5.71 The summary statistics were computed for each bank as follows:

	Range	$s = \sqrt{\dfrac{\sum x^2 - \dfrac{(\sum x)^2}{n}}{n-1}}$	Coefficient of Variation $(s/\bar{x})$
Bank 1	$31 - 10 = 21$	$\sqrt{\dfrac{2554 - \dfrac{116^2}{6}}{5}} = 7.8909$	$\dfrac{7.8909}{19.3333} = .408$
Bank 2	$70 - 54 = 16$	$\sqrt{\dfrac{22974 - \dfrac{370^2}{6}}{5}} = 5.6095$	$\dfrac{5.6095}{61.6667} = .091$
Bank 3	$129 - 81 = 48$	$\sqrt{\dfrac{64791 - \dfrac{617^2}{6}}{5}} = 16.3880$	$\dfrac{16.3880}{102.8333} = .159$

The rankings (from high to low) based on each measure are:

(a) Range: Bank 3, Bank 1, Bank 2
(b) Standard deviation: Bank 3, Bank 1, Bank 2
(c) Coefficient of variation: Bank 1, Bank 3, Bank 2

5.73 Let x equal the annual sales per sales person. The random variable x has an approximately normal distribution with $\mu = \$180,000$ and $\sigma = \$50,000$.

a. p_1 = proportion of sales people who receive a $1,000 bonus
   = proportion of sales people who sell less than $100,000 per year

$$z = \frac{x - \mu}{\sigma} = \frac{100,000 - 180,000}{50,000} = -1.6$$

$$P(x < 100,000) = P(z < -1.6)$$
$$= .5000 - .4452$$
$$= .0548$$
(Table IV, Appendix B)

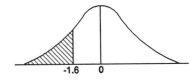

b. p_2 = proportion of sales people who receive a $5,000 bonus
   = proportion of sales people who sell from $100,000 to $200,000

$$z = \frac{x - \mu}{\sigma} = \frac{200,000 - 180,000}{50,000} = .4$$

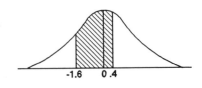

$$P(100,000 \leq x \leq 200,000) = P(-1.6 \leq z \leq .4)$$

(Refer to part **a.**)

$$= P(-1.6 \leq z \leq 0) + P(0 \leq z \leq .4)$$
$$= P(0 \leq z \leq 1.6) + P(0 \leq z \leq .4)$$
$$= .4452 + .1554$$
$$= .6006$$

(Table IV, Appendix B)

c. p_3 = proportion of sales people who receive a $10,000 bonus
      = proportion of sales people who sell above $200,000

$$P(x > 200,000) = P(z > .4)$$
$$= .5000 - .1554$$
$$= .3446$$

(Refer to part **b.**)

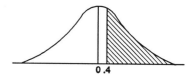

d.   Let y equal the amount of the bonus. The probability distribution for the random variable y is:

y	1,000	5,000	10,000
$p(y)$	.0548	.6006	.3446

(Refers to parts **a, b,** and **c.**)

The mean value of the bonus is

$$\mu = \sum_{\text{All } y} yp(y)$$
$$= 1000(.0548) + 5000(.6006) + 10000(.3446)$$
$$= 54.8 + 3003 + 3446$$
$$= \$6,503.80$$

5.75   Let x equal the difference between the actual weight and recorded weight (the error of measurement). The random variable x is normally distributed with $\mu = 592$ and $\sigma = 628$.

a.   We want to find the probability that the weigh-in-motion equipment understates the actual weight of the truck. This would be true if the error of measurement is positive.

$$z = \frac{x - \mu}{\sigma} = \frac{0 - 592}{628} = -.94$$

$$P(x > 0) = P(z > -.94)$$
$$= .5000 + .3264$$
$$= .8264$$

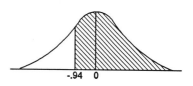

b. $P(\text{overstate the weight}) = 1 - P(\text{understate the weight})$
$$= 1 - .8264$$
$$= .1736 \quad (\text{Refer to part } \mathbf{a.})$$

For 100 measurements, approximately $100(.1736) = 17.36$ or 17 times the weight would be overstated.

c. $z = \dfrac{x - \mu}{\sigma} = \dfrac{400 - 592}{628} = -.31$

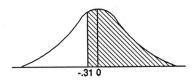

$$P(x > 400) = P(z > -.31)$$
$$= .5000 + .1217$$
$$= .6217$$

d.   We want $P(\text{understate the weight}) = .5$

To understate the weight, $x > 0$. Thus, we want to find μ so that $P(x > 0) = .5$

$$P(x > 0) = P\left[z > \dfrac{0 - \mu}{628}\right] = .5$$

From Table IV, Appendix B, $z_0 = 0$. To find μ, substitute into the z-score formula:

$$z = \dfrac{x - \mu}{\sigma}$$

$$\Rightarrow 0 = \dfrac{0 - \mu}{628} \Rightarrow \mu = 0$$

Thus, the mean error should be set at 0.

We want $P(\text{understate the weight}) = .4$

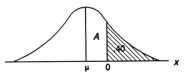

To understate the weight, $x > 0$. Thus, we want to find μ so that $P(x > 0) = .4$

$A = .5 - .40 = .1$. Look up the area .1000 in the body of Table IV, Appendix B, $z_0 = .25$.

To find μ, substitute into the z-score formula:

$$z = \dfrac{x - \mu}{\sigma}$$
$$\Rightarrow .25 = \dfrac{0 - \mu}{628} \Rightarrow \mu = 0 - (.25)628 = -157$$

Thus, the mean error should be set at -157.

CHAPTER SIX

. . . .
Sampling Distributions

6.1   **a.**   "The sampling distribution of the sample statistic A" is the probability distribution of the variable A.

       b.   "A" is an unbiased estimator of α if the mean of the sampling distribution of A is α.

       c.   If both A and B are unbiased estimators of α, then the statistic whose standard deviation is smaller is a better estimator of α.

       d.   No. The Central Limit Theorem applies only to the sample mean. If A is the sample mean, $\bar{x}$, and n is sufficiently large, then the Central Limit Theorem will apply. However, both A and B cannot be sample means. Thus, we cannot apply the Central Limit Theorem to both A and B.

6.3   **a.** $\mu_{\bar{x}} = \mu = 8$ $\sigma_{\bar{x}} = \dfrac{\sigma}{\sqrt{n}} = \dfrac{3}{\sqrt{16}} = .75$

       b. $\mu_{\bar{x}} = \mu = 200$ $\sigma_{\bar{x}} = \dfrac{\sigma}{\sqrt{n}} = \dfrac{15}{\sqrt{15}} = 3.873$

       c. $\mu_{\bar{x}} = \mu = 50$ $\sigma_{\bar{x}} = \dfrac{\sigma}{\sqrt{n}} = \dfrac{4}{\sqrt{100}} = .4$

       d. $\mu_{\bar{x}} = \mu = -10$ $\sigma_{\bar{x}} = \dfrac{\sigma}{\sqrt{n}} = \dfrac{17}{\sqrt{110}} = 1.621$

6.5   We know that the sampling distribution of $\bar{x}$ will be normal since the sampled population is normal. We also know that the sampling distribution will have mean and standard deviation

$$\mu_{\bar{x}} = \mu = 10 \qquad \sigma_{\bar{x}} = \dfrac{\sigma}{\sqrt{n}} = \dfrac{12}{\sqrt{36}} = 2$$

       a. $P(\bar{x} > 11) = P\left(z > \dfrac{11 - \mu}{\sigma_{\bar{x}}}\right) = P\left(z > \dfrac{11 - 10}{2}\right) = P(z > .5)$

$$= .5 - P(0 < z < .5) = .5 - .1915 = .3085$$

b. $P(\bar{x} < 11) = P\left(z < \dfrac{11 - \mu}{\sigma_{\bar{x}}}\right) = P\left(z < \dfrac{11 - 10}{2}\right) = P(z < .5)$

$$= .5 + P(0 < z < .5) = .5 + .1915 = .6915$$

Note: $\bar{x} < 11$ is the complement of $\bar{x} > 11$;

therefore, $P(\bar{x} < 11) = 1 - P(\bar{x} > 11) = 1 - .3085 = .6915$

c. $P(\bar{x} > 13.1) = P\left(z > \dfrac{13.1 - \mu}{\sigma_{\bar{x}}}\right) = P\left(z > \dfrac{13.1 - 10}{2}\right) = P(z > 1.55)$

$$= .5 - P(0 < z < 1.55) = .5 + P(0 < z < 1.55)$$
$$= .5 - .4394 = .0606$$

d. $P(9 < \bar{x} < 11) = P\left(\dfrac{9 - \mu}{\sigma_{\bar{x}}} < z < \dfrac{11 - \mu}{\sigma_{\bar{x}}}\right) = P\left(\dfrac{9 - 10}{2} < z < \dfrac{11 - 10}{2}\right)$

$$= P(-.50 < z < .50) = 2P(0 < z < .50)$$
$$= 2(.1915) = .3830$$

e. $P(\bar{x} < 6.7) = P\left(z < \dfrac{6.7 - \mu}{\sigma_{\bar{x}}}\right) = P\left(z < \dfrac{6.7 - 10}{2}\right) = P(z < -1.65)$

$$= .5 - P(-1.65 < z < 0) = .5 - P(0 < z < 1.65)$$
$$= .5 - .4505 = .0495$$

6.7 The probability distribution for choosing an observation, x, from the population is:

x	1	2	3	4
$p(x)$	.30	.20	.20	.30

$\mu = \displaystyle\sum_{\text{All } x} xp(x) = 1(.30) + 2(.20) + 3(.20) + 4(.30)$

$$= .30 + .40 + .60 + 1.20$$
$$= 2.5$$

$\sigma^2 = \displaystyle\sum_{\text{All } x} (x - \mu)^2 p(x) = (1 - 2.5)^2(.3) + (2 - 2.5)^2(.2) + (3 - 2.5)^2(.2) + (4 - 2.5)^2(.3)$

$$= .675 + .05 + .05 + .675$$
$$= 1.45$$

$\sigma = \sqrt{\sigma^2} = \sqrt{1.45} = 1.2042$

a. $\mu_{\bar{x}} = \mu = 2.5$

$\sigma_{\bar{x}} = \dfrac{\sigma}{\sqrt{n}} = \dfrac{1.2042}{\sqrt{40}} = .1904$

b. We know by the Central Limit Theorem that the sampling distribution of $\bar{x}$ will be approximately normal since $n \geq 30$. Therefore, the answer does depend on the sample size.

6.11 a. The number of samples of size $n = 2$ that could be selected without replacement from a population of size $N = 4$ is:

$$\begin{pmatrix} N \\ n \end{pmatrix} = \begin{pmatrix} 4 \\ 2 \end{pmatrix} = \frac{4!}{2!(4-2)!} = 6$$

The samples are:

chip 1, chip 2	chip 2, chip 3
chip 1, chip 3	chip 2, chip 4
chip 1, chip 4	chip 3, chip 4

where chip 1 is marked 1, chips 2 and 3 are marked 2, and chip 4 is marked 3.

b. Each of the six outcomes is equally likely since the chips are chosen at random; therefore, each has probability 1/6.

c.

Sample (Number Marked on Chip)	$\bar{x}$
1, 2	1.5
1, 2	1.5
1, 3	2
2, 2	2
2, 3	2.5
2, 3	2.5

d.

$\bar{x}$	$p(\bar{x})$
1.5	$\dfrac{2}{6} = \dfrac{1}{3}$
2	$\dfrac{2}{6} = \dfrac{1}{3}$
2.5	$\dfrac{2}{6} = \dfrac{1}{3}$

e. Population Probability Distribution

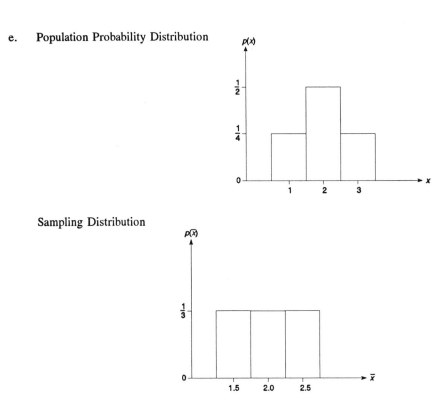

Sampling Distribution

6.13 Given: $\mu = 100$ and $\sigma = 10$

n	1	5	10	20	30	40	50
$\dfrac{\sigma}{\sqrt{n}}$	10	4.472	3.162	2.236	1.826	1.581	1.414

The graph of $\sigma/\sqrt{n}$ against n is given here.

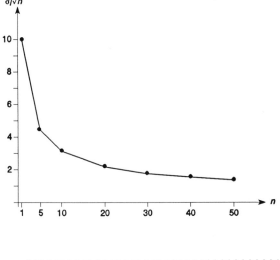

6.15 a. $\sigma_{\bar{x}} = \dfrac{\sigma}{\sqrt{n}} = \dfrac{200}{\sqrt{30}} = 36.515$

b. To reduce the standard deviation of the sampling distribution of $\bar{x}$ to 1/2 of the value in part **a**, we need 4 times as many observations.

(Therefore, $30 \times 4 = 120$ observations are needed.)

c. To reduce the standard deviation to 3/4 of the value in part **a**, we need $16/9 = 1.7778$ times as many observations.

(Therefore, $30 \times 16/9 = 53.33$ (54) observations are needed.)

6.17 a. By the Central Limit Theorem, the sampling distribution of $\bar{x}$ is approximately normal (since $n \geq 30$) with

$$\mu_{\bar{x}} = \mu = 10 \text{ and } \sigma_{\bar{x}} = \dfrac{\sigma}{\sqrt{n}} = \dfrac{16}{\sqrt{61}} = 2.0486$$

b. Find the probability of getting a sample mean of 13.49 or larger.

$$P(\bar{x} \geq 13.49) = P\left(z \geq \dfrac{13.49 - 10}{2.0486}\right)$$
$$= P(z \geq 1.70) = .5 - P(0 \leq z \leq 1.70)$$
$$= .5 - .4554 = .0446$$

This is not a very likely result since the probability is not very large.

c. The random variable x is the difference between the actual number of units sold and the forecast (forecast error). If $\bar{x} = 0$ and $s^2 = 1$, then the forecasts were very accurate since the mean error is 0 and the variance is only 1.

6.19 The mean, μ, of the length of the steel rods is 3 meters with a standard deviation, σ, of .03. By the Central Limit Theorem, the sampling distribution of $\bar{x}$ is approximately normal since $n \geq 30$, and

$$\mu_{\bar{x}} = \mu = 3 \text{ and } \sigma_{\bar{x}} = \dfrac{\sigma}{\sqrt{n}} = \dfrac{.03}{\sqrt{100}} = .003$$

a. Since the lots are accepted if the sample mean is 3.005 meters or more,

$$P(\bar{x} < 3.005) = P\left(z < \dfrac{3.005 - 3}{.003}\right)$$
$$= P(z < 1.67) = .5 + P(0 < z < 1.67)$$
$$= .5 + .4525 = .9525$$

Thus, $.9525 \times 100 = 95.25\%$ of the lots will be returned to the vendor.

b.   We will only accept the lot if the sample mean is 3.005 meters or more. If all the rods are between 2.999 and 3.004 meters in length, then the sample mean must also be between 2.999 and 3.004 meters. Therefore, all of the lots (100%) will be returned to the vendor since the sample mean will never be 3.005 meters or more.

6.21   a.   By the Central Limit Theorem, the sampling distribution of $\bar{x}$ is approximately normal since $n > 30$ and

$$\mu_{\bar{x}} = \mu = 840 \qquad \sigma_{\bar{x}} = \frac{\sigma}{\sqrt{n}} = \frac{15}{\sqrt{50}} = 2.1213$$

b. $P(\bar{x} \leq 830) = P\left[z \leq \frac{830 - 840}{2.1213} \right] = P(z \leq -4.71) \approx .5 - .5 = 0$

c.   Since the probability of observing a mean of 830 or less is extremely small (≈ 0) if the true mean is 840, we would tend to believe that the mean is not 840, but something less.

d.   By the Central Limit Theorem, the sampling distribution of $\bar{x}$ is approximately normal since $n > 30$ and

$$\mu_{\bar{x}} = \mu = 840 \qquad \sigma_{\bar{x}} = \frac{\sigma}{\sqrt{n}} = \frac{45}{\sqrt{50}} = 6.3640$$

$$P(\bar{x} \leq 830) = P\left[z \leq \frac{830 - 840}{6.3640} \right] = P(z \leq -1.57) \approx .5 - .4418 = .0582$$

6.23   The mean, μ, diameter of the bearings is unknown with a standard deviation, σ, of .001 inch. By the Central Limit Theorem, the sampling distribution of $\bar{x}$ is approximately normal since $n \geq 30$, with

$$\mu_{\bar{x}} = \mu \qquad \sigma_{\bar{x}} = \frac{\sigma}{\sqrt{n}} = \frac{.001}{\sqrt{36}} = .000167$$

Having the sample mean fall within .0001 inch of μ implies $|\bar{x} - \mu| \leq .0001$ or $-.0001 \leq \bar{x} - \mu \leq .0001$.

$$P(-.0001 \leq \bar{x} - \mu \leq .0001)$$

$$= P\left[\frac{-.0001}{.000167} \leq z \leq \frac{.0001}{.000167} \right] = P(-.60 \leq z \leq .60)$$

$$= 2P(0 \leq z \leq .60) = 2(.2257) = .4514$$

6.25   a.   We know by the Central Limit Theorem that the sampling distribution of $\bar{x}$ is approximately normal since $n \geq 30$ with the mean, $\mu_{\bar{x}}$, unknown and

$$\sigma_{\bar{x}} = \frac{\sigma}{\sqrt{n}} = \frac{1.2}{\sqrt{280}} = .0717$$

b.   We know by the Central Limit Theorem that the sampling distribution of $\bar{y}$ is approximately normal since $n \geq 30$ with the mean, $\mu_{\bar{y}}$, unknown and

$$\sigma_{\bar{y}} = \frac{\sigma_y}{\sqrt{n}} = \frac{1}{\sqrt{348}} = .0536$$

c. In order for $\sigma_{\bar{x}}^2 = \sigma_{\bar{y}}^2$,

$$\frac{1.2^2}{n} = \frac{1^2}{348} \Rightarrow \frac{1.44}{n} = \frac{1}{348} \Rightarrow n = 501.12 \approx 502$$

Thus, we would need to sample approximately 502 households.

6.27 The mean, μ, of the percentage of alkali in a test specimen of soap is 2% with a standard deviation, σ, of 1%. The sampling distribution of $\bar{x}$ is approximately normal since x is approximately normal and if $n = 4$,

$$\mu_{\bar{x}} = \mu = 2 \qquad \sigma_{\bar{x}} = \frac{\sigma}{\sqrt{n}} = \frac{1}{\sqrt{4}} = .5$$

a. The control limits are located $3\sigma_{\bar{x}}$ above and below μ.

$$3\sigma_{\bar{x}} = 3(.5) = 1.5$$

Therefore, the upper and lower control limits are located 1.5% above and below μ.

b. $\bar{x}$ will fall outside the control limits if it is smaller than $\mu - 3\sigma_{\bar{x}}$ or larger than $\mu + 3\sigma_{\bar{x}}$. If the process is in control,

$$P(\bar{x} < \mu - 3\sigma_{\bar{x}}) + P(\bar{x} > \mu + 3\sigma_{\bar{x}})$$

$$= P\left(z < \frac{\mu - 3\sigma_{\bar{x}} - \mu}{\sigma_{\bar{x}}}\right) + P\left(z > \frac{\mu + 3\sigma_{\bar{x}} - \mu}{\sigma_{\bar{x}}}\right)$$

$$= P(z < -3) + P(z > 3)$$
$$= 2(z > 3)$$
$$= 2(.5 - P(0 < z < 3))$$
$$= 2(.5 - .4987)$$
$$= 2(.0013)$$
$$= .0026$$

c. The process is deemed to be out of control if $\bar{x}$ is outside the control limits. The control limits are located at $\mu \pm 3\sigma_{\bar{x}} \Rightarrow 2 \pm 1.5 \Rightarrow (.5, 3.5)$. If the mean shifts to $\mu = 3\%$,

$$P(\bar{x} < .5) + P(\bar{x} > 3.5)$$

$$= P\left(z < \frac{.5 - \mu}{\sigma_{\bar{x}}}\right) + P\left(z > \frac{3.5 - \mu}{\sigma_{\bar{x}}}\right)$$

$$= P\left(z < \frac{.5 - 3}{.5}\right) + P\left(z > \frac{3.5 - 3}{.5}\right)$$

$$= P(z < -5) + P(z > 1)$$
$$= .5 - P(-5 < z < 0) + .5 - P(0 < z < 1)$$
$$= .5 - .5 + .5 - .3413$$
$$= .1587$$

6.29   Referring to Exercise 6.27, the mean, μ, of the percentage of alkali in a test specimen of soap is 2% with a standard deviation, σ, of 1%. The sampling distribution of $\bar{x}$ is approximately normal since x is approximately normal and if $n = 4$,

$$\mu_{\bar{x}} = \mu = 2 \qquad \sigma_{\bar{x}} = \frac{\sigma}{\sqrt{n}} = \frac{1}{\sqrt{4}} = .5$$

a. $\bar{x}$ will fall outside the warning limits if it is smaller than $\mu - 1.96\sigma_{\bar{x}}$ or larger than $\mu + 1.96\sigma_{\bar{x}}$. If the process is in control,

$$P(\bar{x} < \mu - 1.96\sigma_{\bar{x}}) + P(\bar{x} > \mu + 1.96\sigma_{\bar{x}})$$

$$= P\left(z < \frac{\mu - 1.96\sigma_{\bar{x}}}{\sigma_{\bar{x}}}\right) + P\left(z > \frac{\mu + 1.96\sigma_{\bar{x}}}{\sigma_{\bar{x}}}\right)$$

$$= P(z < -1.96) + P(z > 1.96)$$
$$= 2P(z > 1.96)$$
$$= 2((.5 - P(0 < z < 1.96))$$
$$= 2(.5 - .4750)$$
$$= 2(.025)$$
$$= .05$$

b. $P(\bar{x} > \mu + 1.96\sigma_{\bar{x}}) = P(z > 1.96) = .025$ (Refer to part a.)

Therefore, $40 \times .025 = 1$ of the next 40 values of $\bar{x}$ is expected to fall above the upper warning limit.

c. P(next two values of $\bar{x}$ fall below the lower warning limit)
$$= P(\bar{x} < \mu - 1.96\sigma_{\bar{x}})P(\bar{x} < \mu - 1.96\sigma_{\bar{x}}) \text{ (by independence)}$$
$$= P(z < -1.96)P(z < -1.96)$$
$$= .025(.025) \text{ (Refer to part a).}$$
$$= .000625$$

6.31   For $n = 35$, the Central Limit Theorem indicates the sampling distribution of $\bar{x}$ is approximately normal with a mean of $\mu_{\bar{x}} = \mu = 2.5$ and a standard deviation of

$$\sigma_{\bar{x}} = \frac{\sigma}{\sqrt{n}} = \frac{\sqrt{2.5}}{\sqrt{35}} = \sqrt{.0714} = .2673$$

The contractor will not purchase the load if $\bar{x} > 2.1$.

$$P(\bar{x} > 2.1) = P\left(z > \frac{2.1 - 2.5}{.2673}\right) = P(z > -1.50) = .5 + .4332 = .9332$$

CHAPTER SEVEN

. . . .

Inferences Based on a Single Sample: Estimation

7.1   a. $\alpha = .10$, $\alpha/2 = .05$, then $z_{.05}$ is the z value that locates .05 in one tail of the standard normal distribution. Since the total area to the right of the mean is .5, $z_{.05}$ will be the z value corresponding to the tabulated area to the right of the mean equal to $.5 - .05 = .4500$. Using Table IV of Appendix B, this z value is 1.645.

        b. $\alpha = .01$, $\alpha/2 = .005$. Then $z_{.005}$ is the z value that locates .005 in one tail of the standard normal distribution. Looking up an area of $.5 - .005 = .4950$ in Table IV, Appendix B, $z_{.005}$ is 2.575.

        c. $\alpha = .05$, $\alpha/2 = .025$. Looking up an area of $.5 - .025 = .4750$ in Table IV, Appendix B, $z_{.025}$ is 1.96.

        d. $\alpha = .20$, $\alpha/2 = .10$. Looking up an area of $.5 - .10 = .4000$ in Table IV, Appendix B, $z_{.10}$ is 1.28.

7.3   a.   First, compute the mean and standard deviation of the sample.

$$\bar{x} = \frac{\sum x}{n} = \frac{850}{50} = 17$$

$$s^2 = \frac{\sum (x - \bar{x})^2}{n - 1} = \frac{4720}{49} = 96.3265$$

$$s = \sqrt{s^2} = 9.8146$$

For confidence coefficient .95, $\alpha = 1 - .95 = .05$ and $\alpha/2 = .05/2 = .025$. From Table IV, Appendix B, $z_{.025} = 1.96$.

The 95% confidence interval is:

$$\bar{x} \pm z_{.025}\sigma_{\bar{x}} \Rightarrow \bar{x} \pm 1.96\frac{\sigma}{\sqrt{n}} \Rightarrow 17 \pm 1.96\frac{9.8146}{\sqrt{50}} \Rightarrow 17 \pm 2.720 \Rightarrow (14.28, 19.72)$$

        b.   We are 95% confident that μ lies between 14.28 and 19.72.

7.5   a.   For confidence coefficient .95, $\alpha = 1 - .95 = .05$ and $\alpha/2 = .05/2 = .025$. From Table IV, Appendix B, $z_{.025} = 1.96$. The 95% confidence interval for μ is:

$$\bar{x} \pm z_{.025}\sigma_{\bar{x}} \Rightarrow \bar{x} \pm 1.96\frac{\sigma}{\sqrt{n}} \Rightarrow 26.2 \pm 1.96\frac{4.1}{\sqrt{40}} \Rightarrow 26.2 \pm 1.271 \Rightarrow (24.929, 27.471)$$

b. Sample size is inversely related to the variability and, hence, the width of the confidence interval. As the sample size is reduced, the confidence interval becomes wider. As the sample size is increased, the confidence interval becomes narrower.

7.7 A point estimator is a single value used to estimate the parameter, μ. An interval estimator is two values, an upper and lower bound, which define an interval with which we attempt to enclose the parameter, μ. An interval estimate also has a measure of confidence associated with it.

7.9 As the sample size increases, the width of the confidence interval decreases. As the sample size decreases, the width of the confidence interval increases.

7.11 a. The population of interest is the set of amounts credit card customers spent on their first visit to the chain store's new store.

b. The population parameter the chain wishes to estimate is the average or mean amount of money spent by the chain's credit card customers, μ.

c. For confidence coefficient .90, $\alpha = .10$ and $\alpha/2 = .05$. From Table IV, Appendix B, $z_{.05} = 1.645$. The 90% confidence interval is:

$$\bar{x} \pm z_{.05}\sigma_{\bar{x}} \Rightarrow \bar{x} \pm 1.645\frac{\sigma}{\sqrt{n}} \Rightarrow 62.56 \pm 1.645\frac{\sqrt{400}}{\sqrt{50}} \Rightarrow 62.56 \pm 4.65$$
$$\Rightarrow (\$57.91, \$67.21)$$

7.13 a. The population of interest to the auditor is the set of differences between the actual amounts owed the company and the amounts indicated on the invoices.

b. The auditor measured the difference between the actual amount owed the company and the amount indicated on the invoice for 35 randomly selected invoices from all invoices produced by the company's new billing system.

c. For confidence coefficient .98, $\alpha = .02$ and $\alpha/2 = .01$. From Table IV, Appendix B, $z_{.01} = 2.33$. The 98% confidence interval is:

$$\bar{x} \pm z_{.01}\sigma_{\bar{x}} \Rightarrow \bar{x} \pm 2.33\frac{\sigma}{\sqrt{n}} \Rightarrow 1 \pm 2.33\frac{124}{\sqrt{35}} \Rightarrow 1 \pm 48.84 \Rightarrow (\$-47.84, \$49.84)$$

d. We are 98% confident that the mean difference between the actual amount owed the company and the amount indicated on the invoice is between $\$-47.84$ and $\$49.84$.

e. The interval constructed in part d is fairly wide. It appears the new billing system is not very accurate.

7.15 a. For confidence coefficient .90, $\alpha = 1 - .90 = .10$ and $\alpha/2 = .10/2 = .05$. From Table IV, Appendix B, $z_{.05} = 1.645$. The 90% confidence interval for μ is:

$$\bar{x} \pm z_{.05}\sigma_{\bar{x}} \Rightarrow \bar{x} \pm 1.645\frac{\sigma}{\sqrt{n}} \Rightarrow 23.43 \pm 1.645\frac{10.82}{\sqrt{96}} \Rightarrow 23.43 \pm 1.817$$
$$\Rightarrow (21.613, 25.247)$$

b.   We are 90% confident the true mean number of years of service lies between 21.613 years and 25.247 years.

c.   That you have chosen a random sample where n is sufficiently large to use the normal distribution.

d.   Yes, $\bar{x}$ is an unbiased estimate of μ; therefore, $E(\bar{x}) = \mu = E(x)$.

7.17   To compute the necessary sample size, use

$$n = \frac{4(z_{\alpha/2})^2 \sigma^2}{W^2} \text{ where } \alpha = 1 - .95 = .05 \text{ and } \alpha/2 = .05/2 = .025.$$

From Table IV, Appendix B, $z_{.025} = 1.96$.   Thus,

$$n = \frac{4(1.96)^2(7.2)}{.3^2} = 1229.312 \approx 1230$$

In Exercise 7.16, the bound is .3 which means the width is .6.   The confidence interval resulting from the sample will be wider in Exercise 7.16 than in this exercise.   Therefore, the sample size in Exercise 7.16 is smaller.   As the sample size decreases, the width of the confidence interval increases.

7.19   a.   An estimate of σ is obtained from:

range $\approx 4s$

$$s \approx \frac{\text{range}}{4} = \frac{34 - 30}{4} = 1$$

To compute the necessary sample size, use

$$n = \frac{(z_{\alpha/2})^2 \sigma^2}{B^2} \text{ where } \alpha = 1 - .90 = .10 \text{ and } \alpha/2 = .05.$$

From Table IV, Appendix B, $z_{.05} = 1.645$.   Thus,

$$n = \frac{(1.645)^2(1)^2}{.2^2} = 67.65 \approx 68$$

b.   A less conservative estimate of σ is obtained from:

range $\approx 6s$

$$s \approx \frac{\text{range}}{6} = \frac{34 - 30}{6} = .6667$$

Thus, $n = \dfrac{(z_{\alpha/2})^2 \sigma^2}{B^2} = \dfrac{(1.645)^2(.6667)^2}{.2^2} = 30.07 \approx 31$

· ·

7.21 a. To compute the needed sample size, use

$$n = \frac{(z_{\alpha/2})^2 \sigma^2}{B^2} \text{ where } \alpha = 1 - .80 = .20 \text{ and } \alpha/2 = .10.$$

From Table IV, Appendix B, $z_{.10} = 1.28$. Thus,

$$n = \frac{(1.28)^2 (2)^2}{.1^2} = 655.36 \approx 656$$

b. As the sample size decreases, the width of the confidence interval increases. Therefore, if we sample 100 parts instead of 656, the confidence interval would be wider.

c. To compute the maximum confidence level that could be attained meeting the management's specifications,

$$n = \frac{(z_{\alpha/2})^2 \sigma^2}{B^2} \Rightarrow 100 = \frac{(z_{\alpha/2})^2 (2)^2}{.1^2} \Rightarrow (z_{\alpha/2})^2 = \frac{100(.01)}{4} = .25 \Rightarrow z_{\alpha/2} = .5$$

Using Table IV, Appendix B, $P(0 \le z \le .5) = .1915$. Thus, $\alpha/2 = .5000 - .1915 = .3085$, $\alpha = 2(.3085) = .617$, and $1 - \alpha = 1 - .617 = .383$.

The maximum confidence level would be 38.3%.

7.23 A conservative estimate of σ is obtained from:

$$\text{range} \approx 4s$$
$$s \approx \frac{\text{range}}{4} = \frac{45 - 15}{4} = 7.5$$

To compute the needed sample size, use

$$n = \frac{(z_{\alpha/2})^2 \sigma^2}{B^2} \text{ where } \alpha = 1 - .95 = .05 \text{ and } \alpha/2 = .025.$$

From Table IV, Appendix B, $z_{.025} = 1.96$. Thus,

$$n = \frac{(1.96)^2 (7.5)^2}{2^2} = 54.02 \approx 55$$

7.25 To compute the necessary sample size, use

$$n = \frac{4(z_{\alpha/2})^2 \sigma^2}{W^2} \text{ where } \alpha = 1 - .90 = .10 \text{ and } \alpha/2 = .05.$$

From Table IV, Appendix B, $z_{.05} = 1.645$. Thus,

$$n = \frac{4(1.645)^2 (10)^2}{2^2} = 270.6 \approx 271$$

7.27 a. For confidence coefficient .80, $\alpha = 1 - .80 = .20$ and $\alpha/2 = .20/2 = .10$. From Table IV, Appendix B, $z_{.10} = 1.28$. From Table VI, with df $= n - 1 = 5 - 1 = 4$, $t_{.10} = 1.533$.

b. For confidence coefficient .90, $\alpha = 1 - .90 = .05$ and $\alpha/2 = .10/2 = .05$. From Table IV, Appendix B, $z_{.05} = 1.645$. From Table VI, with df $= n - 1 = 5 - 1 = 4$, $t_{.05} = 2.132$.

c. For confidence coefficient .95, $\alpha = 1 - .95 = .05$ and $\alpha/2 = .05/2 = .025$. From Table IV, Appendix B, $z_{.025} = 1.96$. From Table VI, with df $= n - 1 = 5 - 1 = 4$, $t_{.025} = 2.776$.

d. For confidence coefficient .98, $\alpha = 1 - .98 = .02$ and $\alpha/2 = .02/2 = .01$. From Table IV, Appendix B, $z_{.01} = 2.33$. From Table VI, with df $= n - 1 = 5 - 1 = 4$, $t_{.01} = 3.747$.

e. For confidence coefficient .99, $\alpha = 1 - .99 = .02$ and $\alpha/2 = .02/2 = .005$. From Table IV, Appendix B, $z_{.005} = 2.575$. From Table VI, with df $= n - 1 = 5 - 1 = 4$, $t_{.005} = 4.604$.

f. Both the t and z distributions are symmetric around 0 and mound shaped. The t distribution is more spread out than the z distribution.

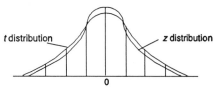

7.29 a. $P(-t_0 < t < t_0) = .95$ where df = 10

Because of symmetry, the statement can be written

$$P(0 < t < t_0) = .475 \text{ where df} = 10$$
$$\Rightarrow P(t \geq t_0) = .025$$
$$t_0 = 2.228$$

b. $P(t \leq -t_0 \text{ or } t \geq t_0) = .05$ where df = 10

$$\Rightarrow 2P(t \geq t_0) = .05$$
$$\Rightarrow P(t \geq t_0) = .025 \text{ where df} = 10$$
$$t_0 = 2.228$$

c. $P(t \leq t_0) = .05$ where df = 10

Because of symmetry, the statement can be written

$$\Rightarrow P(t \geq -t_0) = .05 \text{ where df} = 10$$
$$t_0 = -1.812$$

d. $P(t \leq -t_0 \text{ or } t \geq t_0) = .05$ where df = 20
$$\Rightarrow 2P(t \geq t_0) = .05$$
$$\Rightarrow P(t \geq t_0) = .025 \text{ where df} = 20$$
$$t_0 = 2.086$$

e. $P(t \le -t_0 \text{ or } t \ge t_0) = .01$ where df $= 5$

$\Rightarrow 2P(t \ge t_0) = .01$

$\Rightarrow P(t \ge t_0) = .005$ where df $= 5$

$t_0 = 4.032$

7.31 For this sample,

$$\bar{x} = \frac{\sum x}{n} = \frac{2400}{24} = 100$$

$$s^2 = \frac{\sum x^2 - \frac{(\sum x)^2}{n}}{n-1} = \frac{244{,}382 - \frac{2400^2}{24}}{24 - 1} = 190.5217$$

$$s = \sqrt{s^2} = 13.8030$$

a. For confidence coefficient, .80, $\alpha = 1 - .80 = .20$ and $\alpha/2 = .20/2 = .10$. From Table VI, Appendix B, with df $= n - 1 = 24 - 1 = 23$, $t_{.10} = 1.319$. The 80% confidence interval for μ is:

$$\bar{x} \pm t_{.10}\frac{s}{\sqrt{n}} \Rightarrow 100 \pm 1.319\frac{13.803}{\sqrt{24}} \Rightarrow 100 \pm 3.72 \Rightarrow (96.28, 103.72)$$

b. For confidence coefficient, .95, $\alpha = 1 - .95 = .05$ and $\alpha/2 = .05/2 = .025$. From Table VI, Appendix B, with df $= n - 1 = 24 - 1 = 23$, $t_{.025} = 2.069$. The 95% confidence interval for μ is:

$$\bar{x} \pm t_{.025}\frac{s}{\sqrt{n}} \Rightarrow 100 \pm 2.069\frac{13.803}{\sqrt{24}} \Rightarrow 100 \pm 5.83 \Rightarrow (94.17, 105.83)$$

The 95% confidence interval for μ is wider than the 80% confidence interval for μ found in part a.

c. For part a:

We are 80% confident that the true population mean lies in the interval 96.28 to 103.72.

For part b:

We are 95% confident that the true population mean lies in the interval 94.17 to 105.83.

The 95% confidence interval is wider than the 80% confidence interval because the more confident you want to be that μ lies in the interval, the more numbers must be in the interval. The more confident you are, the larger the t-value, which makes the interval wider.

7.33 a. First we make some preliminary calculations:

$$\bar{x} = \frac{\sum x}{n} = \frac{862}{11} = 78.364$$

$$s^2 = \frac{\sum x^2 - \frac{(\sum x)^2}{n}}{n-1} = \frac{67,980 - \frac{862^2}{11}}{11-1} = 43.0545$$

$$s = \sqrt{43.0545} = 6.56161$$

For confidence coefficient, .95, $\alpha = .05$ and $\alpha/2 = .025$. From Table VI, Appendix B, with df $= n - 1 = 11 - 1 = 10$, $t_{.025} = 2.228$. The 95% confidence interval is:

$$\bar{x} \pm t_{.025}\frac{s}{\sqrt{n}} \Rightarrow 788.364 \pm 2.228\frac{6.5616}{\sqrt{11}} \Rightarrow 78.364 \pm 4.408 \Rightarrow (73.956, 82.772)$$

b. We are 95% confident that the mean participation rate for Fortune 500 companies is between 73.956% and 82.772%.

c. We must assume that the population of Fortune 500 participation rates is normal.

d. If we constructed a 60% confidence interval rather than a 95% confidence interval, the center would be exactly the same, but the width would be narrower.

7.35   a. The population of interest is all the costs of hiring secretaries in the corporation for the last 2 years.

b. First, compute the sample mean and standard deviation:

$$\bar{x} = \frac{\sum x}{n} = \frac{15215}{8} = 1901.875$$

$$s^2 = \frac{\sum x^2 - \frac{(\sum x)^2}{n}}{n-1} = \frac{26,649,475 - \frac{(15,215)^2}{8}}{7} = 101,778.12$$

$$s = \sqrt{s^2} = 319.027$$

For confidence coefficient, .90, $\alpha = .10$ and $\alpha/2 = .05$. From Table VI, Appendix B, with df $= n - 1 = 8 - 1 = 7$, $t_{.05} = 1.895$. The 95% confidence interval is:

$$\bar{x} \pm t_{.05}\frac{s}{\sqrt{n}} \Rightarrow 1901.875 \pm 1.895\frac{319.027}{\sqrt{8}} \Rightarrow 1901.875 \pm 213.743$$

$$\Rightarrow (1,688.132, \ 2,115.618)$$

c. The width of the confidence interval is:

$$2115.618 - 1688.132 = 427.486 = 2(213.743)$$

A 95% confidence interval would be wider because increasing the confidence level causes the use of a larger $t_{\alpha/2}$ value in every case; thus, increasing the width of the confidence interval.

7.37 **a.** For confidence coefficient .95, $\alpha = 1 - .95 = .05$ and $\alpha/2 = .05/2 = .025$. From Table VI, Appendix B, with df $= n - 1 = 23 - 1 = 22$, $t_{.025} = 2.074$. The 95% confidence interval is:

$$\bar{x} \pm t_{.025}\frac{s}{\sqrt{n}} \Rightarrow 135 \pm 2.074\frac{32}{\sqrt{23}} \Rightarrow 135 \pm 13.839 \Rightarrow (121.161, 148.839)$$

b. We must assume a random sample was selected and that the population of all health insurance costs per worker per month is normally distributed.

c. "95% confidence interval" means that if repeated samples of size 23 were selected from the population and 95% confidence intervals formed, 95% of all confidence intervals will contain the true value of μ.

7.39 An unbiased estimator is one in which the mean of the sampling distribution is the parameter of interest, i.e., $E(\hat{p}) = p$.

7.41 **a.** The sample size is large enough if the interval $\hat{p} \pm 3\sigma_{\hat{p}}$ does not include 0 or 1.

$$\hat{p} \pm 3\sigma_{\hat{p}} \Rightarrow \hat{p} \pm 3\sqrt{\frac{pq}{n}} \Rightarrow \hat{p} \pm 3\sqrt{\frac{\hat{p}\hat{q}}{n}} \Rightarrow .88 \pm 3\sqrt{\frac{.88(1 - .88)}{121}} \Rightarrow .88 \pm .089$$
$$\Rightarrow (.791, .969)$$

Since the interval lies within the interval $(0, 1)$, the normal approximation will be adequate.

b. For confidence coefficient .90, $\alpha = .10$ and $\alpha/2 = .05$. From Table IV, Appendix B, $z_{.05} = 1.645$. The 90% confidence interval is:

$$\hat{p} \pm z_{.05}\sqrt{\frac{pq}{n}} \Rightarrow \hat{p} \pm 1.645\sqrt{\frac{\hat{p}\hat{q}}{n}} \Rightarrow .88 \pm 1.645\sqrt{\frac{.88(.12)}{121}} \Rightarrow .88 \pm .049$$
$$\Rightarrow (.831, .929)$$

c. We must assume that the sample is a random sample from the population of interest.

7.43 **a.** Since p is the proportion of consumers who like the snack food, $\hat{p}$ will be:

$$\hat{p} = \frac{\text{number of 1's in sample}}{n} = \frac{15}{50} = 0.3$$

Check to see if the normal approximation is appropriate.

$$\hat{p} \pm 3\sigma_{\hat{p}} \Rightarrow \hat{p} \pm 3\sqrt{\frac{pq}{n}} \Rightarrow \hat{p} \pm 3\sqrt{\frac{\hat{p}\hat{q}}{n}} \Rightarrow .3 \pm 3\sqrt{\frac{.3(.7)}{50}} \Rightarrow .3 \pm .194$$
$$\Rightarrow (.106, .494)$$

This interval lies within the interval $(0, 1)$, so the normal approximation is adequate.

For confidence coefficient .80, $\alpha = .20$ and $\alpha/2 = .10$. From Table IV, Appendix B, $z_{.10} = 1.28$. The 80% confidence interval is:

$$\hat{p} \pm z_{.10}\sqrt{\frac{pq}{n}} \Rightarrow \hat{p} \pm 1.28\sqrt{\frac{\hat{p}\hat{q}}{n}} \Rightarrow .3 \pm 1.28\sqrt{\frac{.3(.7)}{50}} \Rightarrow .3 \pm .083 \Rightarrow (.217, .383)$$

b.   We are 80% confident the true proportion of consumers who like the snack food lies between .217 to .383.

7.45 $\hat{p} = \dfrac{x}{n} = \dfrac{26}{46} = .565$

First, we check to see if the sample size is sufficiently large.

$$\hat{p} \pm 3\sigma_{\hat{p}} \Rightarrow \hat{p} \pm 3\sqrt{\frac{\hat{p}\hat{q}}{n}} \Rightarrow .565 \pm 3\sqrt{\frac{.565(.435)}{46}} \Rightarrow .565 \pm .219 \Rightarrow (.346, .784)$$

Since the interval is wholly contained in the interval $(0, 1)$, we may conclude that the normal approximation is reasonable.

For confidence coefficient .90, $\alpha = .10$ and $\alpha/2 = .05$. From Table IV, Appendix B, $z_{.05} = 1.645$. The 90% confidence interval is:

$$\hat{p} \pm z_{.05}\sqrt{\frac{\hat{p}\hat{q}}{n}} \Rightarrow .565 \pm 1.645\sqrt{\frac{.565(.435)}{46}} \Rightarrow .565 \pm .120 \Rightarrow (.445, .685)$$

7.47   a.   The population of interest is all owners and managers of small U.S. businesses.

b.   The sample size is large enough if $\hat{p} \pm 3\sigma_{\hat{p}}$ lies within the interval $(0, 1)$.

$$\hat{p} \pm 3\sigma_{\hat{p}} \Rightarrow \hat{p} \pm 3\sqrt{\frac{pq}{n}} \Rightarrow \hat{p} \pm 3\sqrt{\frac{\hat{p}\hat{q}}{n}} \Rightarrow .4 \pm 3\sqrt{\frac{.4(.6)}{258}} \Rightarrow .4 \pm .091$$
$$\Rightarrow (.309, .491)$$

Since the interval is wholly contained in the interval $(0, 1)$, we may conclude that the normal approximation is reasonable.

c.   For confidence coefficient .95, $\alpha = .05$ and $\alpha/2 = .025$. From Table IV, Appendix B, $z_{.025} = 1.96$. The 95% confidence interval is:

$$\hat{p} \pm z_{.025}\sqrt{\frac{pq}{n}} \Rightarrow \hat{p} \pm 1.96\sqrt{\frac{\hat{p}\hat{q}}{n}} \Rightarrow .4 \pm 1.96\sqrt{\frac{.4(.6)}{258}} \Rightarrow .4 \pm .06 \Rightarrow (.34, .46)$$

d.   The interval would be narrower for $\hat{p} = .30$ rather than $\hat{p} = .40$. As $\hat{p}$ approaches .5, the standard error, $\sigma_{\hat{p}}$, gets larger; therefore, a smaller $\hat{p}$ produces a smaller $\sigma_{\hat{p}}$ and a narrower confidence interval.

7.49 a. To compute the needed sample size, use

$$n = \frac{(z_{\alpha/2})^2 pq}{B^2} \text{ where } z_{.05} = 1.645 \text{ from Table IV, Appendix B.}$$

Thus, $n = \dfrac{(1.645)^2(.8)(.2)}{.03^2} = 481.07 \approx 482$

b. To compute the needed sample size, use

$$n = \frac{(z_{\alpha/2})^2 pq}{B^2} = \frac{(1.645)^2(.5)(.5)}{.03^2} = 751.67 \approx 752$$

7.51 To find the sample size used:

$$n = \frac{4(z_{\alpha/2})^2 (pq)}{W^2} \text{ where } z_{.05} = 1.645 \text{ from Table IV, Appendix B.}$$

$W = .54 - .26 = .28$ and $p \approx \hat{p} = \dfrac{.26 + .54}{2} = .4$

Thus, $n = \dfrac{4(1.645)^2(.4)(.6)}{.28^2} = 33.135 \approx 34$

A sample size $n = 34$ was used to compute the 90% confidence interval given.

7.53 From the problem, $\hat{p} = \dfrac{x}{n} = \dfrac{190}{532} = .357$. To compute the necessary sample size, use

$$n = \frac{(z_{\alpha/2})^2 (pq)}{B^2} \text{ where } z_{.025} = 1.96 \text{ from Table IV, Appendix B.}$$

Thus, $n = \dfrac{(1.96)^2(.357)(.643)}{.03^2} = 979.8 \approx 980$

Since the sample was for $n = 532$ managers, it was not large enough to meet the specifications. We needed to sample 980 managers.

7.55 a. To compute the necessary sample size, use:

$$n = \frac{(z_{\alpha/2})^2 (pq)}{B^2} \text{ where } z_{.05} = 1.645 \text{ from Table IV, Appendix B}$$

Thus, $n = \dfrac{(1.645)^2(.05)(.95)}{.02^2} = 321.3 \approx 322$

You would need to observe 322 customers.

7.57 a. For a small sample from a normal distribution with unknown standard deviation, we use the t statistic. For confidence coefficient .95, $\alpha = 1 - .95 = .05$ and $\alpha/2 = .05/2 = .025$. From Table VI, Appendix B, with df $= n - 1 = 23 - 1 = 22$, $t_{.025} = 2.074$.

　　　b. For a large sample from a distribution with an unknown standard deviation, we can estimate the population standard deviation with s and use the z statistic. For confidence coefficient .95, $\alpha = 1 - .95 = .05$ and $\alpha/2 = .05/2 = .025$. From Table IV, Appendix B, $z_{.025} = 1.96$.

　　　c. For a small sample from a normal distribution with known standard deviation, we use the z statistic. For confidence coefficient .95, $\alpha = 1 - .95 = .05$ and $\alpha/2 = .05/2 = .025$. From Table IV, Appendix B, $z_{.025} = 1.96$.

　　　d. For a large sample from a distribution about which nothing is known, we can estimate the population standard deviation with s and use the z statistic. For confidence coefficient .95, $\alpha = 1 - .95 = .05$ and $\alpha/2 = .05/2 = .025$. From Table IV, Appendix B, $z_{.025} = 1.96$.

　　　e. For a small sample from a distribution about which nothing is known, we can use neither z nor t.

7.59 a. The average number of checks the bank processed per week for the 50 randomly sampled weeks.

　　　b. The proportion of weeks the bank processed more than 100,000 checks in the 50 randomly sampled weeks.

　　　c. The true standard deviation of the number of checks the bank processes each week.

　　　d. The true average number of checks the bank processes per week.

　　　e. The number of weeks that are randomly sampled ($n = 50$).

　　　f. The true standard deviation of the average number of checks the bank processed each week for the 50 randomly selected weeks.

　　　g. The true proportion of weeks the bank processes more than 100,000 checks.

　　　h. The standard deviation of the number of checks the bank processed each week in the 50 randomly sampled weeks.

7.63 a. For confidence coefficient .95, $\alpha = .05$ and $\alpha/2 = .025$. From Table IV, Appendix B, $z_{.025} = 1.96$. The 95% confidence interval is:

$$\bar{x} \pm z_{.025}\frac{\sigma}{\sqrt{n}} \Rightarrow \bar{x} \pm 1.96\frac{s}{\sqrt{n}} \Rightarrow 12,522 \pm 1.96\frac{4000}{\sqrt{100}} \Rightarrow 12,522 \pm 784$$

$$\Rightarrow (11,738, 13,306)$$

b.   The fact that both values are positive indicates that the actual income was larger than reported income. Also, we can be 95% confident the true mean error in reported income lies in the interval $11,738 and $13,306.

7.65 $\hat{p} = \dfrac{x}{n} = \dfrac{212}{346} = .613$

For confidence coefficient .95, $\alpha = .05$ and $\alpha/2 = .05/2 = .025$. From Table IV, Appendix B, $z_{.025} = 1.96$. The 95% confidence interval is:

$$\hat{p} \pm z_{.025}\sqrt{\dfrac{\hat{p}\hat{q}}{n}} \Rightarrow .613 \pm 1.96\sqrt{\dfrac{(.613)(.387)}{346}} \Rightarrow .613 \pm .051 \Rightarrow (.562, .664)$$

We are 95% confident the proportion of households in the city who are using available recycling facilities is between .562 and .664.

7.67   a.   The point estimate for the mean wear is $\bar{x} = 42,250$ miles.

b.   For confidence coefficient .90, $\alpha = 1 - .90 = .10$ and $\alpha/2 = .10/2 = .05$. From Table IV, Appendix B, $z_{.05} = 1.645$. The 90% confidence interval is:

$$\bar{x} \pm z_{.05}\sigma_{\bar{x}} \Rightarrow \bar{x} \pm 1.645\dfrac{\sigma}{\sqrt{n}} \Rightarrow 42,250 \pm 1.645\dfrac{4355}{\sqrt{200}} \Rightarrow 42,250 \pm 506.570$$

$$\Rightarrow (41,743.430, 42,756.570)$$

c.   Interval estimation is better. The chance of estimating the true mean exactly with a single number is 0. With a confidence interval, we can estimate the true mean using a range of values. We are then fairly confident the true mean will fall in this range.

The interval in this problem is smaller than the one in Exercise 7.66, because the sample size is larger. In Exercise 7.66, we had to use the t statistic, while in this problem, we used the z statistic.

7.69   To compute the needed sample size, use:

$$n = \dfrac{4(z_{\alpha/2})^2 pq}{w^2} \text{ where } z_{.025} = 1.96 \text{ from Table IV, Appendix B.}$$

Thus, $n = \dfrac{4(1.96)^2(.094)(.906)}{.04^2} \approx 817.9 \approx 818$

7.71   a.   To compute the necessary sample size, use:

$$n = \dfrac{(z_{\alpha/2})^2 \sigma^2}{B^2} \text{ where } \alpha = 1 - .99 = .01 \text{ and } \alpha/2 = .01/2 = .005$$

From Table IV, Appendix B, $z_{.005} = 2.575$.

Thus, $n = \dfrac{(2.575)^2 11.34^2}{1^2} = 852.7 \approx 853$.

b.   We would have to assume the sample was a random sample.

CHAPTER EIGHT

Inferences Based on a Single Sample: Tests of Hypotheses

8.1   The null hypothesis is the "status quo" hypothesis, while the alternative hypothesis is the research hypothesis.

8.3   The "level of significance" of a test is α. This is the probability that the test statistic will fall in the rejection region when the null hypothesis is true.

8.5   The four possible results are:

1.   Rejecting the null hypothesis when it is true. This would be a Type I error.
2.   Accepting the null hypothesis when it is true. This would be a correct decision.
3.   Rejecting the null hypothesis when it is false. This would be a correct decision.
4.   Accepting the null hypothesis when it is false. This would be a Type II error.

8.7   When you reject the null hypothesis in favor of the alternative hypothesis, this does not prove the alternative hypothesis is correct. We are $100(1 - \alpha)\%$ confident that there is sufficient evidence to conclude that the alternative hypothesis is correct.

If we were to repeatedly draw samples from the population and perform the test each time, approximately $100(1 - \alpha)\%$ of the tests performed would yield the correct decision.

8.9   a.   Since the company must give proof the drug is safe, the null hypothesis would be the drug is unsafe. The alternative hypothesis would be the drug is safe.

b.   A Type I error would be concluding the drug is safe when it is not safe. A Type II error would be concluding the drug is not safe when it is. α is the probability of concluding the drug is safe when it is not. β is the probability of concluding the drug is not safe when it is.

c.   In this problem, it would be more important for α to be small. We would want the probability of concluding the drug is safe when it is not to be as small as possible.

8.11   a.

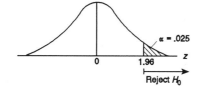

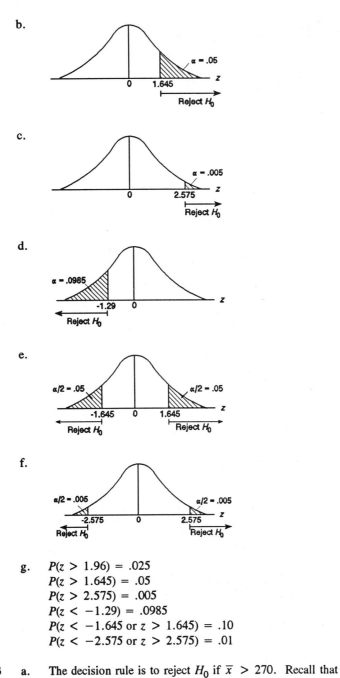

b.

$\alpha = .05$

0   1.645 z

Reject H_0

c.

$\alpha = .005$

0   2.575 z

Reject H_0

d.

$\alpha = .0985$

-1.29   0 z

Reject H_0

e.

$\alpha/2 = .05$ $\alpha/2 = .05$

-1.645   0   1.645 z

Reject H_0 Reject H_0

f.

$\alpha/2 = .005$ $\alpha/2 = .005$

-2.575   0   2.575 z

Reject H_0 Reject H_0

g. $P(z > 1.96) = .025$
   $P(z > 1.645) = .05$
   $P(z > 2.575) = .005$
   $P(z < -1.29) = .0985$
   $P(z < -1.645 \text{ or } z > 1.645) = .10$
   $P(z < -2.575 \text{ or } z > 2.575) = .01$

8.13 a. The decision rule is to reject H_0 if $\bar{x} > 270$. Recall that

$$z = \frac{\bar{x} - \mu_0}{\sigma_{\bar{x}}}$$

Therefore, reject H_0 if $\bar{x} > 270$

can be written reject H_0 if $z > \dfrac{\bar{x} - \mu_0}{\sigma_{\bar{x}}}$

$$z > \frac{270 - 255}{63/\sqrt{81}}$$

$$z > 2.14$$

The decision rule in terms of z is to reject H_0 if $z > 2.14$.

b. $P(z > 2.14) = .5 - P(0 < z < 2.14)$
$$= .5 - .4838$$
$$= .0162$$

8.15 The mean, μ, of the amount paid for long-distance calls is \$23.14 per month with a standard deviation, σ, of \$10.48. The sampling distribution of $\bar{x}$ is approximately normal by the Central Limit Theorem if $n \geq 30$.

a. $P(\bar{x} > 25) = P\left[z > \dfrac{25 - \mu}{\sigma_{\bar{x}}}\right] = P\left[z > \dfrac{25 - 23.14}{10.48/\sqrt{50}}\right]$

$$= P(z > 1.25) = .5 - P(0 \leq z < 1.25)$$
$$= .5 - .3944 = .1056$$

b. $P(\bar{x} > 25) = P\left[z > \dfrac{25 - \mu}{\sigma_{\bar{x}}}\right] = P\left[z > \dfrac{25 - 23.14}{10.48/\sqrt{100}}\right]$

$$= P(z > 1.77) = .5 - P(0 \leq z < 1.77)$$
$$= .5 - .4616 = .0384$$

c. To determine if the mean amount paid per month for long distance calls has increased, we test:

H_0: $\mu = 23.14$
H_a: $\mu > 23.14$

The test statistic is $z = \dfrac{\bar{x} - \mu_0}{\sigma_{\bar{x}}} = \dfrac{27.21 - 23.14}{\dfrac{10.48}{\sqrt{100}}} = 3.88$

The rejection region requires $\alpha = .05$ in the upper tail of the z distribution. From Table IV, Appendix B, $z_{.05} = 1.645$. The rejection region is $z > 1.645$.

Since the observed value of the test statistic falls in the rejection region ($z = 3.88 > 1.645$), H_0 is rejected. There is sufficient evidence to indicate the mean level of the amounts billed per month for long distance telephone calls has increased from \$23.14 at $\alpha = .05$.

8.17   a.   The null hypothesis is H_0: $\mu = 35$ mpg. The alternative hypothesis is H_a: $\mu > 35$ mpg.

     b.   We want to test:

$$H_0: \ \mu = 35$$
$$H_a: \ \mu > 35$$

The test statistic is $z = \dfrac{\bar{x} - \mu_0}{\sigma_{\bar{x}}} \approx \dfrac{\bar{x} - \mu_0}{s/\sqrt{n}} = \dfrac{37.3 - 35}{6.4/\sqrt{36}} = 2.16$

The rejection region requires $\alpha = .05$ in the upper tail of the z distribution. From Table IV, Appendix B, $z_{.05} = 1.645$. The rejection region is $z > 1.645$.

Since the observed value of the test statistic falls in the rejection region ($z = 2.16 > 1.645$), H_0 is rejected. There is sufficient evidence to support the auto manufacturer's claim that the mean miles per gallon for the car exceeds the EPA estimate of 35 at $\alpha = .05$.

8.19   a.   To determine if the mean lifetime of the new cartridges exceeds that of the old, we test:

$$H_0: \ \mu = 1502.5$$
$$H_a: \ \mu > 1502.5$$

     b.   The test statistic is $z = \dfrac{\bar{x} - \mu_0}{\sigma_{\bar{x}}} \approx \dfrac{\bar{x} - \mu_0}{s/\sqrt{n}} = \dfrac{1511.4 - 1502.5}{35.7/\sqrt{225}} = 3.74$

The rejection region requires $\alpha = .005$ in the upper tail of the z distribution. From Table IV, Appendix B, $z_{.005} = 2.575$. The rejection region is $z > 2.575$.

Since the observed value of the test statistic falls in the rejection region ($z = 3.74 > 2.575$), H_0 is rejected. There is sufficient evidence to indicate the mean lifetime of the new cartridges exceeds that of the old at $\alpha = .005$.

8.21   We will reject the null hypothesis if the observed significance level (p-value) is less than α.

     a.   .10 ≮ .05, do not reject H_0

     b.   .05 < .10, reject H_0

     c.   .001 < .01, reject H_0

     d.   .05 ≮ .025, do not reject H_0

     e.   .45 ≮ .10, do not reject H_0

8.23   We reject the null hypothesis (H_0) if the p-value $< \alpha$. Since the p-value is .083, we would reject H_0 if α is greater than .083.

8.25 p-value $= P(z \le -1.77 \text{ or } z \ge 1.77)$
$= 2P(z \ge 1.77)$
$= 2[.5 - P(0 < z \le 1.77)]$
$= 2[.5 - .4616]$
$= 2(.0384) = .0768$

8.27   First, find the value of the test statistic:

$$z = \frac{\bar{x} - \mu_0}{\sigma_{\bar{x}}} \approx \frac{\bar{x} - \mu_0}{s/\sqrt{n}} = \frac{35.7 - 35}{3.1/\sqrt{100}} = 2.26$$

p-value $= P(z \ge 2.26) = .5 - P(0 < z < 2.26) = .5 - .4881 = .0119$

The results are highly significant. We would reject H_0 for any $\alpha > .0119$.

8.29   a.   To determine if the manufacturer's claim is correct, we test:

H_0: $\mu = 3.5$
H_a: $\mu < 3.5$

The test statistic is $z = \dfrac{\bar{x} - \mu_0}{\sigma_{\bar{x}}} \approx \dfrac{\bar{x} - \mu_0}{s/\sqrt{n}} = \dfrac{3.3 - 3.5}{1.1/\sqrt{50}} = -1.29$

where $s = \dfrac{66}{60} = 1.1$

The rejection region requires $\alpha = .05$ in the lower tail of the z distribution. From Table IV, Appendix B, $z_{.05} = 1.645$. The rejection region is $z < -1.645$.

Since the observed value of the test statistic does not fall in the rejection region ($z = -1.29 \not< -1.645$), H_0 is not rejected. There is insufficient evidence to indicate the pain reliever brings relief to headache sufferers in less than 35 minutes. The data do not support the manufacturer's claim at $\alpha = .05$.

b. p-value $= P(z \le -1.29) = .5 - P(0 < z \le 1.29)$
$= .5 - .4015$
$= .0985$

c.   Since we reject H_0 (support the manufacturer's claim) when the p-value is less than α, small p-values support the manufacturer's claim.

8.31   a. H_0: $\mu = \$9,083$
H_a: $\mu > \$9,083$

b.   First, find the value of the test statistic:

$$z = \frac{\bar{x} - \mu_0}{\sigma_{\bar{x}}} \approx \frac{\bar{x} - \mu_0}{s/\sqrt{n}} = \frac{9,667 - 9,083}{1,721/\sqrt{30}} = 1.86$$

p-value $= P(z \geq 1.86)$
$= .5 - P(0 < z < 1.86)$
$\approx .5 - .4686$
$= .0314$

The smallest value of α for which we could reject H_0 is just greater than .0314. The results are statistically significant.

 c. The mean amount for tuition and fees in $1992-1993$ is significantly larger than it was in $1990-1991$, statistically speaking. The observed significance level was .0314. Had the observed significance level been something like .13, the results would not be statistically significant. However, practically speaking, the results might be significant enough for the purposes of the students or the private colleges.

8.33 a. The test statistic is $z = 1.37$ with a p-value of .0853.

   There is no evidence to reject H_0 for $\alpha = .05$. There is insufficient evidence to indicate the mean lifetime of the new cartridges is more than 1502.5 for $\alpha = .05$.

   There is evidence to reject H_0 for $\alpha = .10$. There is sufficient evidence to indicate the mean lifetime of the new cartridges is more than 1502.5 for $\alpha = .10$.

 b. If the test had been two-tailed, the p-value would double. The p-value would be $2(.0853) = .1706$.

   There is no evidence to reject H_0 for $\alpha \leq .10$. There is insufficient evidence to indicate the mean lifetime of the new cartridges is different from 1502.5 for $\alpha \leq .10$.

8.35 We may use the t distribution in testing a hypothesis about a population mean when the relative frequency distribution of the sampled population is approximately normal and σ is unknown.

8.37 $\alpha = P(\text{Type I error}) = P(\text{Reject } H_0 \text{ when } H_0 \text{ is true})$

 a. $\alpha = P(t > 1.440)$ where df $= 6$
   $= .10$ Table VI, Appendix B

 b. $\alpha = P(t < -1.782)$ where df $= 12$
   $= P(t > 1.782)$
   $= .05$ Table VI, Appendix B

 c. $\alpha = P(t < -2.060 \text{ or } t > 2.060)$ where df $= 25$
   $= 2P(t > 2.060)$
   $= 2(.025)$
   $= .05$ Table VI, Appendix B

8.39   From Exercise 8.38, $t = -1.76$ with df $= 5$.

a.   To test:

$$H_0: \mu = 11$$
$$H_a: \mu < 11$$

p-value $= P(t \leq -1.76)$ where df $= 5$
$\qquad\quad\;\; = P(t \geq 1.76)$
$\qquad\qquad .05 < p\text{-value} < .10$

b.   To test:

$$H_0: \mu = 11$$
$$H_a: \mu \neq 11$$

p-value $= P(t \leq -1.76 \text{ or } t \geq 1.76)$ where df $= 5$
$\qquad\quad\;\; = 2P(t \geq 1.76)$
$\qquad\qquad 2(.05) < p\text{-value} < 2(.10)$
$\qquad\quad \Rightarrow .10 < p\text{-value} < .20$

8.41   a.   We must assume that we have a random sample from a normal population.

b.   The test statistic is $t = 1.894$ and the p-value $= .0382$. There is evidence to reject H_0 for $\alpha > .0382$. There is sufficient evidence to indicate that the mean is greater than 1000 for $\alpha > .0382$.

c.   For a two-tailed test, the p-value would double. Thus, p-value $= 2(.0382) = .0764$. There is no evidence to reject H_0 for $\alpha = .05$. There is insufficient evidence to indicate that the mean is different from 1000 for $\alpha = .05$. There is evidence to reject H_0 for $\alpha = .10$. There is sufficient evidence to indicate that the mean is different from 1000 for $\alpha = .10$.

8.43   To determine if the mean fill is less than 50 gallons, we test:

$$H_0: \mu = 50$$
$$H_a: \mu < 50$$

The test statistic is $t = \dfrac{\bar{x} - \mu_0}{s/\sqrt{n}} = \dfrac{49.70 - 50}{.32/\sqrt{20}} = -4.193$

The rejection region requires $\alpha = .10$ in the lower tail of the t distribution with df $= n - 1 = 20 - 1 = 19$. From Table VI, Appendix B, $t_{.10} = 1.328$. The rejection region is $t < -1.328$.

Since the observed value of the test statistic falls in the rejection region, ($t = -4.193 < -1.328$), H_0 is rejected. There is sufficient evidence to indicate the mean fill per 50-gallon drum is less than 50 gallons for $\alpha = .10$.

Assumption: The sample is a random sample selected from a population with a relative frequency distribution that is approximately normal.

8.45 **a.** Notice that perceiving yourself as older than your chronological age implies that (perceived age − chronological age) should be a positive value. Then to test the hypothesis, compute the difference between perceived age and chronological age, and use this as a new set of data to test:

H_0: $\mu = 0$
H_a: $\mu > 0$ (where μ is the mean difference between perceived and chronological age)

Chronological Age	Perceived Age	(Perceived Age − Chronological Age)
20	22	2
19	21	2
25	30	5
22	25	3
26	22	−4
19	19	0
18	20	2
20	18	−2
20	21	1
21	21	0

Treating the set of differences as a new data set, the sample mean and standard deviation are:

$$\bar{x} = \frac{\sum x}{n} = \frac{9}{10} = 0.9$$

$$s^2 = \frac{\sum x^2 - \frac{(\sum x)^2}{n}}{n - 1} = \frac{67 - \frac{(9)^2}{10}}{9} = 6.544$$

$$s = \sqrt{s^2} = 2.558$$

H_0: $\mu = 0$
H_a: $\mu > 0$

The test statistic is $t = \dfrac{\bar{x} - \mu_0}{s/\sqrt{n}} = \dfrac{0.9 - 0}{2.558/\sqrt{10}} = 1.113$

The rejection region requires $\alpha = .10$ in the upper tail of the t distribution with df $= n - 1 = 10 - 1 = 9$. From Table VI, Appendix B, $t_{.10} = 1.383$. The rejection region is $t > 1.383$.

Since the observed value of the test statistic does not fall in the rejection region, ($t = 1.113 \not> 1.383$), H_0 is not rejected. There is insufficient evidence to indicate the mean difference is greater than 0. That is, there is insufficient evidence to indicate that people under the age of 30 perceived themselves as older than their actual age.

b. The differences must be a random sample from a normally distributed population of differences.

8.47 The sample size is large enough if the interval $p_0 \pm 3\sigma_{\hat{p}}$ is contained in the interval $(0, 1)$.

a. $p_0 \pm 3\sqrt{\dfrac{p_0 q_0}{n}} \Rightarrow .975 \pm 3\sqrt{\dfrac{(.975)(.025)}{900}} \Rightarrow .975 \pm .016 \Rightarrow (.959, .991)$

Since the interval is contained in the interval $(0, 1)$, the sample size is large enough.

b. $p_0 \pm 3\sqrt{\dfrac{p_0 q_0}{n}} \Rightarrow .01 \pm 3\sqrt{\dfrac{(.01)(.99)}{125}} \Rightarrow .01 \pm .027 \Rightarrow (-.017, .037)$

Since the interval is not contained in the interval $(0, 1)$, the sample size is not large enough.

c. $p_0 \pm 3\sqrt{\dfrac{p_0 q_0}{n}} \Rightarrow .75 \pm 3\sqrt{\dfrac{(.75)(.25)}{40}} \Rightarrow .75 \pm .205 \Rightarrow (.545, .955)$

Since the interval is contained in the interval $(0, 1)$, the sample size is large enough.

d. $p_0 \pm 3\sqrt{\dfrac{p_0 q_0}{n}} \Rightarrow .75 \pm 3\sqrt{\dfrac{(.75)(.25)}{15}} \Rightarrow .75 \pm .335 \Rightarrow (.415, 1.085)$

Since the interval is not contained in the interval $(0, 1)$, the sample size is not large enough.

e. $p_0 \pm 3\sqrt{\dfrac{p_0 q_0}{n}} \Rightarrow .62 \pm 3\sqrt{\dfrac{(.62)(.38)}{12}} \Rightarrow .62 \pm .420 \Rightarrow (.120, 1.040)$

Since the interval is not contained in the interval $(0, 1)$, the sample size is not large enough.

8.49 From Exercise 8.48, $n = 100$.

First, check to see if n is large enough.

$$p_0 \pm 3\sigma_{\hat{p}} \Rightarrow p_0 \pm 3\sqrt{\dfrac{pq}{n}} \Rightarrow p_0 \pm 3\sqrt{\dfrac{p_0 q_0}{100}}$$

$$\Rightarrow .9 \pm 3\sqrt{\dfrac{.9(1 - .9)}{100}} \Rightarrow .9 \pm .09 \Rightarrow (.81, .99)$$

Since the interval lies completely in the interval $(0, 1)$, the normal approximation will be adequate.

Inferences Based on a Single Sample: Tests of Hypotheses

a. The test statistic is $z = \dfrac{\hat{p} - p_0}{\sigma_{\hat{p}}} = \dfrac{\hat{p} - p_0}{\sqrt{\dfrac{p_0 q_0}{n}}} = \dfrac{.83 - .9}{\sqrt{\dfrac{.9(1 - .9)}{100}}} = -2.33$

b. The estimated standard deviation of $\hat{p}$ is smaller in this exercise than in Exercise 8.48. Since the standard deviation is smaller, the absolute value of z is larger.

$$\text{In } 8.48 \Rightarrow \sqrt{\dfrac{p_0 q_0}{n}} = \sqrt{\dfrac{.7(1 - .7)}{100}} = .0458$$

$$\text{In } 8.49 \Rightarrow \sqrt{\dfrac{p_0 q_0}{n}} = \sqrt{\dfrac{.9(1 - .9)}{100}} = .03$$

c. H_0: $p = .9$
H_a: $p < .9$

The test statistic is $z = -2.33$.

The rejection region requires $\alpha = .05$ in the lower tail of the z distribution. From Table IV, Appendix B, $z_{.05} = 1.645$. The rejection region is $z < -1.645$.

Since the observed value of the test statistic falls in the rejection region ($z = -2.33 < -1.645$), H_0 is rejected. There is sufficient evidence to indicate p is less than .9 at $\alpha = .05$.

d. p-value $= P(z \le -2.33) = .5 - P(0 < z < 2.33)$
$= .5 - .4901 = .0099$

There is a strong indication that p is less than .9, since we would observe a test statistic this extreme or more extreme only 99 times in 10,000 if $p = .9$. We would reject H_0 in favor of H_a if $\alpha > .0099$.

8.51 Since p is the proportion of consumers who do not like the snack food, $\hat{p}$ will be:

$$\hat{p} = \dfrac{\text{number of 0's in the sample}}{n} = \dfrac{29}{50} = .58$$

First, check to see if the normal approximation will be adequate:

$$p_0 \pm 3\sigma_{\hat{p}} \Rightarrow p_0 \pm 3\sqrt{\dfrac{pq}{n}} \Rightarrow p_0 \pm 3\sqrt{\dfrac{p_0 q_0}{n}}$$

$$\Rightarrow .5 \pm 3\sqrt{\dfrac{.5(1 - .5)}{50}} \Rightarrow .5 \pm .2121 \Rightarrow (.2879, .7121)$$

Since the interval lies completely in the interval $(0, 1)$, the normal approximation will be adequate.

a. H_0: $p = .5$
   H_a: $p > .5$

The test statistic is $z = \dfrac{\hat{p} - p_0}{\sigma_{\hat{p}}} = \dfrac{\hat{p} - p_0}{\sqrt{\dfrac{p_0 q_0}{n}}} = \dfrac{.58 - .5}{\sqrt{\dfrac{.5(1 - .5)}{50}}} = 1.13$

The rejection region requires $\alpha = .10$ in the upper tail of the z distribution. From Table IV, Appendix B, $z_{.10} = 1.28$. The rejection region is $z > 1.28$.

Since the observed value of the test statistic does not fall in the rejection region ($z = 1.13 \not> 1.28$), H_0 is not rejected. There is insufficient evidence to indicate the proportion of customers who do not like the snack food is greater than .5 at $\alpha = .10$.

b. p-value $= P(z \geq 1.13) = .5 - P(0 < z < 1.13)$
   $= .5 - .3780 = .1292$

8.53   a. $\hat{p} = \dfrac{497,584}{1,000,000} = .497584$

To test whether p is less than .5, we test:

H_0: $p = .5$
H_a: $p < .5$

The test statistic is $z = \dfrac{\hat{p} - p_0}{\sigma_{\hat{p}}} = \dfrac{\hat{p} - p_0}{\sqrt{\dfrac{p_0 q_0}{n}}} = \dfrac{.497584 - .5}{\sqrt{\dfrac{.5(.5)}{1,000,000}}} = -4.83$

The rejection region requires $\alpha = .01$ in the lower tail of the z distribution. From Table IV, Appendix B, $z_{.01} = 2.33$. The rejection region is $z < -2.33$.

Since the observed value of the test statistic falls in the rejection region ($z = -4.83 < -2.33$), H_0 is rejected. There is sufficient evidence to indicate the long-run proportion of games won is less than .5 for $\alpha = .01$.

b. Yes. The claim was that the system taught guarantees winning results in the long run. The test shows that, in the long run, the results will be "not winning".

c. To test whether p is less than .5, we test:

H_0: $p = .5$
H_a: $p < .5$

The test statistic is $z = \dfrac{\hat{p} - p_0}{\sigma_{\hat{p}}} = \dfrac{\hat{p} - p_0}{\sqrt{\dfrac{p_0 q_0}{n}}} = \dfrac{.4976 - .5}{\sqrt{\dfrac{.5(.5)}{10,000}}} = -.48$

The rejection region requires $\alpha = .01$ in the lower tail of the z distribution. From Table IV, Appendix B, $z_{.01} = 2.33$. The rejection region is $z < -2.33$.

Since the observed value of the test statistic does not fall in the rejection region ($z = -.48 \not< -2.33$), H_0 is not rejected. There is insufficient evidence to indicate the long-run proportion of games won is less than .5 at $\alpha = .01$.

d. We must assume the sample size is sufficiently large to use the normal approximation.

8.55 a. Let p = proportion of all customers who indicate brand A is softer.

$$\hat{p} = \frac{x}{n} = \frac{146}{250} = .584$$

To test if brand A is perceived as softer than brand B, we test:

H_0: $p = .5$
H_a: $p > .5$

The test statistic is $z = \dfrac{\hat{p} - p_0}{\sigma_{\hat{p}}} = \dfrac{\hat{p} - p_0}{\sqrt{\dfrac{p_0 q_0}{n}}} = \dfrac{.584 - .5}{\sqrt{\dfrac{.5(.5)}{250}}} = 2.66$

The rejection region requires $\alpha = .05$ in the upper tail of the z distribution. From Table IV, Appendix B, $z_{.05} = 1.645$. The rejection region is $z > 1.645$.

Since the observed value of the test statistic falls in the rejection region ($z = 2.66 > 1.645$), H_0 is rejected. There is sufficient evidence to indicate brand A is perceived as softer than brand B for $\alpha = .05$.

b. The p-value for this test is p-value $= P(z \geq 2.66) = .5 - .4961 = .0039$. Since the p-value is less than $\alpha = .05$, there is evidence to reject H_0. There is sufficient evidence to indicate brand A is perceived as softer than brand B for $\alpha = .05$.

8.57 Let p = proportion of all households in the area equipped with VCRs. If at least 5,000 of the 20,000 households are equipped with VCRs, the proportion is $p = 5,000/20,000 = .25$.

$$\hat{p} = \frac{x}{n} = \frac{96}{300} = .32$$

To test if at least 5,000 of the households in the area are equipped with VCRs, we test:

H_0: $p = .25$
H_a: $p > .25$

The test statistic is $z = \dfrac{\hat{p} - p_0}{\sigma_{\hat{p}}} = \dfrac{\hat{p} - p_0}{\sqrt{\dfrac{p_0 q_0}{n}}} = \dfrac{.32 - .25}{\sqrt{\dfrac{.25(.75)}{300}}} = 2.80$

The rejection region requires $\alpha = .01$ in the upper tail of the z distribution. From Table IV, Appendix B, $z_{.01} = 2.33$. The rejection region is $z > 2.33$.

Since the observed value of the test statistic falls in the rejection region ($z = 2.80 > 2.33$), H_0 is rejected. There is sufficient evidence to indicate at least 5,000 of the households in the area are equipped with VCRs for $\alpha = .01$.

8.59   The power of a test increases when:

        1.   The distance between the null and alternative values of μ increases.
        2.   The value of α increases.
        3.   The sample size increases.

8.61   From Exercise 8.60 we want to test $H_0: \mu = 500$ against $H_a: \mu > 500$ using $\alpha = .05$, $\sigma = 100$, $n = 25$, and $\bar{x} = 532.9$.

a. $\beta = P(\bar{x}_0 < 532.9$ when $\mu = 575)$.

$$z = \frac{\bar{x}_0 - \mu_a}{\sigma_{\bar{x}}} \approx \frac{\bar{x}_0 - \mu_a}{s/\sqrt{n}} = \frac{532.9 - 575}{\dfrac{100}{\sqrt{25}}} = -2.11$$

$\beta = P(z < -2.11) = .5 - .4826 = .0174$

b.   Power $= 1 - \beta = 1 - .0174 = .9826$

c.   In Exercise 8.60, $\beta = .1949$ and the power is .8051. The value of β has decreased in this exercise since $\mu = 575$ is further from the hypothesized value than $\mu = 500$. As a result, the power of the test in this exercise has increased (when β decreases, the power of the test increases).

8.63   From Exercise 8.62 we want to test $H_0: \mu = 75$ against $H_a: \mu < 75$ using $\alpha = .10$, $\sigma = 15$, $n = 49$, and $\bar{x}_0 = 72.257$.

Now, find

$\beta = P(\bar{x}_0 > 72.257$ when $\mu = 73)$.

$$z = \frac{\bar{x}_0 - \mu_a}{\sigma_{\bar{x}}} \approx \frac{\bar{x}_0 - \mu_a}{s/\sqrt{n}} = \frac{72.257 - 73}{\dfrac{15}{\sqrt{49}}} = -.35$$

$\beta = P(z > -.35) = .5 + .1368 = .6368$
Power $= 1 - \beta = 1 - .6368 = .3632$

In Exercise 8.62, when $\mu = 70$, the power was .8531. The power of the test decreased with $\mu = 73$ since the distance between the null and alternative values decreased. ($\mu = 73$ is closer to $\mu_0 = 75$ than $\mu = 70$).

8.65   a.   The sampling distribution of $\bar{x}$ will be approximately normal (by the Central Limit Theorem) with $\mu_{\bar{x}} = \mu = 30$ and $\sigma_{\bar{x}} = \dfrac{\sigma}{\sqrt{n}} = \dfrac{1.2}{\sqrt{121}} = .109$.

   b.   The sampling distribution of $\bar{x}$ will be approximately normal (CLT) with
$\mu_{\bar{x}} = \mu = 29.8$ and $\sigma_{\bar{x}} = \dfrac{\sigma}{\sqrt{n}} = \dfrac{1.2}{\sqrt{121}} = .109$.

   c.   First, find

$$\bar{x}_{O,L} = \mu_0 - z_{\alpha/2}\sigma_{\bar{x}} = \mu_0 - z_{\alpha/2}\dfrac{\sigma}{\sqrt{n}} \qquad \text{where } z_{.05/2} = z_{.025} = 1.96 \text{ from Table}$$

IV, Appendix B.

Thus, $\bar{x}_{O,L} = 30 - 1.96\dfrac{1.2}{\sqrt{121}} = 29.79$

$$\bar{x}_{O,U} = \mu_0 + z_{\alpha/2}\sigma_{\bar{x}} = \mu_0 + z_{\alpha/2}\dfrac{\sigma}{\sqrt{n}} = 30 + 1.96\dfrac{1.2}{\sqrt{121}} = 30.21$$

Now, find

$$\beta = P(29.79 < \bar{x} < 30.21 \text{ when } \mu = 29.8)$$

$$z = \dfrac{\bar{x}_{O,L} - \mu_a}{\sigma_{\bar{x}}} \approx \dfrac{\bar{x}_0 - \mu_a}{s/\sqrt{n}} = \dfrac{29.79 - 29.8}{\dfrac{1.2}{\sqrt{121}}} = -.09$$

$$z = \dfrac{\bar{x}_{O,U} - \mu_a}{\sigma_{\bar{x}}} \approx \dfrac{\bar{x}_0 - \mu_a}{s/\sqrt{n}} = \dfrac{30.21 - 29.8}{\dfrac{1.2}{\sqrt{121}}} = 3.76$$

$$\beta = P(-.09 < z < 3.76) = P(0 < z < 3.76) + P(0 < z < .09)$$
$$= .5 + .0359 = .5359$$

   d. $\beta = P(29.79 < \bar{x} < 30.21 \text{ when } \mu = 30.4)$

$$z = \dfrac{\bar{x}_{O,L} - \mu_a}{\sigma_{\bar{x}}} \approx \dfrac{\bar{x}_{O,L} - \mu_a}{s/\sqrt{n}} = \dfrac{29.79 - 30.4}{\dfrac{1.2}{\sqrt{121}}} = -5.59$$

$$z = \dfrac{\bar{x}_{O,U} - \mu_a}{\sigma_{\bar{x}}} \approx \dfrac{\bar{x}_{O,L} - \mu_a}{s/\sqrt{n}} = \dfrac{30.21 - 30.4}{\dfrac{1.2}{\sqrt{121}}} = -1.74$$

$$\beta = P(-5.59 < z < -1.74) = .5 - .4591 = .0409$$

8.67   Referring to Exercise 8.18, we want to test $H_0: \mu = 10$ against $H_a: \mu < 10$.

First, find

$$\bar{x}_0 = \mu_0 - z_\alpha \sigma_{\bar{x}} = \mu_0 - z_\alpha \frac{\sigma}{\sqrt{n}} \qquad \text{where } z_{.05} = 1.645 \text{ from Table IV, Appendix B.}$$

Thus, $\bar{x}_0 = 10 - 1.645 \dfrac{1.2}{\sqrt{48}} = 9.7151$

Now, find

$$\beta = P(\bar{x}_0 > 9.7151 \text{ when } \mu = 9.5)$$

$$z = \frac{\bar{x}_0 - \mu_a}{\sigma_{\bar{x}}} \approx \frac{\bar{x}_0 - \mu_a}{s/\sqrt{n}} = \frac{9.7151 - 9.5}{1.2/\sqrt{48}} = 1.24$$

$$\beta = P(z > 1.24) = .5 - P(0 < z < 1.24) = .5 - .3925 = .1075$$

8.69   First, find:

$$\bar{x}_0 = \mu_0 + z_\alpha \sigma_{\bar{x}} \approx \mu_0 + z_\alpha \frac{\sigma}{\sqrt{n}} \qquad \text{where } z_{.05} = 1.645 \text{ from Table IV, Appendix B.}$$

Thus, $\bar{x}_0 = 35 + 1.645 \dfrac{6}{\sqrt{100}} = 35.987$

Now, find β for each value of μ_a.

$\mu = \mathbf{35.5}$

$$\beta = P(\bar{x}_0 < 35.987 \text{ when } \mu = 35.5)$$

$$z = \frac{\bar{x}_0 - \mu_a}{\sigma_{\bar{x}}} \approx \frac{\bar{x}_0 - \mu_a}{s/\sqrt{n}} = \frac{35.987 - 35.5}{6/\sqrt{100}} = .81$$

$$\beta = P(z < .81) = .5 + .2910 = .7910$$

$$\text{Power} = 1 - \beta = 1 - .7910 = .2090$$

$\mu = \mathbf{36.0}$

$$\beta = P(\bar{x}_0 < 35.987 \text{ when } \mu = 36.0)$$

$$z = \frac{\bar{x}_0 - \mu_a}{\sigma_{\bar{x}}} \approx \frac{\bar{x}_0 - \mu_a}{s/\sqrt{n}} = \frac{35.987 - 36}{6/\sqrt{100}} = -.02$$

$$\beta = P(z < -.02) = .5 - P(0 < z < .02) = .5 - .0080 = .4920$$

$$\text{Power} = 1 - \beta = 1 - .4920 = .5080$$

$\mu = 36.5$

$\beta = P(\bar{x}_0 < 35.987 \text{ when } \mu = 36.5)$

$$z = \frac{\bar{x}_0 - \mu_a}{\sigma_{\bar{x}}} \approx \frac{\bar{x}_0 - \mu_a}{s/\sqrt{n}} = \frac{35.987 - 36.5}{6/\sqrt{100}} = -.86$$

$\beta = P(z < -.86) = .5 - .3051 = .1949$

Power $= 1 - \beta = 1 - .1949 = .8051$

$\mu = 37.0$

$\beta = P(\bar{x}_0 < 35.987 \text{ when } \mu = 37.0)$

$$z = \frac{\bar{x}_0 - \mu_a}{\sigma_{\bar{x}}} \approx \frac{\bar{x}_0 - \mu_a}{s/\sqrt{n}} = \frac{35.987 - 37}{6/\sqrt{100}} = -1.69$$

$\beta = P(z < -1.69) = .5 - .4545 = .0455$

Power $= 1 - \beta = 1 - .0455 = .9545$

$\mu = 37.5$

$\beta = P(\bar{x}_0 < 35.987 \text{ when } \mu = 37.5)$

$$z = \frac{\bar{x}_0 - \mu_a}{\sigma_{\bar{x}}} \approx \frac{\bar{x}_0 - \mu_a}{s/\sqrt{n}} = \frac{35.987 - 37.5}{6/\sqrt{100}} = -2.52$$

$\beta = P(z < -2.52) = .5 - .4941 = .0059$

Power $= 1 - \beta = 1 - .0059 = .9941$

When $n = 100$ instead of $n = 36$, the power curve shifts upward; the power becomes larger in each case.

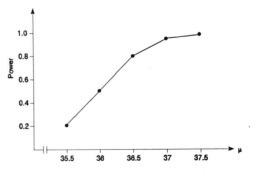

8.71 The smaller the *p*-value associated with a test of hypothesis, the stronger the support for the **alternative** hypothesis. The *p*-value is the probability of observing your test statistic or anything more unusual, given the null hypothesis is true. If this value is small, it would be very unusual to observe this test statistic if the null hypothesis were true. Thus, it would indicate the alternative hypothesis is true.

8.73 a. The hypotheses would be:

H_0: Individual does not have the disease
H_a: Individual does have the disease

b. A Type I error would be: Conclude the individual has the disease when in fact he/she does not. This would be a false positive test.

A Type II error would be: Conclude the individual does not have the disease when in fact he/she does. This would be a false negative test.

c. If the disease is serious, either error would be grave. Arguments could be made for either error being more grave. However, I believe a Type II error would be more grave: Concluding the individual does not have the disease when he/she does. This person would not receive critical treatment, and may suffer very serious consequences. Thus, it is more important to minimize β.

8.75 There is not a direct relationship between α and β. That is, if α is known, it does not mean β is known because β depends on the value of the parameter in the alternative hypothesis and the sample size. However, as α decreases, β increases for a fixed value of the parameter and a fixed sample size.

8.77 a. A Type II error results if we decide not to halt production when we should have since the mean amount of PCB exceeds 3 parts per million.

b. First, find

$$\bar{x}_0 = \mu_0 + z_\alpha \sigma_{\bar{x}} \approx \mu_0 + z_{.01}\frac{s}{\sqrt{n}} \quad \text{where } z_{.01} = 2.33 \text{ from Table IV,}$$

Appendix B.

$$\bar{x}_0 = 3 + 2.33\frac{.5}{\sqrt{50}} = 3.165$$

Then compute

$$\beta = P(\bar{x}_0 < 3.165 \text{ when } \mu = 3.1)$$

$$z = \frac{\bar{x}_0 - \mu_a}{\sigma_{\bar{x}}} \approx \frac{\bar{x}_0 - \mu_a}{s/\sqrt{n}} = \frac{3.165 - 3.1}{.5/\sqrt{50}} = .92$$

$$\beta = P(z < .92) = .5 + P(0 < z < .92) = .5 + .3212 = .8212$$

c. Power $= 1 - \beta = 1 - .8212 = .1788$

d. From part b, $\bar{x}_0 = 3.165$

Then, compute

$$\beta = P(\bar{x}_0 < 3.165 \text{ when } \mu = 3.2)$$

$$z = \frac{\bar{x}_0 - \mu_a}{\sigma_{\bar{x}}} \approx \frac{\bar{x}_0 - \mu_a}{s/\sqrt{n}} = \frac{3.165 - 3.2}{.5/\sqrt{50}} = -.49$$

$$\beta = P(z < -.49) = .5 - .1879 = .3121$$

Power $= 1 - \beta = 1 - .3121 = .6879$

The power becomes larger as the plant's mean PCB departs further from the standard.

8.79 **a.** Some preliminary calculations are:

$$\bar{x} = \frac{\sum x}{n} = \frac{648.8}{6} = 108.133$$

$$s^2 = \frac{\sum x^2 - \frac{(\sum x)^2}{n}}{n - 1} = \frac{90{,}421.16 - \frac{648.8^2}{6}}{6 - 1} = 4{,}052.850666$$

$$s = \sqrt{4{,}052.850666} = 63.662$$

To determine if the annual charitable contributions by millionaires have increased since 1989, we test:

H_0: $\mu = 83.929$
H_a: $\mu > 83.929$

The test statistic is $t = \dfrac{\bar{x} - \mu_0}{s/\sqrt{n}} = \dfrac{108.133 - 83.929}{63.662/\sqrt{6}} = .93$

The rejection region requires $\alpha = .05$ in the upper tail of the t distribution with df $= n - 1 = 6 - 1 = 5$. From Table VI, Appendix B, $t_{.05} = 2.015$. The rejection region is $t > 2.015$.

Since the observed value of the test statistic does not fall in the rejection region, ($t = .93$ $\not> 2.015$), H_0 is not rejected. There is insufficient evidence to indicate the annual charitable contributions by millionaires have increased since 1989 for $\alpha = .05$.

b. We must assume that the population of annual charitable contributions by millionaires is normally distributed. Since a few millionaires are very generous, this distribution may be skewed to the right. Thus, this assumption may not be valid.

8.81 From Exercise 8.80, the test statistic is $t = -1.28$. From Table VI, Appendix B, with df $= n - 1 = 20 - 1 = 19$, $P(t \leq -1.28) > .10$. Thus, the p-value $> .10$.

8.83   From Exercise 8.82, $z = -3.03$ for a one-tailed, lower tail test.

$$p\text{-value} = P(z \le -3.03) = .5 - P(0 < z < 3.03)$$
$$= .5 - .4988$$
$$= .0012$$

Since the p-value is so small, there is strong evidence to reject H_0 for $\alpha > .0012$.

8.85   To determine if the discount chain's tires are more resistant to wear, we test:

H_0: $p = .5$
H_a: $p > .5$ where p = proportion of cars in which the discount tire shows less wear.

First, check to see if the normal approximation will be adequate:

$$p_0 \pm 3\sigma_{\hat{p}} \Rightarrow p_0 \pm 3\sqrt{\frac{pq}{n}} \Rightarrow p_0 \pm 3\sqrt{\frac{p_0 q_0}{n}}$$

$$\Rightarrow .5 \pm 3\sqrt{\frac{.5(.5)}{40}} \Rightarrow .5 \pm .237 \Rightarrow (.263, .737)$$

Since the interval lies completely in the interval $(0, 1)$, the normal approximation will be adequate.

H_0: $p = .5$
H_a: $p > .5$

The test statistic is $z = \dfrac{\hat{p} - p_0}{\sigma_{\hat{p}}} = \dfrac{\hat{p} - p_0}{\sqrt{\dfrac{p_0 q_0}{n}}} = \dfrac{.8 - .5}{\sqrt{\dfrac{.5(.5)}{40}}} = 3.79$ where $\hat{p} = \dfrac{32}{40} = .8$

$$p\text{-value} = P(z \ge 3.79) = .5 - P(0 < z < 3.79)$$
$$\approx .5 - .5$$
$$= 0$$

Reject H_0 if p-value $< \alpha$.

Reject H_0 no matter what α is used. There is sufficient evidence to conclude the discount chain's claim is correct ($p > .5$) for $\alpha > 0$.

8.87   a.   The value of the test statistic is $t = 2.408$. The p-value is .0304, which corresponds to a two-tailed test.

     $P(t \ge 2.408) + P(t \le -2.408) = .0304$. Since the p-value is less than $\alpha = .10$, H_0 is rejected. There is sufficient evidence to indicate the mean beta coefficient of high technology stock is different than 1.

     b.   The p-value would be $.0304/2 = .0152$.

8.89   Using Table IV, Appendix B, the p-value is $P(z > 1.41) = .5 - .4207 = .0793$. The probability of observing a test statistic of 1.41 or anything more unusual is .0793 if H_0 is true. This is not particularly small. There is no evidence to reject H_0 for $\alpha = .05$. There is evidence to reject H_0 for $\alpha = .10$.

8.91   a.   Let p = proportion of residents in the city favoring the increase in property tax.

To determine whether a majority of residents in the city favor the tax, we test:

$$H_0: \ p = .5$$
$$H_a: \ p > .5$$

b.   The test statistic is $z = 1.84$ and the p-value is .0329. Since the p-value is small, there is evidence to reject H_0 for $\alpha > .0329$. There is sufficient evidence to indicate a majority of residents in the city favor the tax at $\alpha > .0329$.

c.   We must assume our sample size was large enough to use the normal approximation.

CHAPTER NINE

....

Inferences Based on Two Samples:
Estimation and Tests of Hypotheses

9.1   a.   From Chapter 6, $\sigma_{\bar{x}_1} = \dfrac{\sigma_1}{\sqrt{n_1}} = \dfrac{\sqrt{900}}{\sqrt{100}} = \dfrac{30}{10} = 3$

$\mu_1 \pm 2\sigma_{\bar{x}_1} \Rightarrow 150 \pm 2(3) \Rightarrow 150 \pm 6 \Rightarrow (144, 156)$

b. $\sigma_{\bar{x}_2} = \dfrac{\sigma_2}{\sqrt{n_2}} = \dfrac{\sqrt{1600}}{\sqrt{100}} = \dfrac{40}{10} = 4$

$\mu_2 \pm 2\sigma_{\bar{x}_2} \Rightarrow 150 \pm 2(4) \Rightarrow 150 \pm 8 \Rightarrow (142, 158)$

c. $\mu_{\bar{x}_1-\bar{x}_2} = \mu_1 - \mu_2 = 150 - 150 = 0$

$\sigma_{\bar{x}_1-\bar{x}_2} = \sqrt{\dfrac{\sigma_1^2}{n_1} + \dfrac{\sigma_2^2}{n_2}} = \sqrt{\dfrac{900}{100} + \dfrac{1600}{100}} = \sqrt{25} = 5$

d. $\mu_{\bar{x}_1-\bar{x}_2} \pm 2\sigma_{\bar{x}_1-\bar{x}_2} \Rightarrow 0 \pm 2(5) \Rightarrow 0 \pm 10 \Rightarrow (-10, 10)$

e.   In general, the variability of the difference between independent sample means is larger than the variability of individual sample means.

9.3   From Exercise 9.2, $\sigma_{\bar{x}_1-\bar{x}_2} = \sqrt{\dfrac{\sigma_1^2}{n_1} + \dfrac{\sigma_2^2}{n_2}} = \sqrt{\dfrac{4^2}{64} + \dfrac{3^2}{64}} = \sqrt{\dfrac{25}{64}} = \dfrac{5}{8} = .625$

$\mu_{\bar{x}_1-\bar{x}_2} = \mu_1 - \mu_2 = 12 - 10 = 2$

a. $z = \dfrac{(\bar{x}_1 - \bar{x}_2) - (\mu_1 - \mu_2)}{\sigma_{\bar{x}_1-\bar{x}_2}} = \dfrac{2.25 - 2}{.625} = .4$

b. $P[(\bar{x}_1 - \bar{x}_2) > 2.25] = P(z > .4) = .5 - .1554 = .3446$ (from Table IV, Appendix B).

c. $P[(\bar{x}_1 - \bar{x}_2) < 2.25] = P(z < .4) = .5 + .1554 = .6554$ (from Table IV, Appendix B).

d. $z = \dfrac{-.8 - 2}{.625} = -4.48$ $\qquad$ $z = \dfrac{.8 - 2}{.625} = -1.92$

$$P[(\bar{x}_1 - \bar{x}_2) < -.8] + P[(\bar{x}_1 - \bar{x}_2) > .8] = P(z < -4.48) + P(z > -1.92)$$
$$= .5 - .5 + (.5 + .4726) = .9726$$

(from Table IV, Appendix B)

9.5 The form of the confidence interval is:

$$(\bar{x}_1 - \bar{x}_2) \pm z_{\alpha/2}\sqrt{\dfrac{\sigma_1^2}{n_1} + \dfrac{\sigma_2^2}{n_2}}$$

For confidence coefficient .95, $\alpha = 1 - .95 = .05$ and $\alpha/2 = .05/2 = .025$. From Table IV, Appendix B, $z_{.025} = 1.96$. The confidence interval is:

$$(15.5 - 26.6) \pm 1.96\sqrt{\dfrac{9}{100} + \dfrac{16}{100}} \Rightarrow -11.1 \pm .98 \Rightarrow (-12.08, -10.12)$$

We are 95% confident that the difference in the two means is between -12.08 and -10.12. The confidence interval gives more information.

9.7 a. The test statistic is $z = -1.576$ and the p-value $= .1150$. Since the p-value is not small, there is no evidence to reject H_0 for $\alpha \le .10$. There is insufficient evidence to indicate the two population means differ for $\alpha \le .10$.

b. If the alternative hypothesis had been one-tailed, the p-value would be half of the value for the two-tailed test. Here, p-value $= .1150/2 = .0575$.

There is no evidence to reject H_0 for $\alpha = .05$. There is insufficient evidence to indicate the mean for population 1 is less than the mean for population 2 at $\alpha = .05$.

There is evidence to reject H_0 for $\alpha > .0575$. There is sufficient evidence to indicate the mean for population 1 is less than the mean for population 2 at $\alpha > .0575$.

9.9 a. To determine if the new method will reduce the mean unloading time, we test:

$$H_0: \mu_1 - \mu_2 = 0$$
$$H_a: \mu_1 - \mu_2 < 0$$

where μ_1 = mean unloading time for the new method, and
μ_2 = mean unloading time for the current method

The test statistic is $z = \dfrac{(\bar{x}_1 - \bar{x}_2) - 0}{\sqrt{\dfrac{\sigma_1^2}{n_1} + \dfrac{\sigma_2^2}{n_2}}} \approx \dfrac{(25.4 - 27.3) - 0}{\sqrt{\dfrac{3.1^2}{50} + \dfrac{3.7^2}{50}}} = -2.78$

The rejection region requires $\alpha = .05$ in the lower tail of the z distribution. From Table IV, Appendix B, $z_{.05} = 1.645$. The rejection region is $z < -1.645$.

Since the observed value of the test statistic falls in the rejection region ($z = -2.78 <$ -1.645), H_0 is rejected. There is sufficient evidence to indicate that the mean unloading time for the new method is less than the mean unloading time for the current method at $\alpha = .05$.

b.   The p-value is $P(z \le -2.78) = .5 - .4973 = .0027$.

9.11   a.   To determine if the decline in the average number of defects per machine between 1979 and 1982 is significant, we test:

H_0: $\mu_1 = \mu_2$
H_a: $\mu_1 > \mu_2$

where μ_1 = average number of defects per machine in 1979, and
    μ_2 = average number of defects per machine in 1982.

The test statistic is $z = \dfrac{(\bar{x}_1 - \bar{x}_2) - 0}{\sqrt{\dfrac{\sigma_1^2}{n_1} + \dfrac{\sigma_2^2}{n_2}}} \approx \dfrac{(4.2 - 1.3) - 0}{\sqrt{\dfrac{2^2}{100} + \dfrac{1.1^2}{100}}} = 12.71$

p-value $= P(z \ge 12.71) = .5 - P(0 < z < 12.71) = .5 - .5 = 0$

This is highly significant. We would reject H_0 for any value of α. There is sufficient evidence to indicate that the decline in the average number of defects per machine between 1979 and 1982 is statistically significant.

b.   A Type I error is to reject H_0 when H_0 is true. In this case you decide there is a significant decline in the average number of defects when there is not.

A Type II error is to accept H_0 when it is false. In this case, you decide there is not a significant decline in the average number of defects when there is a significant decline.

9.13   a.   Population 1 is the number of items produced in every 1 hour time interval for machine 1, and population 2 is the number of items produced in every 1 hour time interval for machine 2.

b.   To determine if machine 1 produces more items than machine 2, on average, we test:

H_0: $\mu_1 - \mu_2 = 0$
H_a: $\mu_1 - \mu_2 > 0$

where μ_1 is the mean production for machine 1 and μ_2 is the mean production for population 2.

The test statistic is $z = \dfrac{(\bar{x}_1 - \bar{x}_2) - 0}{\sqrt{\dfrac{\sigma_1^2}{n_1} + \dfrac{\sigma_2^2}{n_2}}} \approx \dfrac{(51.4 - 49.5) - 0}{\sqrt{\dfrac{2.1^2}{35} + \dfrac{1.8^2}{45}}} = \dfrac{1.9}{.4450} = 4.27$

The rejection region requires $\alpha = .10$ in the upper tail of the z distribution. From Table IV, Appendix B, $z_{.10} = 1.28$. The rejection region is $z > 1.28$.

Since the observed value of the test statistic falls in the rejection region ($z = 4.27 > 1.28$), H_0 is rejected. There is sufficient evidence to indicate the mean number of items produced by machine 1 exceeds that of machine 2 at $\alpha = .10$.

    c.   The p-value is $P(z \geq 4.27) \approx .5 - .5 = 0$ (from Table IV, Appendix B).

9.15   a.   The two populations are the asking prices for all houses in both Edina and St. Louis Park.

    b.   The parameters of interest are the mean asking prices for all houses in both Edina and St. Louis Park.

    c.   To determine if the mean price of a home in Edina is more than \$60,000 greater than the mean price at St. Louis Park, we test:

$$H_0: \mu_1 - \mu_2 = 60,000$$
$$H_a: \mu_1 - \mu_2 > 60,000$$

The test statistic is $z = \dfrac{(\bar{x}_1 - \bar{x}_2) - D_0}{\sqrt{\dfrac{\sigma_1^2}{n_1} + \dfrac{\sigma_2^2}{n_2}}} = \dfrac{(203,142 - 94,300) - 60,000}{\sqrt{\dfrac{112,285^2}{30} + \dfrac{30,687^2}{30}}}$

$$= \dfrac{48,842}{21,252.1499} = 2.30$$

The rejection region requires $\alpha = .05$ in the upper tail of the z distribution. From Table IV, Appendix B, $z_{.05} = 1.645$. The rejection region is $z > 1.645$.

Since the observed value of the test statistic falls in the rejection region ($z = 2.30 > 1.645$), H_0 is rejected. There is sufficient evidence to indicate the mean price of a home in Edina is more than \$60,0000 greater than the mean price at St. Louis Park at $\alpha = .05$.

    d.   For confidence coefficient .95, $\alpha = 1 - .95 = .05$ and $\alpha/2 = .05/2 = .025$. From Table IV, Appendix B, $z_{.025} = 1.96$. The confidence interval is:

$$(\bar{x}_1 - \bar{x}_2) \pm z_{\alpha/2} \sqrt{\dfrac{\sigma_1^2}{n_1} + \dfrac{\sigma_2^2}{n_2}}$$

$$\Rightarrow (203,142 - 94,300) \pm 1.96 \sqrt{\dfrac{112,285^2}{30} + \dfrac{30,687^2}{30}}$$

$$\Rightarrow 108,842 \pm 41,654.21 \Rightarrow (\$67,187.79, \$150,496.21)$$

9.17   a.   No. Both populations must be normal.

    b.   No. Both population variances must be equal.

c. No. Both populations must be normal.

d. Yes.

e. No. Both populations must be normal.

9.19 a. $s_p^2 = \dfrac{(n_1 - 1)s_1^2 + (n_2 - 1)s_2^2}{n_1 + n_2 - 2} = \dfrac{(25 - 1)120 + (25 - 1)100}{25 + 25 - 2} = \dfrac{5280}{48} = 110$

b. $s_p^2 = \dfrac{(20 - 1)12 + (10 - 1)20}{20 + 10 - 2} = \dfrac{408}{28} = 14.5714$

c. $s_p^2 = \dfrac{(6 - 1).15 + (10 - 1).2}{6 + 10 - 2} = \dfrac{2.55}{14} = .1821$

d. $s_p^2 = \dfrac{(16 - 1)3000 + (17 - 1)2500}{16 + 17 - 2} = \dfrac{85,000}{31} = 2741.9355$

e. s_p^2 falls near the variance with the larger sample size.

9.21 a. From Exercise 9.20, $\bar{x}_1 = 2.683$, $s_1^2 = .6057$, $\bar{x}_2 = 3.58$, $s_2^2 = .202$, $s_p^2 = .4263$

From confidence coefficient .90, $\alpha = 1 - .90 = .10$ and $\alpha/2 = .10/2 = .05$. From Table VI, Appendix B, with df $= n_1 + n_2 - 2 = 6 + 5 - 2 = 9$, $t_{.05} = 1.833$. The confidence interval is:

$$(\bar{x}_1 - \bar{x}_2) \pm t_{.05}\sqrt{s_p^2\left[\frac{1}{n_1} + \frac{1}{n_2}\right]} \Rightarrow (2.683 - 3.58) \pm 1.833\sqrt{.4263\left[\frac{1}{6} + \frac{1}{5}\right]}$$

$$\Rightarrow -.897 \pm .725 \Rightarrow (-1.622, -.172)$$

We are 90% confident the difference between the 2 population means falls between -1.622 and $-.172$.

b. The confidence interval gives more information. The test of hypothesis indicated that $\mu_1 < \mu_2$, but the confidence interval gives us an idea of by how much $\mu_1 < \mu_2$.

9.23 From Exercise 9.22, $s_p^2 = 15.3612$. For confidence coefficient .95, $\alpha = 1 - .95 = .05$ and $\alpha/2 = .05/2 = .025$. From Table VI, Appendix B, with df $= n_1 + n_2 - 2 = 12 + 16 - 2 = 26$, $t_{.025} = 2.056$. The confidence interval is:

$$(\bar{x}_1 - \bar{x}_2) \pm t_{.025}\sqrt{s_p^2\left[\frac{1}{n_1} + \frac{1}{n_2}\right]} \Rightarrow (35 - 43) \pm 2.056\sqrt{15.3612\left[\frac{1}{12} + \frac{1}{16}\right]}$$

$$\Rightarrow -8 \pm 3.077 \Rightarrow (-11.077, -4.923)$$

We are 95% confident the difference between the 2 population means falls between -11.077 and -4.923.

9.25   a.   Let μ_1 = mean rate for cable companies with no competition and μ_2 = mean rate for cable companies with competition.

To determine if the mean rate for cable companies with no competition is higher than that for companies with competition, we test:

H_0: $\mu_1 - \mu_2 = 0$
H_a: $\mu_1 - \mu_2 > 0$

b.   Some preliminary calculations are:

$$\bar{x}_1 = \frac{\sum x_1}{n_1} = \frac{144.83}{6} = 24.138$$

$$s_1^2 = \frac{\sum x_1^2 - \frac{\left(\sum x_1\right)^2}{n_1}}{n_1 - 1} = \frac{3552.3193 - \frac{144.83^2}{6}}{5} = 11.2729$$

$$\bar{x}_2 = \frac{\sum x_2}{n_2} = \frac{119.2}{6} = 19.867$$

$$s_2^2 = \frac{\sum x_2^2 - \frac{\left(\sum x_2\right)^2}{n_2}}{n_2 - 1} = \frac{2391.7322 - \frac{119.2^2}{6}}{5} = 4.7251$$

$$s_p^2 = \frac{(n_1 - 1)s_1^2 + (n_2 - 1)s_2^2}{n_1 + n_2 - 2} = \frac{5(11.2729) + 5(4.7251)}{6 + 6 - 2} = 7.999$$

The test statistic is $t = \dfrac{(\bar{x}_1 - \bar{x}_2) - D_0}{\sqrt{s_p^2\left[\dfrac{1}{n_1} + \dfrac{1}{n_2}\right]}} = \dfrac{24.138 - 19.867 - 0}{\sqrt{7.999\left[\dfrac{1}{6} + \dfrac{1}{6}\right]}} = 2.616$

The p-value = $P(t \geq 2.616)$. Using Table VI, Appendix B, with df = $n_1 + n_2 - 2 = 6 + 6 - 2 = 10$, $.01 < p$-value $< .025$. Since the p-value is less than $\alpha = .05$, there is evidence to reject H_0. There is sufficient evidence to indicate the mean rate for cable companies with no competition is higher than that for companies with competition for $\alpha = .05$.

c.   We must assume:

1.   Both populations sampled from are normally distributed.
2.   The variances for the two populations are equal.
3.   Independent random samples were selected from each population.

9.27 Some preliminary calculations are:

$$\bar{x}_1 = \frac{\sum x_1}{n_1} = \frac{796}{20} = 39.8 \qquad \bar{x}_2 = \frac{\sum x_2}{n_2} = \frac{944}{20} = 47.2$$

$$s_1^2 = \frac{\sum x_1^2 - \frac{\left(\sum x_1\right)^2}{n_1}}{n_1 - 1} = \frac{33596 - \frac{796^2}{20}}{20 - 1} = \frac{1915.2}{19} = 100.8$$

$$s_2^2 = \frac{\sum x_2^2 - \frac{\left(\sum x_2\right)^2}{n_2}}{n_2 - 1} = \frac{48082 - \frac{944^2}{20}}{20 - 1} = \frac{3525.2}{19} = 185.537$$

$$s_p^2 = \frac{(n_1 - 1)s_1^2 + (n_2 - 1)s_2^2}{n_1 + n_2 - 2} = \frac{(20 - 1)100.8 + (20 - 1)(18.537)}{20 + 20 - 2} = 143.1684$$

a. To determine if there is a difference in the mean age of purchasers and nonpurchasers, we test:

$$H_0: \mu_1 - \mu_2 = 0$$
$$H_a: \mu_1 - \mu_2 \neq 0$$

The test statistic is $t = \dfrac{(\bar{x}_1 - \bar{x}_2) - D_0}{\sqrt{s_p^2\left[\dfrac{1}{n_1} + \dfrac{1}{n_2}\right]}} = \dfrac{(39.8 - 47.2) - 0}{\sqrt{143.1684\left[\dfrac{1}{20} + \dfrac{1}{20}\right]}} = \dfrac{-7.4}{3.7838}$

$$= -1.96$$

The rejection region requires $\alpha/2 = .10/2 = .05$ in each tail of the t distribution with df $= n_1 + n_2 - 2 = 20 + 20 - 2 = 38$. From Table VI, Appendix B, $t_{.05} \approx 1.684$. The rejection region is $t < -1.684$ or t > 1.684.

Since the observed value of the test statistic falls in the rejection region ($t = -1.96 < -1.684$), H_0 is rejected. There is sufficient evidence to indicate the mean age of purchasers and nonpurchasers differ at $\alpha = .10$.

b. The necessary assumptions are:

1. Both sampled populations are approximately normal.
2. The population variances are equal.
3. The samples are randomly and independently sampled.

c. The observed significance level is $P(t \leq -1.96) + P(t \geq 1.96) = 2P(t \geq 1.96)$

From Table VI, Appendix B, with df $= 38$, $.025 < P(t \geq 1.96) < .05$

Thus, $2(.025) < p\text{-value} < 2(.05)$ or $.05 < p\text{-value} < .10$.

d. For confidence coefficient .90, $\alpha = 1 - .90 = .10$ and $\alpha/2 = .10/2 = .05$. From Table VI, Appendix B, with df = 38, $t_{.05} \approx 1.684$. The confidence interval is:

$$(\bar{x}_1 - \bar{x}_2) \pm t_{.05} \sqrt{s_p^2 \left[\frac{1}{n_1} + \frac{1}{n_2} \right]}$$

$$\Rightarrow (39.8 - 47.2) \pm 1.684 \sqrt{143.1684 \left[\frac{1}{20} + \frac{1}{20} \right]}$$

$$\Rightarrow -7.4 \pm 6.382 \Rightarrow (-13.772, -1.028)$$

9.29 Some preliminary calculations are:

$$\bar{x}_1 = \frac{\sum x_1}{n_1} = \frac{32.81}{5} = 6.562 \qquad\qquad \bar{x}_2 = \frac{\sum x_2}{n_2} = \frac{15.59}{5} = 3.118$$

$$s_1^2 = \frac{\sum x_1^2 - \frac{\left(\sum x_1 \right)^2}{n_1}}{n_1 - 1} = \frac{221.2255 - \frac{32.81^2}{5}}{5 - 1} = \frac{5.92628}{4} = 1.48157$$

$$s_2^2 = \frac{\sum x_2^2 - \frac{\left(\sum x_2 \right)^2}{n_2}}{n_2 - 1} = \frac{54.6337 - \frac{15.59^2}{5}}{5 - 1} = \frac{6.02408}{4} = 1.50602$$

$$s_p^2 = \frac{(n_1 - 1)s_1^2 + (n_2 - 1)s_2^2}{n_1 + n_2 - 2} = \frac{(5 - 1)(1.48157) + (5 - 1)(1.50602)}{5 + 5 - 2} = 1.493795$$

a. To determine if the mean annual percentage turnover for U.S. plants exceeds that for Japanese plants, we test:

$$H_0: \ \mu_1 - \mu_2 = 0$$
$$H_a: \ \mu_1 - \mu_2 > 0$$

The test statistic is $t = \dfrac{(\bar{x}_1 - \bar{x}_2) - D_0}{\sqrt{s_p^2 \left[\dfrac{1}{n_1} + \dfrac{1}{n_2} \right]}} = \dfrac{(6.562 - 3.118) - 0}{\sqrt{1.493795 \left[\dfrac{1}{5} + \dfrac{1}{5} \right]}} = \dfrac{3.444}{.77299} = 4.46$

The rejection region requires $\alpha = .05$ in the upper tail of the t distribution with df $= n_1 + n_2 - 2 = 5 + 5 - 2 = 8$. From Table VI, Appendix B, $t_{.05} = 1.860$. The rejection region is $t > 1.860$.

Since the observed value of the test statistic falls in the rejection region ($t = 4.46 > 1.860$), H_0 is rejected. There is sufficient evidence to indicate the mean annual percentage turnover for U.S. plants exceeds that for Japanese plants at $\alpha = .05$.

b. The observed significance is $P(t \geq 4.46)$. From Table VI, Appendix B, with df = 8, $.001 < P(t \geq 4.46) < .005$.

Since the p-value is so small, there is evidence to reject H_0 for $\alpha > .005$.

c. The necessary assumptions are:

1. Both sampled populations are approximately normal.
2. The population variances are equal.
3. The samples are randomly and independently sampled.

There is no indication that the populations are not normal. Both sample variances are similar, so there is no evidence the population variances are unequal. There is no indication the assumptions are not valid.

9.31 The necessary conditions are:

1. Both sampled populations are normally distributed.
2. The samples are random and independent.

9.33 a. With $\nu_1 = 2$ and $\nu_2 = 30$,
$P(F \geq 5.39) = .01$ (Table X, Appendix B)

b. With $\nu_1 = 24$ and $\nu_2 = 10$,
$P(F \geq 2.74) = .05$ (Table VIII, Appendix B)

Thus, $P(F < 2.74) = 1 - P(F \geq 2.74) = 1 - .05 = .95$.

c. With $\nu_1 = 7$ and $\nu_2 = 1$,
$P(F \geq 236.8) = .05$ (Table VIII, Appendix B)

Thus, $P(F < 236.8) = 1 - P(F \geq 236.8) = 1 - .05 = .95$.

d. With $\nu_1 = 40$ and $\nu_2 = 40$,
$P(F > 2.11) = .01$ (Table X, Appendix B)

9.35 To test $H_0: \sigma_1^2 = \sigma_2^2$ against $H_a: \sigma_1^2 \neq \sigma_2^2$, the rejection region is $F > F_{\alpha/2}$ with $\nu_1 = 10$ and $\nu_2 = 12$.

a. $\alpha = .20, \alpha/2 = .10$
Reject H_0 if $F > F_{.10} = 2.19$ (Table VII, Appendix B)

b. $\alpha = .10, \alpha/2 = .05$
Reject H_0 if $F > F_{.05} = 2.75$ (Table VIII, Appendix B)

c. $\alpha = .05, \alpha/2 = .025$
Reject H_0 if $F > F_{.025} = 3.37$ (Table IX, Appendix B)

d. $\alpha = .02, \alpha/2 = .01$
Reject H_0 if $F > F_{.01} = 4.30$ (Table X, Appendix B)

9.37   a.   Some preliminary calculations are:

$$s_1^2 = \frac{\sum x_1^2 - \frac{\left(\sum x_1\right)^2}{n_1}}{n_1 - 1} = \frac{45.35 - \frac{14.5^2}{6}}{6 - 1} = 2.062$$

$$s_2^2 = \frac{\sum x_2^2 - \frac{\left(\sum x_2\right)^2}{n_2}}{n_2 - 1} = \frac{130.4 - \frac{21.8^2}{5}}{5 - 1} = 8.838$$

To determine if a difference exists between the population variances, we test:

H_0: $\sigma_1^2 = \sigma_2^2$
H_a: $\sigma_1^2 \neq \sigma_2^2$

The test statistic is $F = \dfrac{s_2^2}{s_1^2} = \dfrac{8.838}{2.062} = 4.29$

The rejection region requires $\alpha/2 = .05/2 = .025$ in the upper tail of the F distribution with $\nu_1 = n_2 - 1 = 5 - 1 = 4$ and $\nu_2 = n_1 - 1 = 6 - 1 = 5$. From Table IX, Appendix B, $F_{.025} = 7.39$. The rejection region is $F > 7.39$.

Since the observed value of the test statistic does not fall in the rejection region ($F = 4.29 \not> 7.39$), H_0 is not rejected. There is insufficient evidence to indicate a difference between the population variances at $\alpha = .05$.

b.   The p-value is $2P(F \geq 4.29)$. From Tables VII and VIII, with $\nu_1 = 4$ and $\nu_2 = 5$,

$2(.05) < 2P(F \geq 4.29) < 2(.10)$
$\Rightarrow .10 < p\text{-value} < .20$

There is no evidence to reject H_0 for $\alpha \leq .10$.

9.39   To determine if the variances of the number of units produced per day differ for the two arrangements, we test:

H_0: $\sigma_1^2 = \sigma_2^2$
H_a: $\sigma_1^2 \neq \sigma_2^2$

The test statistic is $F = \dfrac{s_2^2}{s_1^2} = \dfrac{3761}{1432} = 2.63$

The rejection region requires $\alpha/2 = .10/2 = .05$ in the upper tail of the F distribution with $\nu_1 = n_2 - 1 = 21 - 1 = 20$ and $\nu_2 = n_1 - 1 = 21 - 1 = 20$. From Table VIII, Appendix B, $F_{.05} = 2.12$. The rejection region is $F > 2.12$.

Since the observed value of the test statistic falls in the rejection region ($F = 2.63 > 2.12$), H_0 is rejected. There is sufficient evidence to indicate the variances of the number of units produced per day differ for the 2 arrangements at $\alpha = .10$.

Since Assembly line 1 has the smaller variance, you should choose Assembly line 1.

9.41 a. Let σ_1^2 = variance of completion times with the time guideline issued, and
σ_2^2 = variance of completion times without the time guideline issued.

Some preliminary calculations are:

$$s_1^2 = \frac{\sum x_1^2 - \frac{\left(\sum x_1\right)^2}{n_1}}{n_1 - 1} = \frac{7455.34 - \frac{270.8^2}{10}}{10 - 1} = 13.564$$

$$s_2^2 = \frac{\sum x_2^2 - \frac{\left(\sum x_2\right)^2}{n_2}}{n_2 - 1} = \frac{11366.33 - \frac{327.9^2}{10}}{10 - 1} = 68.2766$$

To determine if the variance in completion times for the 2 groups is equal, we test:

$H_0: \sigma_1^2 = \sigma_2^2$
$H_a: \sigma_1^2 \neq \sigma_2^2$

The test statistic is $F = \dfrac{s_2^2}{s_1^2} = \dfrac{68.2766}{13.564} = 5.03$

The rejection region requires $\alpha/2 = .05/2 = .025$ in the upper tail of the F distribution with $\nu_1 = n_2 - 1 = 10 - 1 = 9$ and $\nu_2 = n_1 - 1 = 10 - 1 = 9$. From Table IX, Appendix B, $F_{.025} = 4.03$. The rejection region is $F > 4.03$.

Since the observed value of the test statistic falls in the rejection region ($F = 5.03 > 4.03$), H_0 is rejected. There is sufficient evidence to indicate that the population variances are not equal at $\alpha = .05$.

The two-sample t is not appropriate in this situation. In order to use the two-sample t, you need to assume the population variances are equal. We just showed that there is evidence to indicate that the population variances are not equal.

b. The necessary assumptions are:

1. Both sampled populations are normally distributed.
2. The samples are random and independent.

c. A Type I error is rejecting H_0 when H_0 is true. In this case, you decide there is a difference in the variances of the completion times for the 2 groups, but there is not a difference.

A Type II error is accepting H_0 when it is false. In this case, you decide there is not a difference in the variances of the completion times for the 2 groups, but there is a difference.

d. The p-value is $2P(F \geq 5.03)$. From Tables IX and X, with $\nu_1 = 9$ and $\nu_2 = 9$,

$2(.01) < 2P(F \geq 5.03) < 2(.025)$
$\Rightarrow .02 < p\text{-value} < .05$

9.43 Some preliminary calculations are:

Person	Difference (Before − After)
1	−10
2	−12
3	−2
4	−7
5	−10

a. $\bar{x}_D = \dfrac{\sum x_D}{n_D} = \dfrac{-41}{5} = -8.2$

$s_D^2 = \dfrac{\sum x_D^2 - \dfrac{\left(\sum x_D\right)^2}{n_D}}{n_D - 1} = \dfrac{397 - \dfrac{(-41)^2}{5}}{5 - 1} = 15.2$

$s_D = \sqrt{s_D^2} = 3.8987$

b. $\bar{x}_1 = \dfrac{\sum x_1}{n_1} = \dfrac{373}{5} = 74.6$

$\bar{x}_2 = \dfrac{\sum x_2}{n_2} = \dfrac{414}{5} = 82.8$

$\bar{x}_1 - \bar{x}_2 = 74.6 - 82.8 = -8.2 = \bar{x}_D$

c. $H_0: \mu_1 = \mu_2$
$H_a: \mu_1 \neq \mu_2$

The test statistic is $t = \dfrac{\bar{x}_D - 0}{\dfrac{s_D}{\sqrt{n_D}}} = \dfrac{-8.2 - 0}{\dfrac{3.8987}{\sqrt{5}}} = -4.70$

The rejection region requires $\alpha/2 = .05/2 = .025$ in each tail of the t distribution with df $= n_D - 1 = 5 - 1 = 4$. From Table VI, Appendix B, $t_{.025} = 2.776$. The rejection region is $t < -2.776$ or $t > 2.776$.

Since the observed value of the test statistic falls in the rejection region ($t = -4.70 < -2.776$), H_0 is rejected. There is sufficient evidence to indicate that $\mu_1 \neq \mu_2$ at $\alpha = .05$.

d. p-value $= P(t \leq -4.70 \text{ or } t \geq 4.70)$
$= 2P(t \geq 4.70)$

From Table VI, Appendix B, with df $= 4$,

$2(.001) < 2P(t \geq 3.73) < 2(.005)$
$\Rightarrow .002 < p\text{-value} < .01$

e. The necessary assumptions are:

    1. The population of differences is normal.
    2. The differences are randomly selected.

9.45 a. H_0: $\mu_1 - \mu_2 = 0$
      H_a: $\mu_1 - \mu_2 < 0$

    The rejection region requires $\alpha = .10$ in the lower tail of the t distribution with df $= n_D - 1 = 18 - 1 = 17$. From Table VI, Appendix B, $t_{.10} = 1.333$. The rejection region is $t < -1.333$.

   b. H_0: $\mu_1 - \mu_2 = 0$
      H_a: $\mu_1 - \mu_2 < 0$

    The test statistic is $t = \dfrac{\bar{x}_D - 0}{\dfrac{s_D}{\sqrt{n_D}}} = \dfrac{-3.5 - 0}{\dfrac{\sqrt{21}}{\sqrt{18}}} = -3.24$

    The rejection region is $t < -1.333$. (Refer to part a).

    Since the observed value of the test statistic falls in the rejection region ($t = -3.24 < -1.333$), H_0 is rejected. There is sufficient evidence to indicate $\mu_1 - \mu_2 < 0$ at $\alpha = .10$.

   c. The necessary assumptions are:

    1. The population of differences is normal.
    2. The differences are randomly selected.

   d. For confidence coefficient .90, $\alpha = 1 - .90 = .10$ and $\alpha/2 = .10/2 = .05$. From Table VI, Appendix B, with df $= 17$, $t_{.05} = 1.740$. The confidence interval is:

   $$\bar{x}_D \pm t_{.05}\frac{s_D}{\sqrt{n_D}} \Rightarrow -3.5 \pm 1.740\frac{\sqrt{21}}{\sqrt{18}} \Rightarrow -3.5 \pm 1.88 \Rightarrow (-5.38, -1.62)$$

   e. The confidence interval provides more information since it gives an interval of possible values for the difference between the population means.

9.47 Some preliminary calculations are:

Pair	Difference $x - y$
1	12.1
2	12.0
3	8.8
4	7.9
5	8.7
6	6.5
7	7.4

Inferences Based on Two Samples: Estimation and Tests of Hypotheses

$$\bar{x}_D = \frac{\sum x_D}{n_D} = \frac{63.4}{7} = 9.057$$

$$s_D^2 = \frac{\sum x_D^2 - \frac{\left(\sum x_D\right)^2}{n_D}}{n_D - 1} = \frac{602.96 - \frac{(63.4)^2}{7}}{7 - 1} = 4.7895$$

$$s_D = \sqrt{s_D^2} = 2.1885$$

a. H_0: $\mu_D = 10$
    H_a: $\mu_D \neq 10$

The test statistic is $t = \dfrac{\bar{x}_D - 10}{\dfrac{s_D}{\sqrt{n_D}}} = \dfrac{9.057 - 10}{\dfrac{2.1885}{\sqrt{7}}} = -1.14$

The rejection region requires $\alpha/2 = .05/2 = .025$ in each tail of the t distribution with df $= n_D - 1 = 7 - 1 = 6$. From Table VI, Appendix B, $t_{.025} = 2.447$. The rejection region is $t < -2.447$ or $t > 2.447$.

Since the observed value of the test statistic does not fall in the rejection region ($t = -1.14 \nless -2.447$), H_0 is not rejected. There is insufficient evidence to indicate $\mu_1 - \mu_2 \neq 10$ at $\alpha = .05$.

b. The p-value $= P(t \leq -1.14$ or $t \geq 1.14)$
    $\qquad\qquad\; = 2P(t \geq 1.14)$

From Table VI, Appendix B, with df $= 6$,

$\qquad 2P(t \geq .81) > 2(.10)$
$\qquad \Rightarrow p\text{-value} > .20$

The probability of observing our test statistic or anything more unusual if the difference in the means is 10 is greater than .20. This is not unusual, so H_0 is not rejected for $\alpha \leq .10$.

9.49   Some preliminary calculations are:

Employee ID Number	Difference (1986−1987)
1011	1
0033	−2
0998	−2
0006	−1
1802	1
0246	−3
0777	−
1112	−2

$$\bar{x}_D = \frac{\sum x_D}{n_D} = \frac{-8}{7} = -1.14$$

$$s_D^2 = \frac{\sum x_D^2 - \frac{\left(\sum x_D\right)^2}{n_D}}{n_D - 1} = \frac{24 - \frac{(-8)^2}{7}}{7 - 1} = 2.4762$$

$$s_D = \sqrt{s_D^2} = 1.5736$$

a. To determine if the program has helped increase worker productivity, we test:

$$H_0: \mu_D = 0$$
$$H_a: \mu_D < 0$$

The test statistic is $t = \dfrac{\bar{x}_D - 0}{\dfrac{s_D}{\sqrt{n_D}}} = \dfrac{-1.14 - 0}{\dfrac{1.5736}{\sqrt{7}}} = -1.92$

The rejection region requires $\alpha = .10$ in the lower tail of the t distribution with df $= n_D - 1 = 7 - 1 = 6$. From Table VI, Appendix B, $t_{.10} = 1.44$. The rejection region is $t < -1.44$.

Since the observed value of the test statistic falls in the rejection region ($t = -1.92 < -1.44$), H_0 is rejected. There is sufficient evidence to indicate that the program has helped to increase worker productivity at $\alpha = .10$.

The necessary assumptions are:

1. The population of differences is normal.
2. The differences are randomly selected.

b. We could lose some of the workers along the way (promotion, no longer employed, etc.). In this particular problem, we did lose one worker which gave us a smaller sample size.

9.51 Some preliminary calculations are:

Car	Difference (Manufacturer − Competitor)
1	.4
2	.4
3	.5
4	.4
5	.6
6	.2

$$\bar{x}_D = \frac{\sum x_D}{n_D} = \frac{2.5}{6} = .4167$$

$$s_D^2 = \frac{\sum x_D^2 - \frac{\left(\sum x_D\right)^2}{n_D}}{n_D - 1} = \frac{1.13 - \frac{(2.5)^2}{6}}{6 - 1} = .0177$$

$$s_D = \sqrt{.0177} = .1329$$

a. To determine if there is a difference in the mean strength of the two types of shocks, we test:

$$H_0: \ \mu_D = 0$$
$$H_a: \ \mu_D \neq 0$$

The test statistic is $t = \dfrac{\bar{x}_D - 0}{\dfrac{s_D}{\sqrt{n_D}}} = \dfrac{.4167 - 0}{\dfrac{.1329}{\sqrt{6}}} = 7.68$

The rejection region requires $\alpha/2 = .05/2 = .025$ in each tail of the t distribution with df $= n_D - 1 = 6 - 1 = 5$. From Table VI, Appendix B, $t_{.025} = 2.571$. The rejection region is $t < -2.571$ or $t > 2.571$.

Since the observed value of the test statistic falls in the rejection region ($t = 7.68 > 2.571$), H_0 is rejected. There is sufficient evidence to indicate a difference in mean strength of the two shocks at $\alpha = .05$.

b. The observed significance level is $2P(t \geq 7.68)$. From Table VI, with df $= 5$, $2P(t \geq 7.68) < 2(.005) = .01$.

c. The necessary assumptions are:

  1. The population of differences is normal.
  2. The differences are randomly selected.

d. For confidence coefficient .95, $\alpha = 1 - .95 = .05$ and $\alpha/2 = .05/2 = .025$. From Table VI, Appendix B, with df $= 5$, $t_{.025} = 2.571$. The confidence interval is:

$$\bar{x}_D \pm t_{.025} \frac{s_D}{\sqrt{n_D}} \Rightarrow .4167 \pm 2.571 \left[\frac{.1329}{\sqrt{6}} \right] \Rightarrow .4167 \pm .1395 \Rightarrow (.2772, .5562)$$

We are 95% confident the difference in mean strength between the manufacturer's shock and that of the competitor's shock is between .2772 and .5562.

9.53   a.   Let μ_D = mean difference in pupil dilation between pattern 1 and pattern 2.

To determine if the pupil dilation differs for the two patterns, we test:

H_0: $\mu_D = 0$
H_a: $\mu_D \neq 0$

b.   The test statistic is $t = 5.76$ and the p-value $= .000$. Since the p-value is so small, there is strong evidence to reject H_0. There is evidence to indicate that the pupil dilation differs for the two patterns for $\alpha > .000$.

The p-value is not exactly 0. The p-value $= P(t \leq -5.76) + P(t \geq 5.76)$. Rounded off to 3 decimal places, the p-value is .000.

c.   The 95% confidence interval is $(.15, .328)$. We are 95% confident that the mean difference in pupil dilation is between .15 and .328. Since both values are greater than zero, there is evidence to indicate the mean pupil dilation for pattern 1 is greater than the mean dilation for pattern 2.

d.   The paired difference design is better. There is much variation in pupil dilation from person to person. By using the paired difference design, we can eliminate the person to person differences.

9.55 x_1 and x_2 are both binomial random variables which represent the number of successes in sample 1 and sample 2, respectively.

The sampling distributions of $\hat{p}_1$ and $\hat{p}_2$ are approximately normal with means of p_1 and p_2, respectively, and standard deviations of $\sqrt{\dfrac{p_1 q_1}{n_1}}$ and $\sqrt{\dfrac{p_2 q_2}{n_2}}$, respectively.

9.57   For testing H_0: $p_1 - p_2 = 0$ against H_a: $p_1 - p_2 < 0$, we would reject H_0 in favor of H_a if $z < -z_\alpha$. Using Table IV, Appendix B:

a.   When $\alpha = .01$,
        reject H_0 if $z < -z_{.01} = -2.33$.

b.   When $\alpha = .025$,
        reject H_0 if $z < -z_{.025} = -1.96$.

c.   When $\alpha = .05$,
        reject H_0 if $z < -z_{.05} = -1.645$.

d.   When $\alpha = .10$,
        reject H_0 if $z < -z_{.10} = -1.28$.

9.59 From Exercise 9.58,

$$\hat{p}_1 = \frac{x_1}{n_1} = \frac{47}{200} = .235 \qquad \hat{p}_2 = \frac{x_2}{n_2} = \frac{72}{220} = .327$$

For confidence coefficient .98, $\alpha = 1 - .98 = .02$ and $\alpha/2 = .02/2 = .01$. From Table VI, Appendix B, $z_{.01} = 2.33$. The confidence interval is:

$$(\hat{p}_1 - \hat{p}_2) \pm z_{.01} \sqrt{\frac{\hat{p}_1\hat{q}_1}{n_1} + \frac{\hat{p}_2\hat{q}_2}{n_2}}$$

$$\Rightarrow (.235 - .327) \pm 2.33 \sqrt{\frac{.235(.765)}{200} + \frac{.327(.673)}{220}} \Rightarrow -.092 \pm .1015$$

$$\Rightarrow (-.1935, .0095)$$

We are 98% confident that the difference between p_1 and p_2 is between $-.1935$ and $.0095$.

9.61 From Exercise 9.60,

$$\hat{p}_1 = \frac{x_1}{n_1} = \frac{290}{1000} = .29 \qquad \hat{p}_2 = \frac{x_2}{n_2} = \frac{343}{1000} = .343$$

For confidence coefficient .80, $\alpha = 1 - .80 = .20$ and $\alpha/2 = .20/2 = .10$. From Table VI, Appendix B, $z_{.10} = 1.28$. The confidence interval is:

$$(\hat{p}_1 - \hat{p}_2) \pm z_{.10} \sqrt{\frac{\hat{p}_1\hat{q}_1}{n_1} + \frac{\hat{p}_2\hat{q}_2}{n_2}}$$

$$\Rightarrow (.29 - .343) \pm 1.28 \sqrt{\frac{.29(.71)}{1000} + \frac{.343(.657)}{1000}} \Rightarrow -.053 \pm .0266$$

$$\Rightarrow (-.0796, -.0264)$$

We are 80% confident that p_2 exceeds p_1 by between $.0264$ and $.0796$.

9.63 a. Let p_1 = proportion of new fathers using the program and p_2 = proportion of new fathers that wants to participate.

To determine if there is a difference in the proportion of new fathers using the program and the proportion of new fathers who wants to participate, we test:

$$H_0: \ p_1 - p_2 = 0$$
$$H_a: \ p_1 - p_2 \neq 0$$

b. The sample sizes are large enough if the interval $\hat{p}_i \pm 3\sigma_{\hat{p}_i}$ is contained in the interval $(0, 1)$.

$$\hat{p}_1 = \frac{x_1}{n_1} = \frac{9}{96} = .094 \qquad \hat{p}_2 = \frac{x_2}{n_2} = \frac{35}{100} = .35$$

$$\hat{p}_1 \pm 3\sigma_{\hat{p}_1} \Rightarrow \hat{p}_1 \pm 3\sqrt{\frac{\hat{p}_1\hat{q}_1}{n_1}} \Rightarrow .094 \pm 3\sqrt{\frac{.094(.906)}{96}} \Rightarrow .094 \pm .089$$

$$\Rightarrow (.005, .183)$$

$$\hat{p}_2 \pm 3\sigma_{\hat{p}_2} \Rightarrow \hat{p}_2 \pm 3\sqrt{\frac{\hat{p}_2\hat{q}_2}{n_2}} \Rightarrow .35 \pm 3\sqrt{\frac{.35(.65)}{100}} \Rightarrow .35 \pm .143$$
$$\Rightarrow (.207, .493)$$

Since both intervals are contained in the interval $(0, 1)$, the normal approximation is valid.

c. $\hat{p} = \dfrac{x_1 + x_2}{n_1 + n_2} = \dfrac{9 + 35}{96 + 100} = .224 \qquad \hat{q} = 1 - \hat{p} = 1 - .224 = .776$

The test statistic is $z = \dfrac{(\hat{p}_1 - \hat{p}_2) - 0}{\sqrt{\hat{p}\hat{q}\left(\dfrac{1}{n_1} + \dfrac{1}{n_2}\right)}} = \dfrac{(.094 - .35) - 0}{\sqrt{.224(.776)\left(\dfrac{1}{96} + \dfrac{1}{100}\right)}} = -4.30$

The p-value $= P(z \le -4.30) + P(z \ge 4.30) = 2P(z \ge 4.30) \approx 2(.5 - .5) = 0$.

Since the p-value is less than $\alpha = .05$, there is very strong evidence to reject H_0 for $\alpha = .05$. There is sufficient evidence to indicate that there is a difference in the proportion of new fathers using the program and the proportion of new fathers who wants to participate at $\alpha = .05$.

d. We must assume that the sample sizes are large enough to use the normal approximation and that the two samples are random and independent.

9.65 a. The sample sizes for each of the 3 areas are large enough to use the methods of this section if the intervals

$$\hat{p}_i \pm 3\sigma_{\hat{p}_i} \Rightarrow \hat{p}_i \pm 3\sqrt{\frac{\hat{p}_i\hat{q}_i}{n_i}} \quad \text{do not contain 0 or 1.}$$

For Age:

$$\hat{p}_1 = \frac{x_1}{n_1} = \frac{19}{207} = .092 \qquad \hat{p}_2 = \frac{x_2}{n_2} = \frac{96}{153} = .627$$

$$\hat{p}_1 \pm 3\sigma_{\hat{p}_1} \Rightarrow .092 \pm 3\sqrt{\frac{.092(.908)}{207}} \Rightarrow .092 \pm .06 \Rightarrow (.032, .152)$$

$$\hat{p}_2 \pm 3\sigma_{\hat{p}_2} \Rightarrow .627 \pm 3\sqrt{\frac{.627(.373)}{153}} \Rightarrow .627 \pm .117 \Rightarrow (.51, .744)$$

Since neither interval contains 0 or 1, the sample sizes are large enough to use the methods of this section in the area age.

For Education:

$$\hat{p}_1 = \frac{x_1}{n_1} = \frac{195}{207} = .942 \qquad \hat{p}_2 = \frac{x_2}{n_2} = \frac{116}{153} = .758$$

$$\hat{p}_1 \pm 3\sigma_{\hat{p}_1} \Rightarrow .942 \pm 3\sqrt{\frac{.942(.058)}{207}} \Rightarrow .942 \pm .049 \Rightarrow (.893, .991)$$

$$\hat{p}_2 \pm 3\sigma_{\hat{p}_2} \Rightarrow .758 \pm 3\sqrt{\frac{.758(.242)}{153}} \Rightarrow .758 \pm .104 \Rightarrow (.654, .862)$$

Since neither interval contains 0 or 1, the sample sizes are large enough to use the methods of this section in the area education.

For Employment:

$$\hat{p}_1 = \frac{x_1}{n_1} = \frac{19}{207} = .092 \qquad \hat{p}_2 = \frac{x_2}{n_2} = \frac{47}{153} = .307$$

$$\hat{p}_1 \pm 3\sigma_{\hat{p}_1} \Rightarrow .092 \pm 3\sqrt{\frac{.092(.908)}{207}} \Rightarrow .092 \pm .06 \Rightarrow (.032, .152)$$

$$\hat{p}_2 \pm 3\sigma_{\hat{p}_2} \Rightarrow .307 \pm 3\sqrt{\frac{.307(.693)}{153}} \Rightarrow .307 \pm .112 \Rightarrow (.195, .419)$$

Since neither interval contains 0 or 1, the sample sizes are large enough to use the methods of this section in the area employment.

b. To determine if Fortune 500 CEO's and entrepreneurs differ in terms of education, we test:

$$H_0: p_1 - p_2 = 0$$
$$H_a: p_1 - p_2 \neq 0$$

The test statistic is $z = \dfrac{(\hat{p}_1 - \hat{p}_2) - 0}{\sqrt{\hat{p}\hat{q}\left[\dfrac{1}{n_1} + \dfrac{1}{n_2}\right]}}$

where $\hat{p}_1 = .942$, $\hat{p}_2 = .758$, and $\hat{p} = \dfrac{x_1 + x_2}{n_1 + n_2} = \dfrac{195 + 116}{207 + 153} = .864$

Thus, $z = \dfrac{(.942 - .758) - 0}{\sqrt{.864(.136)\left[\dfrac{1}{207} + \dfrac{1}{153}\right]}} = 5.03$

The rejection region requires $\alpha/2 = .05/2 = .025$ in each tail of the z distribution. From Table IV, Appendix B, $z_{.025} = 1.96$. The rejection region is $z < -1.96$ or $z > 1.96$.

Since the observed value of the test statistic falls in the rejection region ($z = 5.03 > 1.96$), H_0 is rejected. There is sufficient evidence to indicate that Fortune 500 CEO's and entrepreneurs differ in terms of education at $\alpha = .05$.

c. We must assume the 2 samples are independent random samples from binomial distributions. Both samples should be large enough that the normal distribution provides an adequate approximation to the sampling distributions of $\hat{p}_1$ and $\hat{p}_2$. This was checked in part a.

9.67 a. To determine if CEO's and entrepreneurs differ in the proportion that have been fired or dismissed from a job, we test:

$$H_0: \ p_1 - p_2 = 0$$
$$H_a: \ p_1 - p_2 \neq 0$$

The test statistic is $z = \dfrac{(\hat{p}_1 - \hat{p}_2) - 0}{\sqrt{\hat{p}\hat{q}\left[\dfrac{1}{n_1} + \dfrac{1}{n_2}\right]}}$

where $\hat{p}_1 = \dfrac{x_1}{n_1} = \dfrac{19}{207} = .092 \qquad \hat{p}_2 = \dfrac{x_2}{n_2} = \dfrac{47}{153} = .307$, and

$$\hat{p} = \dfrac{x_1 + x_2}{n_1 + n_2} = \dfrac{19 + 47}{207 + 153} = .183$$

Thus, $z = \dfrac{(.092 - .307) - 0}{\sqrt{.183(.817)\left[\dfrac{1}{207} + \dfrac{1}{153}\right]}} = -5.22$

The rejection region requires $\alpha/2 = .01/2 = .005$ in each tail of the z distribution. From Table IV, Appendix B, $z_{.005} = 2.575$. The rejection region is $z < -2.575$ or $z > 2.575$.

Since the observed value of the test statistic falls in the rejection region ($z = -5.22 < -2.575$), H_0 is rejected. There is sufficient evidence to indicate there is a difference in the fraction of CEO's and entrepreneurs who have been fired or dismissed from job at $\alpha = .01$.

b. For confidence coefficient .99, $\alpha = 1 - .99 = .01$ and $\alpha/2 = .01/2 = .005$. From Table IV, Appendix B, $z_{.005} = 2.575$. The confidence interval is:

$$(\hat{p}_1 - \hat{p}_2) \pm z_{\alpha/2}\sqrt{\dfrac{\hat{p}_1\hat{q}_1}{n_1} + \dfrac{\hat{p}_2\hat{q}_2}{n_2}}$$

$$\Rightarrow (.092 - .307) \pm 2.575\sqrt{\dfrac{.092(.908)}{207} + \dfrac{.307(.693)}{153}} \Rightarrow -.215 \pm .109$$

$$\Rightarrow (-.324, -.106)$$

c. The confidence interval in part b provides more information about employment records. The confidence interval gives an interval of possible values for the difference in the proportions while the hypothesis test only tells us that they differ.

9.69   a.   A confidence interval reflects the reliability of an estimate.

For confidence coefficient .95, $\alpha = 1 - .95 = .05$ and $\alpha/2 = .05/2 = .025$. From Table IV, Appendix B, $z_{.025} = 1.96$. The confidence interval is:

$$\hat{p}_1 \pm z_{.025}\sqrt{\frac{\hat{p}_1\hat{q}_1}{n_1}} \quad \text{where } \hat{p}_1 = \frac{x_1}{n_1} = \frac{1653}{9542} = .1732$$

$$\Rightarrow .1732 \pm 1.96\sqrt{\frac{.1732(.8268)}{9542}} \Rightarrow .1732 \pm .0076 \Rightarrow (.1656, .1808)$$

b.   A 95% confidence interval for p_2 is:

$$\hat{p}_2 \pm z_{.025}\sqrt{\frac{\hat{p}_2\hat{q}_2}{n_2}} \quad \text{where } \hat{p}_2 = \frac{x_2}{n_2} = \frac{501}{6631} = .0756$$

$$\Rightarrow .0756 \pm 1.96\sqrt{\frac{.0756(.9244)}{6631}} \Rightarrow .0756 \pm .0064 \Rightarrow (.0692, .082)$$

c.   A 95% confidence interval for $p_1 - p_2$ is:

$$(\hat{p}_1 - \hat{p}_2) \pm z_{.025}\sqrt{\frac{\hat{p}_1\hat{q}_1}{n_1} + \frac{\hat{p}_2\hat{q}_2}{n_2}}$$

$$\Rightarrow (.1732 - .0756) \pm 1.96\sqrt{\frac{.1732(.8268)}{9542} + \frac{.0756(.9244)}{6631}}$$

$$\Rightarrow .0976 \pm .0099 \Rightarrow (.0877, .1075)$$

9.71   In Exercise 9.70, we used $\alpha = .05$. If α were set at .01 instead, we would be less likely to reject the null hypothesis if in fact it is true. Recall $\alpha = P(\text{Type I error}) = P(\text{Reject } H_0 \text{ when } H_0 \text{ is true})$. Therefore, the smaller α is, the less likely we are of committing a Type I error.

9.73   For confidence coefficient .90, $\alpha = 1 - .90 = .10$ and $\alpha/2 = .10/2 = .05$. From Table IV, Appendix B, $z_{.05} = 1.645$. We estimate $p_1 = p_2 = .5$ to be conservative.

$$n_1 = n_2 = \frac{4(z_{\alpha/2})^2(p_1q_1 + p_2q_2)}{W^2} = \frac{4(1.645)^2(.5(.5) + .5(.5))}{.04^2} = 3382.5 \approx 3383$$

9.75   First, find the sample sizes needed for width 4:

For confidence coefficient .9, $\alpha = 1 - .9 = .1$ and $\alpha/2 = .1/2 = .05$. From Table IV, Appendix B, $z_{.05} = 1.645$.

$$n_1 = n_2 = \frac{4(z_{\alpha/2})^2(\sigma_1^2 + \sigma_2^2)}{W^2} = \frac{4(1.645)^2(10^2 + 10^2)}{4^2} = 135.30 \approx 136$$

Thus, the necessary sample size from each population is 136. Therefore, sufficient funds have not been allocated to meet the specifications since $n_1 = n_2 = 100$ are not large enough samples.

9.77 a. For confidence coefficient .80, $\alpha = 1 - .80 = .20$ and $\alpha/2 = .20/2 = .10$. From Table IV, Appendix B, $z_{.10} = 1.28$.

$$n_1 = n_2 = \frac{4(z_{\alpha/2})^2(p_1 q_1 + p_2 q_2)}{W^2}$$

Since we have no information on p_1 and p_2, we will use $p_1 = p_2 = .5$ to be conservative.

$$n_1 = n_2 = \frac{4(1.28)^2(.5(.5) + .5(.5))}{.06^2} = 910.2 \approx 911$$

The total number included should be $2(911) = 1822$.

b. For confidence coefficient .9, $\alpha = 1 - .9 = .1$ and $\alpha/2 = .1/2 = .05$. From Table IV, Appendix B, $z_{.05} = 1.645$.

$$n_1 = \frac{(z_{\alpha/2})^2(p_1 q_1)}{B^2} = \frac{(1.645)^2(.5(.5))}{.02^2} = 1691.3 \approx 1692$$

Thus, the necessary sample size is 1692 for both men and women. This is much greater than the sample size of 911 each in part **a**.

9.79 For confidence coefficient .90, $\alpha = 1 - .90 = .10$ and $\alpha/2 = .10/2 = .05$. From Table IV, Appendix B, $z_{.05} = 1.645$. We estimate $p_1 = p_2 = .5$ to be conservative.

$$n_1 = n_2 = \frac{(z_{\alpha/2})^2(p_1 q_1 + p_2 q_2)}{B^2} = \frac{(1.645)^2(.5(.5) + .5(.5))}{.05^2} = 541.2 \approx 542$$

9.81 For confidence coefficient .95, $\alpha = 1 - .95 = .05$ and $\alpha/2 = .025$. From Table IV, Appendix B, $z_{.025} = 1.96$.

$$n_1 = n_2 = \frac{(z_{\alpha/2})^2(\sigma_1^2 + \sigma_2^2)}{B^2} = \frac{(1.645)^2(35^2 + 80^2)}{10^2} = 292.9 \approx 293$$

9.83 For confidence coefficient .95, $\alpha = .05$ and $\alpha/2 = .025$. From Table IV, Appendix B, $z_{.025} = 1.96$. The required sample sizes are:

From Exercise 9.82, $\hat{p}_1 = \frac{133}{337} = .395$ and $\hat{p}_4 = \frac{109}{372} = .293$

$$n_1 = n_4 = \frac{(z_{\alpha/2})^2(p_1 q_1 + p_4 q_4)}{B^2} = \frac{1.96^2((.395)(.605) + (.293)(.707))}{.03^2} = 1904.26$$

$$\approx 1905$$

9.85 From Exercise 9.84, $\bar{x}_D = .3$, $s_D^2 = 41.4086$, and $s_D = 6.4349$. For confidence coefficient .95, $\alpha = 1 - .95 = .05$ and $\alpha/2 = .05/2 = .025$. From Table VI, Appendix B, with df $= n_D - 1 = 8 - 1 = 7$, $t_{.025} = 2.365$. The confidence interval is:

$$\bar{x}_D \pm t_{.025} \frac{s_D}{\sqrt{n_D}} \Rightarrow .3 \pm 2.365 \left[\frac{6.4349}{\sqrt{8}} \right] \Rightarrow .3 \pm 5.381 \Rightarrow (-5.081, 5.681)$$

We are 95% confident the difference between 1989 and 1988 R&D expenditures is between -5.081 and 5.681 million dollars.

9.87 Let p_1 = proportion of shoppers that are exposed to the commercial that purchase XYZ, and p_2 = proportion of shoppers that are not exposed to the commercial that purchase XYZ.

Some preliminary calculations are:

$$\hat{p}_1 = \frac{x_1}{n_1} = \frac{84}{387} = .2171 \qquad \hat{p}_2 = \frac{x_2}{n_2} = \frac{57}{392} = .1454$$

$$\hat{p} = \frac{x_1 + x_2}{n_1 + n_2} = \frac{84 + 57}{387 + 392} = .181$$

a. To determine if the new commercial motivates shoppers to purchase the XYZ brand, we test:

$$H_0: p_1 = p_2$$
$$H_a: p_1 > p_2$$

The test statistic is $z = \dfrac{(\hat{p}_1 - \hat{p}_2) - 0}{\sqrt{\hat{p}\hat{q}\left[\dfrac{1}{n_1} + \dfrac{1}{n_2}\right]}} = \dfrac{(.2171 - .1454) - 0}{\sqrt{.181(.819)\left[\dfrac{1}{387} + \dfrac{1}{392}\right]}} = 2.60$

The rejection region requires $\alpha = .05$ in the upper tail of the z distribution. From Table IV, Appendix B, $z_{.05} = 1.645$. The rejection region is $z > 1.645$.

Since the observed value of the test statistic falls in the rejection region ($z = 2.60 > 1.645$), H_0 is rejected. There is sufficient evidence to indicate the new XYZ commercial motivates shoppers to purchase the XYZ brand at $\alpha = .05$.

b. The p-value is $P(z \geq 2.60) = .5 - P(0 < z < 2.60)$
$$= .5 - .4953$$
$$= .0047 \text{ (Table IV, Appendix B)}$$

This is highly significant. The probability of observing a test statistic of 2.60 or higher is .0047 if the proportions are equal. We would reject H_0 for $\alpha > .0047$.

9.89 For confidence coefficient .90, $\alpha = 1 - .90 = .10$ and $\alpha/2 = .10/2 = .05$. From Table IV, Appendix B, $z_{.05} = 1.645$. We estimate $p_1 = p_2 = .5$.

$$n_1 = n_2 = \frac{(z_{\alpha/2})^2(p_1q_1 + p_2q_2)}{B^2} = \frac{(1.645)^2(.5(.5) + .5(.5))}{.05^2} = 541.205 \approx 542$$

9.91 a. Let p_1 = proportion of consumers who agree with the action and p_2 = proportion of managers who agree with the action.

$$\hat{p}_1 = \frac{x_1}{n_1} = \frac{89}{100} = .89 \qquad \hat{p}_2 = \frac{x_2}{n_2} = \frac{37}{100} = .37$$

For confidence coefficient .95, $\alpha = .05$ and $\alpha/2 = .025$. From Table IV, Appendix B, $z_{.025} = 1.96$. The confidence interval is:

$$(\hat{p}_1 - \hat{p}_2) \pm z_{.025}\sqrt{\frac{\hat{p}_1\hat{q}_1}{n_1} + \frac{\hat{p}_2\hat{q}_2}{n_2}}$$

$$\Rightarrow (.89 - .37) \pm 1.96\sqrt{\frac{.89(.11)}{100} + \frac{.37(.63)}{100}}$$

$$\Rightarrow .52 \pm .113 \Rightarrow (.407, .633)$$

b. The intervals $\hat{p}_i \pm 3\sigma_{\hat{p}_i}$ must no contain 0 or 1.

$$\hat{p}_1 \pm 3\sigma_{\hat{p}_1} \Rightarrow .89 \pm 3\sqrt{\frac{.89(.11)}{100}} \Rightarrow .89 \pm .094 \Rightarrow (.796, .984)$$

$$\hat{p}_2 \pm 3\sigma_{\hat{p}_2} \Rightarrow .37 \pm 3\sqrt{\frac{.37(.63)}{100}} \Rightarrow .37 \pm .145 \Rightarrow (.225, .515)$$

Since neither interval contains 0 or 1, the results are valid.

9.93 From Exercise 9.92, $\hat{p}_1 = .2$, $\hat{p}_2 = .1$, $n_1 = 387$, and $n_2 = 311$.

$$\hat{p} = \frac{n_1\hat{p}_1 + n_2\hat{p}_2}{n_1 + n_2} = \frac{387(.2) + 311(.1)}{387 + 311} = .1554$$

To determine if a difference exists between the proportions of the two types of executives who do not know how much poor quality costs their company, we test:

$H_0: p_1 - p_2 = 0$
$H_a: p_1 - p_2 \neq 0$

The test statistic is $z = \dfrac{(\hat{p}_1 - \hat{p}_2) - 0}{\sqrt{\hat{p}\hat{q}\left[\dfrac{1}{n_1} + \dfrac{1}{n_2}\right]}} = \dfrac{(.2 - .1) - 0}{\sqrt{.1554(.8446)\left[\dfrac{1}{387} + \dfrac{1}{311}\right]}} = 3.62$

The rejection region requires $\alpha/2 = .10/2 = .05$ in each tail of the z distribution. From Table IV, Appendix B, $z_{.05} = 1.645$. The rejection region is $z < -1.645$ or $z > 1.645$.

Since the observed value of the test statistic falls in the rejection region ($z = 3.62 > 1.645$), H_0 is rejected. There is sufficient evidence to indicate there is a difference in the proportions of the two types of executives who do not know how much poor quality costs their company at $\alpha = .10$.

9.95   a.   Let μ_1 = mean level of concern about product tampering for men and μ_2 = mean level of concern about product tampering for women.

To determine if a difference exists in the mean level of concern about product tampering between men and women, we test:

$$H_0: \mu_1 - \mu_2 = 0$$
$$H_a: \mu_1 - \mu_2 \neq 0$$

    b.   The test statistic is $z = -2.69$ and the p-value $= .0072$. Since the p-value is very small, there is strong evidence to reject H_0 for $\alpha > .0072$. There is sufficient evidence to indicate there is a difference in the mean level of concern about product tampering between men and women for $\alpha > .0072$.

    c.   We must assume that the 2 samples are random samples from the 2 populations and that the samples are independent.

9.97   If the p-value is less than α, reject H_0. Otherwise, do not reject H_0.

    a. p-value $= .0429 < .05 \Rightarrow$ Reject H_0

    b. p-value $= .1984 \nless .05 \Rightarrow$ Do not reject H_0

    c. p-value $= .0001 < .05 \Rightarrow$ Reject H_0

    d. p-value $= .0344 < .05 \Rightarrow$ Reject H_0

    e. p-value $= .0545 \nless .05 \Rightarrow$ Do not reject H_0

    f. p-value $= .9633 \nless .05 \Rightarrow$ Do not reject H_0

    g.   We must assume:

        1.   Both sampled populations are normal.
        2.   Both population variances are equal.
        3.   Samples are random and independent.

9.99   a.   For confidence coefficient .95, $\alpha = .05$ and $\alpha/2 = .025$. From Table IV, Appendix B, $z_{.025} = 1.96$. The required sample sizes are:

From Exercise 9.98, $\hat{p}_1 = .155$ and $\hat{p}_3 = .108$

$$n_1 = n_3 = \frac{(z_{\alpha/2})^2 (p_1 q_1 + p_3 q_3)}{B^2} = \frac{1.96^2((.155)(.845) + (.108)(.892))}{.04^2} = 545.77 \approx 546$$

b. H_0: $p_1 - p_2 = 0$
H_a: $p_1 - p_2 > 0$

$$\hat{p} = \frac{p_1 + p_2}{n_1 + n_2} = \frac{.155(546) + .108(546)}{546 + 546} = .1315$$

$$\hat{q} = 1 - \hat{p} = 1 - .1315 = .8685$$

The test statistic is $z = \dfrac{(\hat{p}_1 - \hat{p}_2) - 0}{\sqrt{\hat{p}\hat{q}\left[\dfrac{1}{n_1} + \dfrac{1}{n_2}\right]}} = \dfrac{(.155 - .108) - 0}{\sqrt{.1315(.8685)\left[\dfrac{1}{546} + \dfrac{1}{546}\right]}} = 2.30$

The p-value $= P(z \geq 2.30) = .5 - .4893 = .0107$

(From Table IV, Appendix B.)

Since the p-value is so small, there is evidence to reject H_0 for $\alpha > .0107$. There is sufficient evidence to indicate a greater proportion of employees in the poor fitness category show signs of stress than those in the good fitness category for $\alpha > .0107$.

9.101 Some preliminary calculations are:

$$\hat{p}_1 = \frac{x_1}{n_1} = \frac{12}{100} = .12 \qquad \hat{p}_2 = \frac{x_2}{n_2} = \frac{15}{90} = .1667$$

For confidence coefficient .90, $\alpha = 1 - .90 = .10$ and $\alpha/2 = .10/2 = .05$. From Table IV, Appendix B, $z_{.05} = 1.645$. The confidence interval is:

$$(\hat{p}_1 - \hat{p}_2) \pm z_{.05}\sqrt{\frac{\hat{p}_1\hat{q}_1}{n_1} + \frac{\hat{p}_2\hat{q}_2}{n_2}}$$

$$\Rightarrow (.12 - .1667) \pm 1.645\sqrt{\frac{.12(.88)}{100} + \frac{.1667(.8333)}{90}} \Rightarrow -.0467 \pm .0839$$

$$\Rightarrow (-.1306, .0372)$$

We are 90% confident the difference in proportion of defective smoke detectors between brand A and brand B is between $-.1306$ and $.0372$. Since 0 is in the interval, there is no evidence to indicate these proportions are different.

9.103 a. Let $\mu_1 =$ mean hours worked for Japanese workers and $\mu_2 =$ mean hours worked for American workers.

To determine if the gap in average hours worked differed in 1993 from 1991, we test:

H_0: $\mu_1 - \mu_2 = 180$
H_a: $\mu_1 - \mu_2 \neq 180$

b. The test statistic is $z = \dfrac{(\bar{x}_1 - \bar{x}_2) - D_0}{\sqrt{\dfrac{\sigma_1^2}{n_1} + \dfrac{\sigma_2^2}{n_2}}} = \dfrac{(2097 - 1933) - 180}{\sqrt{\dfrac{43^2}{60} + \dfrac{38^2}{50}}} = -2.07$

The p-value $= P(z \leq -2.07) + P(z \geq 2.07) = 2P(z \geq 2.07) = 2(.5 - .4808)$
$= .0384$.

Since the p-value is so small, there is evidence to reject H_0 for $\alpha > .0384$. There is sufficient evidence to indicate the gap in average hours worked differed in 1993 from 1991 for $\alpha > .0384$.

c. We must assume the 2 samples were independently and randomly selected from the 2 populations.

d. For confidence coefficient .95, $\alpha = .05$ and $\alpha/2 = .025$. From Table IV, Appendix B, $z_{.025} = 1.96$. The confidence interval is:

$$(\bar{x}_1 - \bar{x}_2) \pm z_{.025} \sqrt{\dfrac{\sigma_1^2}{n_1} + \dfrac{\sigma_2^2}{n_2}} \Rightarrow (2097 - 1933) \pm 1.96 \sqrt{\dfrac{43^2}{60} + \dfrac{38^2}{50}}$$

$$\Rightarrow 164 \pm 15.144 \Rightarrow (148.856, 179.144)$$

The hypothesized value of $D_0 = 180$. This interval does not cover this value.

CHAPTER TEN
. . . .
Simple Linear Regression

10.1 a.   b.

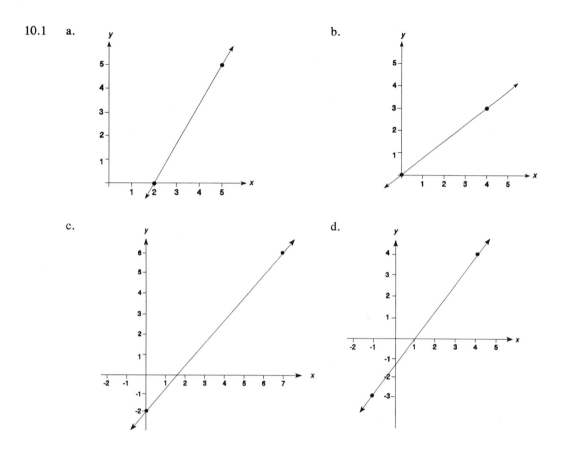

c.   d.

10.3 a.   The equation for a straight line (deterministic) is $y = \beta_0 + \beta_1 x$

Of the line passes through (2, 0), then $0 = \beta_0 + \beta_1(2)$

Likewise, through (5, 5), $5 = \beta_0 + \beta_1(5)$

Solving for these 2 equations

$$0 = \beta_0 + \beta_1(2)$$
$$-(5 = \beta_0 + \beta_1(5))$$
$$\overline{}$$
$$-5 = \qquad -3\beta_1 \quad \Rightarrow \beta_1 = \frac{5}{3}$$

Solving for β_0, $0 = \beta_0 + \frac{5}{3}(2)$ or $\beta_0 = \frac{-10}{3}$

Thus, $y = \frac{-10}{3} + \frac{5}{3}x$

b. The equation for a straight line is $y = \beta_0 + \beta_1 x$. If the line passes through $(0, 0)$, then $0 = \beta_0 + \beta_1(0)$, which implies $\beta_0 = 0$. Likewise, through the point $(4, 3)$, then $3 = \beta_0 + 4\beta_1$ or $\beta_0 = 3 - 4\beta_1$. Substituting $\beta_0 = 0$, we get $0 = 3 - 4\beta_1$, or $\beta_1 = \frac{3}{4}$. Therefore, the line passing through $(0, 0)$ and $(4, 3)$ is $y = \frac{3}{4}x$.

c. The equation for a straight line is $y = \beta_0 + \beta_1 x$. If the line passes through $(0, -2)$, then $-2 = \beta_0 + \beta_1(0)$ or $\beta_0 = -2$. Likewise through the point $(7, 6)$, $6 = \beta_0 + \beta_1(7)$.

Substituting $\beta_0 = -2$, we get $6 = -2 + \beta_1(7)$ or $\beta_1 = \frac{8}{7}$.

Therefore, the line passing through $(0, -2)$ and $(7, 6)$ is $y = -2 + \frac{8}{7}x$.

d. The equation for a straight line is $y = \beta_0 + \beta_1 x$. If the line passes through $(-1, -3)$, then $-3 = \beta_0 - \beta_1$. Likewise, through the point $(4, 4)$, $4 = \beta_0 + \beta_1 4$. Solving these equations simultaneously,

$$
\begin{aligned}
4 &= \beta_0 + \beta_1 4 \\
-[(-3) &= \beta_0 - \beta_1] \\
\hline
7 &= \qquad 5\beta_1 \quad \text{or } \beta_1 = \frac{7}{5}
\end{aligned}
$$

Solving for β_0, $4 = \beta_0 + \frac{7}{5}(4) \Rightarrow 4 - \frac{28}{5} = \beta_0$ or $\beta_0 = \frac{-8}{5}$

Therefore, $y = \frac{-8}{5} + \frac{7}{5}x$

10.5 a. $y = 4 + x$. The slope is the value for β_1 or $\beta_1 = 1$. The intercept is that value for β_0 or $\beta_0 = 4$.

b. $y = -4 + x$. The slope is the value for β_1, which is 1; the intercept is the value for β_0 or -4.

c. $y = 4 + 2x$. The slope is the value for β_1, which is 2; the intercept is the value β_0 or 4.

d. $y = -2x$. We note $\beta_0 = 0$, so the intercept is 0. The slope, $\beta_1 = -2$.

e. $y = x$. We note $\beta_0 = 0$, so the intercept is 0. The slope, $\beta_1 = 1$.

f. $y = .5 + .75x$. The slope, $\beta_1 = .75$, and the intercept $\beta_0 = .5$.

10.7   The "line of means" is the deterministic component of the probabilistic model, because the mean of y, $E(y)$, is equal to the straight-line component of the model. That is, $E(y) = \beta_0 + \beta_1 x$.

10.9   a.

x_i	y_i	x_i^2	$x_i y_i$
7	2	$7^2 = 49$	$7(2) = 14$
4	4	$4^2 = 16$	$4(4) = 16$
6	2	$6^2 = 36$	$6(2) = 12$
2	5	$2^2 = 4$	$2(5) = 10$
1	7	$1^2 = 1$	$1(7) = 7$
1	6	$1^2 = 1$	$1(6) = 6$
3	5	$3^2 = 9$	$3(5) = 15$
Totals $\sum x_i = 24$	$\sum y_i = 31$	$\sum x_i^2 = 116$	$\sum x_i y_i = 80$

Totals: $\sum x_i = 7 + 4 + 6 + 2 + 1 + 1 + 3 = 24$

$\sum y_i = 2 + 4 + 2 + 5 + 7 + 6 + 5 = 31$

$\sum x_i^2 = 49 + 16 + 36 + 4 + 1 + 1 + 9 = 116$

$\sum x_i y_i = 14 + 16 + 12 + 10 + 7 + 6 + 15 = 80$

b. $SS_{xy} = \sum_{i=1}^{n} x_i y_i - \dfrac{\left[\sum_{i=1}^{n} x_i\right]\left[\sum_{i=1}^{n} y_i\right]}{n} = 80 - \dfrac{(24)(31)}{7} = -26.2857143$

c. $SS_{xx} = \sum x_i^2 - \dfrac{\left(\sum x_i\right)^2}{7} = 116 - \dfrac{(24)^2}{7} = 33.71428571$

d. $\hat{\beta}_1 = \dfrac{SS_{xy}}{SS_{xx}} = \dfrac{-26.28571430}{33.71428571} = -.779661017 \approx -.7797$

e. $\bar{x} = \dfrac{\sum x_i}{n} = \dfrac{24}{7} = 3.428571429 \approx 3.4286$

$\bar{y} = \dfrac{\sum y_i}{n} = \dfrac{31}{7} = 4.428571429 \approx 4.4286$

f. $\hat{\beta}_0 = \bar{y} - \hat{\beta}_1 \bar{x} = 4.428571429 - (-.779661017)(3.428574129) = 7.101694917$
           ≈ 7.1017

g.   The least squares line is $\hat{y} = \hat{\beta}_0 + \hat{\beta}_1 x = 7.1017 - .7797x$

10.11 a.

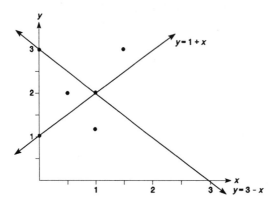

b.   Choose $y = 1 + x$ since it best describes the relation of x and y.

c.

y	x	$\hat{y} = 1 + x$	$y - \hat{y}$
2	.5	1.5	$2 - 1.5 = .5$
1	1	2	$1 - 2 = -1$
3	1.5	2.5	$3 - 2.5 = .5$
			Sum of errors $= 0$

y	x	$\hat{y} = 3 - x$	$y - \hat{y}$
2	.5	$3 - .5 = 2.5$	$2 - 2.5 = .5$
1	1	$3 - 1 = 2$	$1 - 2 = -1$
3	1.5	$3 - 1.5 = 1.5$	$3 - 1.5 = 1.5$
			Sum of errors $= 0$

d.   SSE $= \sum (y - \hat{y})^2$

SSE for 1st model: $y = 1 + x$, SSE $= (.5)^2 + (-1)^2 + (.5)^2 = 1.5$
SSE for 2nd model: $y = 3 - x$, SSE $= (.5)^2 + (-1)^2 + (1.5)^2 = 3.5$

Thus, the first line has the smaller SSE.

e.   The best fitting straight line is the one that has the least squares. The model $y = 1 + x$ has a smaller SSE, and therefore it verifies the visual check in part a. The least squares line,

$$\hat{\beta}_1 = \frac{SS_{xy}}{SS_{xx}} \qquad \sum x_i = 3 \qquad \sum y_i = 6 \qquad \sum x_i y_i = 6.5$$

$$\sum x_i^2 = 3.5 \qquad \sum y_i^2 = 14$$

$$SS_{xy} = \sum x_i y_i - \frac{\left(\sum x_i\right)\left(\sum y_i\right)}{n} = 6.5 - \frac{(3)(6)}{3} = .5$$

$$SS_{xx} = \sum x_i^2 - \frac{\left(\sum x_i\right)^2}{n} = 3.5 - \frac{(3)^2}{3} = .5$$

$$\hat{\beta}_1 = \frac{.5}{.5} = 1 \qquad \bar{x} = \frac{\sum x_i}{3} = \frac{3}{3} = 1 \qquad \bar{y} = \frac{\sum y_i}{3} = \frac{6}{3} = 2$$

$$\hat{\beta}_0 = \bar{y} - \hat{\beta}_1 \bar{x} = 2 - 1(1) = 1 \Rightarrow \hat{y} = \hat{\beta}_0 + \hat{\beta}_1 x = 1 + x$$

10.13   a. $y = \beta_0 + \beta_1 x + \epsilon$. The model implies that the mean sales price and square feet of living space are linearly related.

   b.   The estimate of the y-intercept is $\hat{\beta}_0 = -30,000$ and the estimate of the slope is $\hat{\beta}_1 = 70$.

   c.   The interpretation of the estimate of the y-intercept is not meaningful in this problem because $x = 0$ is not in the range of values for the square footage, x, which is 1500 to 4000.

   d.   For each additional square foot of living space, the mean selling price of a house is estimated to increase by $70.

   e. $\hat{y} = -30,000 + 70(3000) = \$180,000$. This is meaningful because $x = 3000$ is in the observed range of x.

   f. $y = -30,000 + 70(5000) = \$320,000$. This is not meaningful because $x = 5000$ is not in the observed range of x.

10.15   a. $\bar{x} = \dfrac{\sum x}{n} = \dfrac{330.28}{16} = 20.6425 \qquad \bar{y} = \dfrac{\sum y}{n} = \dfrac{1521}{16} = 95.0625$

$$SS_{xy} = \sum xy - \frac{\left(\sum x\right)\left(\sum y\right)}{n} = 34,259.58 - \frac{330.28(1521)}{16} = 2,862.3375$$

$$SS_{xx} = \sum x^2 - \frac{\left(\sum x\right)^2}{n} = 7,824.1822 - \frac{330.28^2}{16} = 1,006.3773$$

$$\hat{\beta}_1 = \frac{SS_{xy}}{SS_{xx}} = \frac{2,862.3375}{1,006.3773} = 2.844199189 \approx 2.8442$$

$$\hat{\beta}_0 = \bar{y} - \hat{\beta}_1 \bar{x} = 95.0625 - (2.844199189)(20.6425) = 36.35111824 \approx 36.3511$$

The least squares line is $\hat{y} = 36.3511 + 2.8442x$.

b. The least squares line fits the data fairly well.

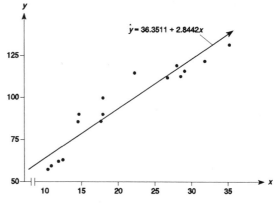

c. Using the least squares line with $x = 15$,

$$\hat{y} = 36.3511 + 2.8442(15) = 79.0141$$

The price of regular gasoline would fall to approximately 79.0141 cents per gallon.

10.17 a. You should expect to have a positive relationship since the higher batting average, the more hits and ultimately the more games won.

b. Yes, it seems there is a positive linear relationship. As batting averages increase, the number of games won also increases.

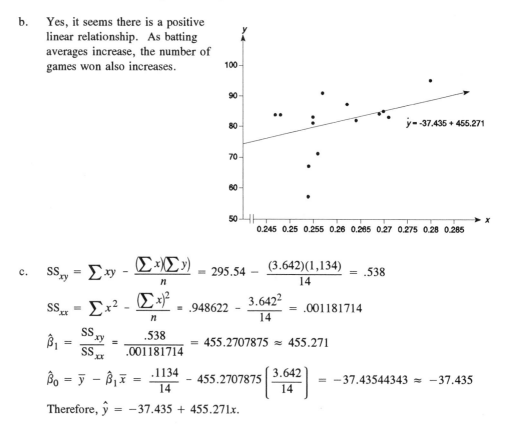

c. $SS_{xy} = \sum xy - \dfrac{(\sum x)(\sum y)}{n} = 295.54 - \dfrac{(3.642)(1,134)}{14} = .538$

$SS_{xx} = \sum x^2 - \dfrac{(\sum x)^2}{n} = .948622 - \dfrac{3.642^2}{14} = .001181714$

$\hat{\beta}_1 = \dfrac{SS_{xy}}{SS_{xx}} = \dfrac{.538}{.001181714} = 455.2707875 \approx 455.271$

$\hat{\beta}_0 = \bar{y} - \hat{\beta}_1\bar{x} = \dfrac{.1134}{14} - 455.2707875\left(\dfrac{3.642}{14}\right) = -37.43544343 \approx -37.435$

Therefore, $\hat{y} = -37.435 + 455.271x$.

d. The line presents an adequate fit.

e. There are more factors to a game than batting, such as pitcher's performance.

10.19 $s^2 = \dfrac{\text{SSE}}{n-2} = \dfrac{8.34}{26-2} = \dfrac{8.34}{24} = .3475$

10.21 $\text{SSE} = \text{SS}_{yy} - \hat{\beta}_1 \text{SS}_{xy}$

where $\text{SS}_{yy} = \sum y_i^2 - \dfrac{\left(\sum y_i\right)^2}{n}$

From Exercise 10.9,

$\sum y_i^2 = 159 \quad \sum y_i = 31$

$\text{SS}_{yy} = 159 - \dfrac{31^2}{7} = 159 - 137.285714 = 21.7142857$

$\text{SS}_{xy} = -26.2857143 \qquad \hat{\beta}_1 = -.779661017$

Therefore, $\text{SSE} = 21.7142857 - (-.779661017)(-26.2857143) = 1.22033896$

$s^2 = \dfrac{\text{SSE}}{n-2} = \dfrac{1.22033896}{5} = .2441, s = \sqrt{.2441} = .4940$

From Exercise 10.12,

$\text{SS}_{xx} = 23.4286, \text{SS}_{xy} = 23.2857, \bar{y} = 3.8571, \text{ and } \bar{x} = 4.7143$

$\sum y = 27 \text{ and } \sum y^2 = 133$

$\text{SS}_{yy} = \sum y_i^2 - \dfrac{\left(\sum y_i\right)^2}{n} = 133 - \dfrac{27^2}{7} = 28.8571$

$\hat{\beta}_1 = \dfrac{\text{SS}_{xy}}{\text{SS}_{xx}} = \dfrac{-23.2857}{23.4286} = -.993900617$

$\text{SSE} = \text{SS}_{yy} - \hat{\beta}_1 \text{SS}_{xy} = 28.8571 - (-.993900617)(-23.2857) = 5.7134284$

$s^2 = \dfrac{\text{SSE}}{n-2} = \dfrac{5.7134284}{7-2} = 1.14269 \qquad s = \sqrt{1.14269} = 1.0690$

10.23 a. $\hat{\beta}_1 = \dfrac{\text{SS}_{xy}}{\text{SS}_{xx}} = \dfrac{1,419,492.796}{3,809,368.452} = .372632055 \approx .373$

$\hat{\beta}_0 = \bar{y} - \hat{\beta}_1 \bar{x} = 302.52 - .372632055(792.04) = 7.3805068 \approx 7.381$

The least squares line is $\hat{y} = 7.381 + .373x$

The graph of the data is:

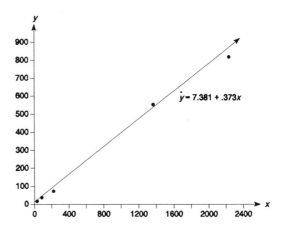

b. For x = \$1,600 billion, $\hat{y}$ = 7.381 + .373(1,600) = \$604.181 billion.

c. SSE = SS_{yy} − $\hat{\beta}_1 SS_{xy}$ = 531,174.148 − .372632055(1,419,492.796) = 2225.6298

$$s^2 = \frac{SSE}{n - 2} = \frac{2225.6298}{5 - 2} = 741.8766$$

d. We would expect almost all of the observed values of y to fall within $2s$ or 2(741.8766) or 1483.7532 of their least squares predicted values.

10.25 a. First, we compute percentage of channels watched by dividing the number of channels watched by the number of channels available and multiplying by 100. The percentages are:

Household	% Channels Watched	Household	% Channels Watched
1	50	11	40
2	34.5	12	75
3	75	13	80
4	40	14	40
5	30	15	56.25
6	60	16	100
7	83.3	17	20
8	100	18	28.9
9	57.1	19	14.3
10	30	20	20

$$\sum x_i = 357 \qquad \sum x_i^2 = 10535 \qquad \sum x_i y_i = 13100.7$$

$$\sum y_i = 1034.35 \qquad \sum y_i^2 = 66743.3125$$

$$SS_{xy} = \sum x_i y_i - \frac{\sum x_i \sum y_i}{n} = 13100.7 - \frac{357(1034.35)}{20} = -5362.4475$$

$$SS_{xx} = \sum x_i^2 - \frac{\left(\sum x_i\right)^2}{n} = 10535 - \frac{357^2}{20} = 4162.55$$

$$SS_{yy} = \sum y_i^2 - \frac{\left(\sum y_i\right)^2}{n} = 66743.3125 - \frac{1034.35^2}{20} = 13249.3164$$

$$\hat{\beta}_1 = \frac{SS_{xy}}{SS_{xx}} = \frac{-5362.4475}{4162.55} = -1.288260201 \approx -1.288$$

$$\hat{\beta}_0 = \bar{y} - \hat{\beta}_1 \bar{x} = \frac{1034.35}{20} - (-1.288260201)\left(\frac{357}{20}\right) = 74.7129446 \approx 74.713$$

The least squares line is $\hat{y} = 74.713 - 1.288x$. Because our estimate for β_1 is negative, -1.288, this supports the Nielson findings.

b.

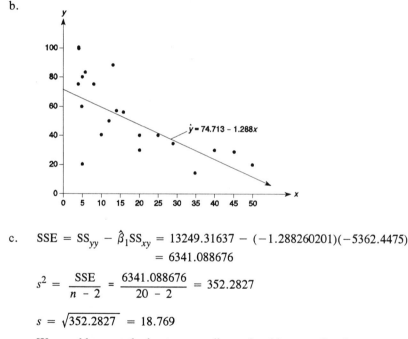

$\hat{y} = 74.713 - 1.288x$

c. $$SSE = SS_{yy} - \hat{\beta}_1 SS_{xy} = 13249.31637 - (-1.288260201)(-5362.4475)$$
$$= 6341.088676$$

$$s^2 = \frac{SSE}{n-2} = \frac{6341.088676}{20-2} = 352.2827$$

$$s = \sqrt{352.2827} = 18.769$$

We would expect the least squares line to be able to predict the percentage of channels watched to within 2 standard deviations or $\pm 2(18.769) = \pm 37.538$.

10.27   a.   For confidence coefficient .95, $\alpha = 1 - .95 = .05$ and $\alpha/2 = .05/2 = .025$. From Table VI, Appendix B, with df $= n - 2 = 10 - 2 = 8$, $t_{.025} = 2.306$.

The 95% confidence interval for β_1 is:

$$\hat{\beta}_1 \pm t_{.025} s_{\hat{\beta}_1} \text{ where } s_{\hat{\beta}_1} = \frac{s}{\sqrt{SS_{xx}}} = \frac{3}{\sqrt{35}} = .5071$$

$$\Rightarrow 31 \pm 2.306(.5071) \Rightarrow 31 \pm 1.17 \Rightarrow (29.83, 32.17)$$

For confidence coefficient .90, $\alpha = 1 - .90 = .10$ and $\alpha/2 = .10/2 = .05$. From Table VI, Appendix B, with df $= 8$, $t_{.05} = 1.860$.

The 90% confidence interval for β_1 is:

$$\hat{\beta}_1 \pm t_{.05}s_{\hat{\beta}_1} \Rightarrow 31 \pm 1.860(.5071) \Rightarrow 31 \pm .94 \Rightarrow (30.06, 31.94)$$

b. $s^2 = \dfrac{\text{SSE}}{n-2} = \dfrac{1,960}{14-2} = 163.3333, s = \sqrt{s^2} = 12.7802$

For confidence coefficient .95, $\alpha = 1 - .95 = .05$ and $\alpha/2 = .05/2 = .025$. From Table VI, Appendix B, with df $= n - 2 = 14 - 2 = 12$, $t_{.025} = 2.179$. The 95% confidence interval for β_1 is:

$$\hat{\beta}_1 \pm t_{.025}s_{\hat{\beta}_1} \text{ where } s_{\hat{\beta}_1} = \frac{s}{\sqrt{\text{SS}_{xx}}} = \frac{12.7802}{\sqrt{30}} = 2.3333$$

$$\Rightarrow 64 \pm 2.179(2.3333) \Rightarrow 64 \pm 5.08 \Rightarrow (58.92, 69.08)$$

For confidence coefficient .90, $\alpha = 1 - .90 = .10$ and $\alpha/2 = .10/2 = .05$. From Table VI, Appendix B, with df $= 12$, $t_{.05} = 1.782$.

The 90% confidence interval for β_1 is:

$$\hat{\beta}_1 \pm t_{.05}s_{\hat{\beta}_1} \Rightarrow 64 \pm 1.782(2.3333) \Rightarrow 64 \pm 4.16 \Rightarrow (59.84, 68.16)$$

c. $s^2 = \dfrac{\text{SSE}}{n-2} = \dfrac{146}{20-2} = 8.1111, s = \sqrt{s^2} = 2.848$

For confidence coefficient .95, $\alpha = 1 - .95 = .05$ and $\alpha/2 = .05/2 = .025$. From Table VI, Appendix B, with df $= n - 2 = 20 - 2 = 18$, $t_{.025} = 2.101$. The 95% confidence interval for β_1 is:

$$\hat{\beta}_1 \pm t_{.025}s_{\hat{\beta}_1} \text{ where } s_{\hat{\beta}_1} = \frac{s}{\sqrt{\text{SS}_{xx}}} = \frac{2.848}{\sqrt{64}} = .356$$

$$\Rightarrow -8.4 \pm 2.101(.356) \Rightarrow -8.4 \pm .75 \Rightarrow (-9.15, -7.65)$$

For confidence coefficient .90, $\alpha = 1 - .90 = .10$ and $\alpha/2 = .10/2 = .05$. From Table VI, Appendix B, with df $= 18$, $t_{.05} = 1.734$.

The 90% confidence interval for β_1 is:

$$\hat{\beta}_1 \pm t_{.05}s_{\hat{\beta}_1} \Rightarrow -8.4 \pm 1.734(.356) \Rightarrow -8.4 \pm .62 \Rightarrow (-9.02, -7.78)$$

10.29   From Exercise 10.28 $\hat{\beta}_1 = .8214$, $s = 1.1922$, $\text{SS}_{xx} = 28$, and $n = 7$.

For confidence coefficient .80, $\alpha = 1 - .80 = .20$ and $\alpha/2 = .20/2 = .10$. From Table VI, Appendix B, with df $= n - 2 = 7 - 2 = 5$, $t_{.10} = 1.476$. The 80% confidence interval for β_1 is:

$$\hat{\beta}_1 \pm t_{.025} s_{\hat{\beta}_1} \text{ where } s_{\hat{\beta}_1} = \frac{s}{\sqrt{SS_{xx}}} = \frac{1.1922}{\sqrt{28}} = .2253$$

$$\Rightarrow .8214 \pm 1.476(.2253) \Rightarrow .8214 \pm .3325 \Rightarrow (.4889, 1.1539)$$

For confidence coefficient .98, $\alpha = 1 - .98 = .02$ and $\alpha/2 = .02/2 = .01$. From Table VI, Appendix B, with df $= 5$, $t_{.01} = 3.365$.

The 98% confidence interval for β_1 is:

$$\hat{\beta}_1 \pm t_{.01} s_{\hat{\beta}_1} \Rightarrow .8214 \pm 3.365(.2253) \Rightarrow .8214 \pm .7581 \Rightarrow (.0633, 1.5795)$$

10.31   a.   For $n = 100$, df $= n - 2 = 100 - 2 = 98$. From Table VI with df $= 98$, p-value $= 2P(t \geq 6.572) < 2 \,(.0005) = .0010$. Since this p-value is very small, we would reject H_0. There is sufficient evidence to indicate the slope is not 0. Thus, there is evidence to indicate sales price and square feet of living space are linearly related.

         b.   We are 95% confident the slope is between 49.1 and 90.9. For each additional square foot of living space, the price of the house is estimated to increase from \$49.1 to \$90.9. This is valid only for houses with square footage between 1,500 and 4,000. The interval could be made narrower by decreasing the level of confidence.

10.33   a.   Some preliminary calculations are:

$$\sum x_i = 226.7 \qquad \sum x_i^2 = 4552.45 \qquad \sum x_i y_i = 9716.62$$

$$\sum y_i = 492.9 \qquad \sum y_i^2 = 20{,}872.01$$

$$SS_{xy} = \sum x_i y_i - \frac{\sum x_i \sum y_i}{n} = 9716.62 - \frac{226.7(492.9)}{12} = 404.9175$$

$$SS_{xx} = \sum x_i^2 - \frac{\left(\sum x_i\right)^2}{n} = 4552.45 - \frac{226.7^2}{12} = 269.7091667$$

$$SS_{yy} = \sum y_i^2 - \frac{\left(\sum y_i\right)^2}{n} = 20872.01 - \frac{492.9^2}{12} = 626.1425$$

$$\hat{\beta}_1 = \frac{SS_{xy}}{SS_{xx}} = \frac{404.9175}{269.7091667} = 1.501311598 \approx 1.50$$

$$\hat{\beta}_0 = \bar{y} - \hat{\beta}_1 \bar{x} = \frac{492.9}{12} - (1.501311598)\left[\frac{226.7}{12}\right] = 12.71272173 \approx 12.71$$

The fitted line is $\hat{y} = 12.71 + 1.50x$.

$$SSE = SS_{yy} - \hat{\beta}_1 SS_{xy} = 626.1425 - (1.501311598)(404.9175) = 18.235161$$

$$s^2 = \frac{SSE}{n - 2} = \frac{18.235161}{12 - 2} = 1.8235161, \; s = \sqrt{1.8235161} = 1.3504$$

b. To determine whether the straight-line model contributes information for the prediction of overhead costs, we test:

H_0: $\beta_1 = 0$
H_a: $\beta_1 \neq 0$

The test statistic is $t = \dfrac{\hat{\beta}_1 - 0}{s_{\hat{\beta}_1}} = \dfrac{1.501 - 0}{\dfrac{1.3504}{\sqrt{269.7091667}}} = 18.25$

The rejection region requires $\alpha/2 = .05/2 = .025$ in each tail of the t distribution with df $= n - 2 = 12 - 2 = 10$. From Table VI, Appendix B, $t_{.025} = 2.228$. The rejection region is $t > 2.228$ or $t < -2.228$.

Since the observed value of the test statistic falls in the rejection region ($t = 18.25 > 2.228$), H_0 is rejected. There is sufficient evidence to indicate the straight-line model contributes information for the prediction of overhead costs at $\alpha = .05$.

c. The assumption of independence of error terms may be inappropriate since the data are collected sequentially.

10.35 From Exercise 10.14,

$SS_{xx} = 4,362,209,330$ $\hat{\beta}_1 = -.002186456$
$SS_{xy} = -9,537,780$

$\sum y_i = 7497$ $\sum y_i^2 = 4,061,063$

$SS_{yy} = \sum y_i^2 - \dfrac{\left(\sum y_i\right)^2}{n} = 4,061,063 - \dfrac{7497^2}{15} = 314062.4$

$SSE = SS_{yy} - \hat{\beta}_1 SS_{xy} = 314062.4 - (-.002186456)(-9,537,780) = 293208.4637$

$s^2 = \dfrac{SSE}{n - 2} = \dfrac{293208.4637}{15 - 2} = 22554.49721$ $s = \sqrt{22554.49721} = 150.1815$

To determine if extent of retaliation is related to whistle blower's power, we test:

H_0: $\beta_1 = 0$
H_a: $\beta_1 \neq 0$

The test statistic is $t = \dfrac{\hat{\beta}_1 - 0}{s_{\hat{\beta}_1}} = \dfrac{-.0022}{\dfrac{150.1815}{\sqrt{4362209330}}} = -.96$

The rejection region requires $\alpha/2 = .05/2 = .025$ in each tail of the t distribution with df $= n - 2 = 15 - 2 = 13$. From Table VI, Appendix B, $t_{.025} = 2.160$. The rejection region is $t > 2.160$ or $t < -2.160$.

Since the observed value of the test statistic does not fall in the rejection region ($t = -.96 \not< -2.160$), H_0 is not rejected. There is insufficient evidence to indicate the extent of retaliation is related to the whistle blower's power at $\alpha = .05$. This agrees with Near and Miceli.

10.37   From Exercise 10.25, $n = 20$, $SS_{xx} = 4162.55$, $s = 18.769$, and $\hat{\beta}_1 = -1.28826$.

To determine if the percentage of channels watched for 10 minutes or more decreases as the number of channels available increases, we test:

H_0: $\beta_1 = 0$
H_a: $\beta_1 < 0$

The test statistic is $t = \dfrac{\hat{\beta}_1 - 0}{s_{\hat{\beta}_1}} = \dfrac{-1.28826 - 0}{\dfrac{18.769}{\sqrt{4162.55}}} = -4.43$

The rejection region requires $\alpha = .10$ in the lower tail of the t distribution with df $= n - 2$ $= 20 - 2 = 18$. From Table VI, Appendix B, $t_{.10} = 1.330$. The rejection region is $t < -1.330$.

Since the observed value of the test statistic falls in the rejection region ($t = -4.43 < -1.330$), H_0 is rejected. There is sufficient evidence to indicate the percentage of channels watched decreases as the number of channels available increases at $\alpha = .10$.

The necessary assumptions are:

1.   The mean of the probability distribution of ϵ is 0.
2.   The variance of the probability distribution of ϵ is constant for all values of x.
3.   The probability distribution of ϵ is normal
4.   The errors associated with any 2 different observations are independent.

10.39   a.   If $r = 1$, there is a perfect positive linear relationship between x and y. As x increases, y increases.

b.   If $r = -1$, there is a perfect negative linear relationship between x and y. As x increases, y decreases.

c.   If $r = 0$, there is no linear relationship between x and y.

d.   If $r = .90$, there is a strong positive linear relationship between x and y. As x increases, y tends to increase.

e.   If $r = .10$, there is a weak positive linear relationship between x and y. As x increases, y tends to increase.

f.   If $r = -.88$, there is a strong negative linear relationship between x and y. As x increases, y tends to decrease.

10.41   a.

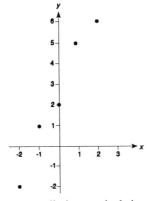

Some preliminary calculations are:

$$\sum x_i = 0 \qquad \sum x_i^2 = 10 \qquad \sum x_iy_i = 20$$

$$\sum y_i = 12 \qquad \sum y_i^2 = 70$$

$$SS_{xy} = \sum x_iy_i - \frac{\sum x_i \sum y_i}{n} = 20 - \frac{0(12)}{5} = 20$$

$$SS_{xx} = \sum x_i^2 - \frac{\left(\sum x_i\right)^2}{n} = 10 - \frac{0^2}{5} = 10$$

$$SS_{yy} = \sum y_i^2 - \frac{\left(\sum y_i\right)^2}{n} = 70 - \frac{12^2}{5} = 41.2$$

$$r = \frac{SS_{xy}}{\sqrt{SS_{xx}SS_{yy}}} = \frac{20}{\sqrt{10(41.2)}} = .985$$

There is a strong positive linear relationship between x and y.

$r^2 = .985^2 = .971$

97.1% of the sample variability around $\bar{y}$ is explained by the linear relationship between x and y.

b.

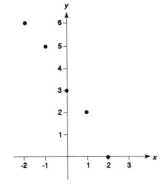

Some preliminary calculations are:

$$\sum x_i = 0 \qquad \sum x_i^2 = 10 \qquad \sum x_i y_i = -15$$

$$\sum y_i = 16 \qquad \sum y_i^2 = 74$$

$$SS_{xy} = \sum x_i y_i - \frac{\sum x_i \sum y_i}{n} = -15 - \frac{0(16)}{5} = -15$$

$$SS_{xx} = \sum x_i^2 - \frac{\left(\sum x_i\right)^2}{n} = 10 - \frac{0^2}{5} = 10$$

$$SS_{yy} = \sum y_i^2 - \frac{\left(\sum y_i\right)^2}{n} = 74 - \frac{16^2}{5} = 22.8$$

$$r = \frac{SS_{xy}}{\sqrt{SS_{xx}SS_{yy}}} = \frac{-15}{\sqrt{10(22.8)}} = -.9934$$

There is a strong negative linear relationship between x and y.

$$r^2 = (-.993)^2 = .987$$

98.7% of the sample variability around $\bar{y}$ is explained by the linear relationship between x and y.

c.

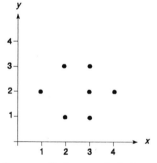

Some preliminary calculations are:

$$\sum x_i = 18 \qquad \sum x_i^2 = 52 \qquad \sum x_i y_i = 36$$

$$\sum y_i = 14 \qquad \sum y_i^2 = 32$$

$$SS_{xy} = \sum x_i y_i - \frac{\sum x_i \sum y_i}{n} = 36 - \frac{18(14)}{7} = 0$$

$$SS_{xx} = \sum x_i^2 - \frac{\left(\sum x_i\right)^2}{n} = 52 - \frac{18^2}{7} = 5.7143$$

$$SS_{yy} = \sum y_i^2 - \frac{\left(\sum y_i\right)^2}{n} = 32 - \frac{14^2}{7} = 4$$

$$r = \frac{SS_{xy}}{\sqrt{SS_{xx}SS_{yy}}} = \frac{0}{\sqrt{5.7143(4)}} = 0$$

There is no linear relationship between x and y.

$r^2 = 0^2 = 0$

None of the sample variability around $\bar{y}$ is explained by the linear relationship between x and y.

d.

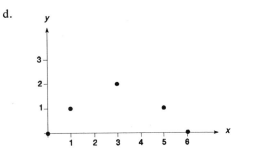

Some preliminary calculations are:

$$\sum x_i = 15 \qquad \sum x_i^2 = 71 \qquad \sum x_i y_i = 12$$

$$\sum y_i = 4 \qquad \sum y_i^2 = 6$$

$$SS_{xy} = \sum x_i y_i - \frac{\sum x_i \sum y_i}{n} = 12 - \frac{15(4)}{5} = 0$$

$$SS_{xx} = \sum x_i^2 - \frac{\left(\sum x_i\right)^2}{n} = 71 - \frac{15^2}{5} = 26$$

$$SS_{yy} = \sum y_i^2 - \frac{\left(\sum y_i\right)^2}{n} = 6 - \frac{4^2}{5} = 2.8$$

$$r = \frac{SS_{xy}}{\sqrt{SS_{xx}SS_{yy}}} = \frac{0}{\sqrt{26(2.8)}} = 0$$

There is no linear relationship between x and y.

$r^2 = 0^2 = 0$

None of the sample variability around $\bar{y}$ is explained by the linear relationship between x and y.

10.43   Some preliminary calculations are:

$$\sum x_i = 246.9 \qquad \sum x_i^2 = 6815.25 \qquad \sum x_i y_i = 276.378$$

$$\sum y_i = 9.98 \qquad \sum y_i^2 = 11.2626$$

$$SS_{xy} = \sum x_i y_i - \frac{\sum x_i \sum y_i}{n} = 276.378 - \frac{246.9(9.98)}{9} = 2.5933333$$

$$SS_{xx} = \sum x_i^2 - \frac{\left(\sum x_i\right)^2}{n} = 6815.25 - \frac{246.9^2}{9} = 41.96$$

$$SS_{yy} = \sum y_i^2 - \frac{\left(\sum y_i\right)^2}{n} = 11.2626 - \frac{9.98^2}{9} = .19588889$$

$$r = \frac{SS_{xy}}{\sqrt{SS_{xx}SS_{yy}}} = \frac{2.5933333}{\sqrt{41.96(.19588889)}} = .9046$$

The relationship between x and y is positive because $r > 0$. Since r is close to 1, the relationship between the number of hunting licenses sold in the U.S. and the number of divorces in the U.S. is very strong.

$$r^2 = .9046^2 = .8183$$

81.83% of the sample variability around the sample mean number of divorces is explained by the linear relationship between the number of hunting licenses sold and number of divorces in the U.S.

This does not mean that there is a causal relationship between the number of hunting licenses sold and the number of divorces. There are many factors that probably have contributed to both of these variables.

10.45   To determine which variable, x_1 or x_2, provides more information about y, find the correlation coefficient for each variable with y and compare them.

Some preliminary calculations are:

$$\sum x_1 = 4.16 \qquad \sum x_1^2 = 3.5026 \qquad \sum x_1 y = 179.25$$

$$\sum x_2 = 32 \qquad \sum x_2^2 = 211.5 \qquad \sum x_2 y = 1402.5$$

$$\sum y = 220 \qquad \sum y^2 = 10,050$$

$$SS_{x_1 x_1} = \sum x_1^2 - \frac{\left(\sum x_1\right)^2}{n} = 3.5026 - \frac{4.16^2}{5} = .04148$$

$$SS_{x_2 x_2} = \sum x_2^2 - \frac{\left(\sum x_2\right)^2}{n} = 211.5 - \frac{32^2}{5} = 6.7$$

$$SS_{x_1y} = \sum x_1y - \frac{(\sum x_1)(\sum y)}{n} = 179.25 - \frac{4.16(220)}{5} = -3.79$$

$$SS_{x_2y} = \sum x_2y - \frac{(\sum x_2)(\sum y)}{n} = 1402.5 - \frac{32(220)}{5} = -5.5$$

$$SS_{yy} = \sum y^2 - \frac{(\sum y)^2}{n} = 10,050 - \frac{220^2}{5} = 370$$

$$r_1 = \frac{SS_{x_1y}}{\sqrt{SS_{x_1x_1}SS_{yy}}} = \frac{-3.79}{\sqrt{.04148(370)}} = -.967$$

$$r_2 = \frac{SS_{x_2y}}{\sqrt{SS_{x_2x_2}SS_{yy}}} = \frac{-5.5}{\sqrt{6.7(370)}} = -.110$$

Since r_1 is closer to -1, the price (x_1) provides more information about the sales than does amount spent on advertising (x_2).

10.47   a.

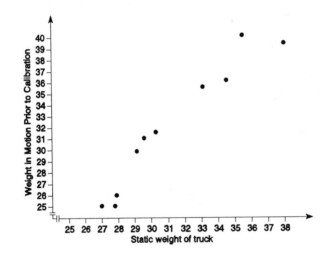

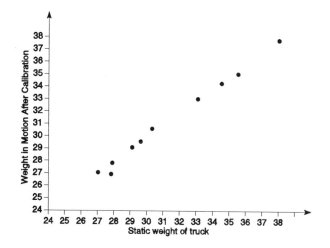

b. We note there is a much stronger linear relationship after calibration than prior because the points fall closer to a straight line.

c. Some preliminary calculations are:

$$\sum x_i = 312.8 \quad \sum y_{i1} = 320.2 \quad \sum x_i y_{i1} = 10201.41 \quad \sum y_{i2} = 311.2$$

$$\sum x_i^2 = 9911.42 \quad \sum y_{i1}^2 = 10543.68 \quad \sum x_i y_{i2} = 9859.84 \quad \sum y_{i2}^2 = 9809.52$$

$$SS_{xy1} = \sum x_i y_{i1} - \frac{\sum x_i \sum y_{i1}}{n} = 10201.41 - \frac{312.8(320.2)}{10} = 185.554$$

$$SS_{xx} = \sum x_i^2 - \frac{\left(\sum x_i\right)^2}{n} = 9911.42 - \frac{312.8^2}{10} = 127.036$$

$$SS_{y_1 y_1} = \sum y_{i1}^2 - \frac{\left(\sum y_{i1}\right)^2}{n} = 10543.68 - \frac{320.2^2}{10} = 290.876$$

$$SS_{xy2} = \sum x_i y_{i2} - \frac{\sum x_i \sum y_{i2}}{n} = 9859.84 - \frac{312.8(311.2)}{14} = 125.504$$

$$SS_{y_2 y_2} = \sum y_{i2}^2 - \frac{\left(\sum y_{i2}\right)^2}{n} = 9809.52 - \frac{311.2^2}{14} = 124.976$$

For prior adjustment,

$$r = \frac{SS_{xy_1}}{\sqrt{SS_{xx} SS_{y_1 y_1}}} = \frac{185.554}{\sqrt{127.036(290.876)}} = .965$$

The correlation between weight and weight-in-motion after adjustment is .965. This relationship is positive since $r > 0$. Since $r = .965$ is close to 1, the relationship is very strong.

For after adjustment,

$$r = \frac{SS_{xy_2}}{\sqrt{SS_{xx}SS_{y_2y_2}}} = \frac{125.504}{\sqrt{127.036(124.976)}} = .996$$

The correlation between weight and weight-in-motion after adjustment is .996. This relationship is positive since $r > 0$. Since $r = .996$ is almost 1, the relationship is almost a perfect relationship.

d. Yes, it could happen. If the difference between the static weights and weight-in-motion readings was constant for every pair, the correlation would be 1.

10.49 a. Some preliminary calculations are:

$$\sum x_i = 10,166 \qquad \sum x_i^2 = 37,987,830 \quad \sum x_iy_i = 45,986,410$$

$$\sum y_i = 10,595 \qquad \sum y_i^2 = 57,142,963$$

$$SS_{xy} = \sum x_iy_i - \frac{\sum x_i \sum y_i}{n} = 45,986,410 - \frac{10166(10595)}{10} = 35,215,533$$

$$SS_{xx} = \sum x_i^2 - \frac{\left(\sum x_i\right)^2}{n} = 37,987,830 - \frac{10166^2}{10} = 27,653,074.4$$

$$SS_{yy} = \sum y_i^2 - \frac{\left(\sum y_i\right)^2}{n} = 57,142,963 - \frac{10595^2}{10} = 45,917,560.5$$

$$\hat{\beta}_1 = \frac{SS_{xy}}{SS_{xx}} = \frac{35,215,533}{27,653,074.4} = 1.273476232 \approx 1.273$$

$$\hat{\beta}_0 = \bar{y} - \hat{\beta}_1\bar{x} = \frac{10595}{10} - (1.273476232)\left(\frac{10166}{10}\right) = -235.115937$$
$$\approx -235.1$$

$$SSE = SS_{yy} - \hat{\beta}_1 SS_{xy} = 45,917,560.5 - 1.273476232(35,215,533)$$
$$= 1,071,416.23$$

$$s^2 = \frac{SSE}{n-2} = \frac{1071416.23}{10-2} = 133,927.0288$$

$$s = \sqrt{133,927.0288} = 365.9604$$

The fitted line is $\hat{y} = -235.1 + 1.273x_2$

b. To determine if the number of arrests increase as the number of law enforcement employees increase, we test:

$$H_0: \beta_1 = 0$$
$$H_a: \beta_1 > 0$$

The test statistic is $t = \dfrac{\hat{\beta}_1 - 0}{s_{\hat{\beta}_1}} = \dfrac{1.2735}{\dfrac{365.9604}{\sqrt{27,653,074.4}}} = 18.30$

The rejection region requires $\alpha/2 = .05$ in the upper tail of the t distribution with df $= n - 2 = 10 - 2 = 8$. From Table VI, Appendix B, $t_{.05} = 1.860$. The rejection region is $t > 1.860$.

Since the observed value of the test statistic falls in the rejection region ($t = 18.30 > 1.860$), H_0 is rejected. There is sufficient evidence to indicate that as the number of law enforcement employees increases, the number of arrests increases at $\alpha = .05$.

c. $r^2 = 1 - \dfrac{\text{SSE}}{\text{SS}_{yy}} = 1 - \dfrac{1,071,416.23}{45,917,560.5} = .977$

97.7% of the variability in the number of arrests is explained by the linear relationship between number of arrests and number of law enforcement employees.

d. SSE from Exercise 10.48 is 19,082,078.13. SSE from Exercise 10.49 is 1,071,416.23. Thus, SSE from Exercise 10.49 is smaller.

e. $r^2 = .584$ from Exercise 10.48; $r^2 = .977$ from Exercise 10.49. Thus, x_2 explains more of the variation in y because its r^2 is larger.

10.51 Some preliminary calculations are:

$$\sum x_i = 230 \qquad \sum x_i^2 = 12150 \qquad \sum x_i y_i = 9850$$

$$\sum y_i = 215 \qquad \sum y_i^2 = 12781$$

$$SS_{xy} = \sum x_i y_i - \frac{\sum x_i \sum y_i}{n} = 9850 - \frac{230(215)}{5} = -40$$

$$SS_{xx} = \sum x_i^2 - \frac{\left(\sum x_i\right)^2}{n} = 12150 - \frac{230^2}{5} = 1570$$

$$SS_{yy} = \sum y_i^2 - \frac{\left(\sum y_i\right)^2}{n} = 12781 - \frac{215^2}{5} = 3536$$

$$\hat{\beta}_1 = \frac{SS_{xy}}{SS_{xx}} = \frac{-40}{1570} = -.025477707$$

$$\hat{\beta}_0 = \bar{y} - \hat{\beta}_1 \bar{x} = \frac{215}{5} - (-.025477707)\left[\frac{230}{5}\right] = 44.17197452$$

$$SSE = SS_{yy} - \hat{\beta}_1 SS_{xy} = 3536 - (-.025477707)(-40) = 3534.980892$$

$$s^2 = \frac{SSE}{n-2} = \frac{3534.980892}{5-2} = 1178.326964 \quad s = \sqrt{1178.326964} = 34.3268$$

a. The fitted line is $\hat{y} = 44.17 - .0255x$

b.

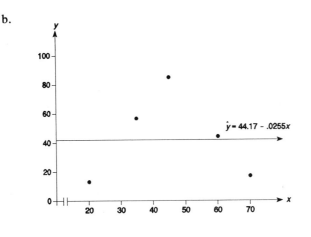

c. $H_0: \beta_1 = 0$
    $H_a: \beta_1 \neq 0$

The test statistic is $t = \dfrac{\hat{\beta}_1 - 0}{s_{\hat{\beta}_1}} = \dfrac{-.0255}{\dfrac{34.3268}{\sqrt{1570}}} = -.03$

The rejection region requires $\alpha/2 = .05/2 = .025$ in each tail of the t distribution with df $= n - 2 = 5 - 2 = 3$. From Table VI, Appendix B, $t_{.025} = 3.182$. The rejection region is $t > 3.182$ or $t < -3.182$.

Since the observed value of the test statistic does not fall in the rejection region ($t = -.03 \nless -3.182$), H_0 is not rejected. There is insufficient evidence to indicate the number of tires sold and tire price are linearly related.

d. No. It implies tire price and number of tires sold are not **linearly** related.

e. $r^2 = 1 - \dfrac{SSE}{SS_{yy}} = 1 - \dfrac{3534.980892}{3536} = .0003$

.03% of the sample variability in the number of tires sold is explained by the linear relationship between number of tires sold and tire price.

10.53 a. –b.

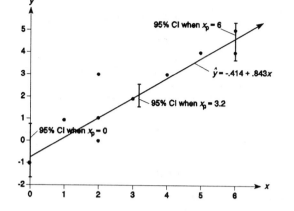

c.   The form of the confidence interval is:

$$\hat{y} \pm t_{\alpha/2}s\sqrt{\frac{1}{n} + \frac{(x_p - \bar{x})^2}{SS_{xx}}}$$

where $s = \sqrt{\dfrac{SSE}{n-2}} = \sqrt{\dfrac{SS_{yy} - \hat{\beta}_1 SS_{xy}}{n-2}} = \sqrt{\dfrac{33.6 - .843(32.8)}{10-2}} = .8624$

For $x_p = 6, \hat{y} = -.414 + .843(6) = 4.644$ and $\bar{x} = \dfrac{\sum x}{n} = \dfrac{31}{10} = 3.1$

For confidence coefficient .95, $\alpha = 1 - .95 = .05$ and $\alpha/2 = .05/2 = .025$. From Table VI, Appendix B, $t_{.025} = 2.036$, with df $= n - 2 = 10 - 2 = 8$. The 95% confidence interval is:

$$4.644 \pm 2.306(.8624)\sqrt{\frac{1}{10} + \frac{(6-3.1)^2}{38.9}} \Rightarrow 4.644 \pm 1.118 \Rightarrow (3.526, 5.762)$$

d.   For $x_p = 3.2, \hat{y} = -.414 + .843(3.2) = 2.284$

The 95% confidence interval is:

$$2.284 \pm 2.306(.8624)\sqrt{\frac{1}{10} + \frac{(3.2-3.1)^2}{38.9}} \Rightarrow 2.284 \pm .630 \Rightarrow (1.654, 2.914)$$

For $x_p = 0, \hat{y} = -.414 + .843(0) = -.414$

The 95% confidence interval is:

$$-.414 \pm 2.306(.8624)\sqrt{\frac{1}{10} + \frac{(0-3.1)^2}{38.9}} \Rightarrow -.414 \pm 1.172 \Rightarrow (-1.586, .758)$$

e. When $x_p = 6$, the width of the confidence interval is $5.762 - 3.526 = 2.236$.

When $x_p = 3.2$, the width of the confidence interval is $2.914 - 1.654 = 1.26$.

When $x_p = 0$, the width of the confidence interval is $.758 - (-1.586) = 2.344$.

The smallest interval will always be when $x_p = \bar{x}$. In this case, $\bar{x} = 3.1$, so the smallest interval is the one for $x_p = 3.2$ and has a width of 1.26. The further x_p is from $\bar{x}$, the larger the interval width will be. This is reflected in this problem.

10.55 a. $\hat{\beta}_1 = \dfrac{SS_{xy}}{SS_{xx}} = \dfrac{28}{32} = .875$

$\hat{\beta}_0 = \bar{y} - \hat{\beta}_1\bar{x} = 4 - .875(3) = 1.375$

The least squares line is $\hat{y} = 1.375 + .875x$

b.

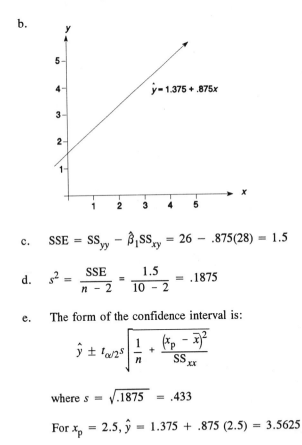

c. $SSE = SS_{yy} - \hat{\beta}_1 SS_{xy} = 26 - .875(28) = 1.5$

d. $s^2 = \dfrac{SSE}{n - 2} = \dfrac{1.5}{10 - 2} = .1875$

e. The form of the confidence interval is:

$$\hat{y} \pm t_{\alpha/2} s \sqrt{\dfrac{1}{n} + \dfrac{(x_p - \bar{x})^2}{SS_{xx}}}$$

where $s = \sqrt{.1875} = .433$

For $x_p = 2.5$, $\hat{y} = 1.375 + .875(2.5) = 3.5625$

For confidence coefficient $.95$, $\alpha = 1 - .95 = .05$ and $\alpha/2 = .05/2 = .025$. From Table VI, Appendix B, $t_{.025} = 2.306$, with df $= n - 2 = 10 - 2 = 8$.

The confidence interval is:

$$3.5625 \pm 2.306(.433)\sqrt{\frac{1}{10} + \frac{(2.5 - 3)^2}{32}} \Rightarrow 3.5625 \pm .3279 \Rightarrow (3.2346, 3.8904)$$

f. The form of the prediction interval is:

$$\hat{y} \pm t_{\alpha/2}s\sqrt{1 + \frac{1}{n} + \frac{(x_p - \bar{x})^2}{SS_{xx}}}$$

For $x_p = 4$, $\hat{y} = 1.375 + .875(4) = 4.875$

The prediction interval is:

$$4.875 \pm 2.306(.433)\sqrt{1 + \frac{1}{10} + \frac{(4 - 3)^2}{32}} \Rightarrow 4.875 \pm 1.062 \Rightarrow (3.813, 5.937)$$

10.57 a. Let x = average wage and y = quit rate.

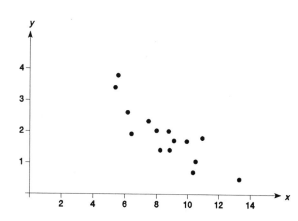

b. Some preliminary calculations are:

$$\sum x_i = 129.05 \qquad \sum x_i^2 = 1{,}178.9601 \qquad \sum x_iy_i = 218.806$$

$$\sum y_i = 28.2 \qquad \sum y_i^2 = 64.34$$

$$SS_{xy} = \sum x_iy_i - \frac{\sum x_i \sum y_i}{n} = 218.806 - \frac{129.05(28.2)}{15} = -23.808$$

$$SS_{xx} = \sum x_i^2 - \frac{\left(\sum x_i\right)^2}{n} = 1{,}178.9601 - \frac{129.05^2}{15} = 68.6999$$

$$SS_{yy} = \sum y_i^2 - \frac{\left(\sum y_i\right)^2}{n} = 64.34 - \frac{28.2^2}{15} = 11.324$$

$$\hat{\beta}_1 = \frac{SS_{xy}}{SS_{xx}} = \frac{-23.808}{68.6999} = -.346550722$$

$$\hat{\beta}_0 = \bar{y} - \hat{\beta}_1\bar{x} = \frac{28.2}{15} - (-.346550722)\left[\frac{129.05}{15}\right] = 4.861491385 \approx 4.861$$

$$SSE = SS_{yy} - \hat{\beta}_1 SS_{xy} = 11.324 - (-.346550722(-23.808) = 3.073320411$$

$$s^2 = \frac{SSE}{n-2} = \frac{3.073320411}{15-2} = .23641 \quad s = \sqrt{.23641} = .48622$$

The least squares line is $\hat{y} = 4.861 - .3466x$

c. To determine if the average hourly wage rate contributes information to predict quit rates, we test:

$H_0: \beta_1 = 0$
$H_a: \beta_1 \neq 0$

The test statistic is $t = \dfrac{\hat{\beta}_1 - 0}{s_{\hat{\beta}_1}} = \dfrac{\hat{\beta}_1 - 0}{\dfrac{s}{\sqrt{SS_{xx}}}} = \dfrac{-.3466 - 0}{\dfrac{.48622}{\sqrt{68.6999}}} = -5.91$

The rejection region requires $\alpha/2 = .05/2 = .025$ in each tail of the t distribution with df $= n - 2 = 15 - 2 = 13$. From Table VI, Appendix B, $t_{.025} = 2.160$. The rejection region is $t > 2.160$ or $t < -2.160$.

Since the observed value of the test statistic falls in the rejection region ($t = -5.91 < -2.160$), H_0 is rejected. There is sufficient evidence to indicate the average hourly wage rate contributes information to predict quit rate at $\alpha = .05$.

Since the slope is negative ($\hat{\beta}_1 = -5.91$), the model suggests that x and y have a negative relationship. As the average hourly wage rate increases, the quit rate tends to decrease.

d. The prediction interval is:

$$\hat{y} \pm t_{\alpha/2}s\sqrt{1 + \frac{1}{n} + \frac{(x_p - \bar{x})^2}{SS_{xx}}}$$

where $x_p = 9$, $\bar{x} = \dfrac{\sum x}{n} = \dfrac{129.05}{15} = 8.6033$, and $\hat{y} = 4.861 - .3466(9) = 1.7416$

For confidence coefficient .95, $\alpha = 1 - .95 = .05$ and $\alpha/2 = .05/2 = .025$. From Table VI, Appendix B, with df $= n - 2 = 15 - 2 = 13$, $t_{.025} = 2.160$. The 95% prediction interval is:

$$\Rightarrow 1.74 \pm 2.160(.48622)\sqrt{1 + \frac{1}{15} + \frac{(9 - 8.6033)^2}{68.6999}} \Rightarrow 1.74 \pm 1.09 \Rightarrow (.65, 2.83)$$

e. The confidence interval is:

$$\hat{y} \pm t_{\alpha/2}s\sqrt{\frac{1}{n} + \frac{(x_p - \bar{x})^2}{SS_{xx}}}$$

where $x_p = 6.50$, and $\hat{y} = 4.861 - .3466(6.5) = 2.6081$

$$\Rightarrow 2.61 \pm 2.160(.48622)\sqrt{\frac{1}{15} + \frac{(6.5 - 8.6033)^2}{68.6999}} \Rightarrow 2.61 \pm .38 \Rightarrow (2.23, 2.99)$$

10.59 a. Some preliminary calculations are:

$$\sum x_i = 540 \qquad \sum x_i^2 = 19,098 \qquad \sum x_iy_i = 14,069$$

$$\sum y_i = 423 \qquad \sum y_i^2 = 10,617$$

$$SS_{xy} = \sum x_iy_i - \frac{\sum x_i \sum y_i}{n} = 14,069 - \frac{540(423)}{20} = 2,648$$

$$SS_{xx} = \sum x_i^2 - \frac{\left(\sum x_i\right)^2}{n} = 19,098 - \frac{540^2}{20} = 4,518$$

$$SS_{yy} = \sum y_i^2 - \frac{\left(\sum y_i\right)^2}{n} = 10,617 - \frac{423^2}{20} = 1,670.55$$

$$\hat{\beta}_1 = \frac{SS_{xy}}{SS_{xx}} = \frac{2,648}{4,518} = .586100044$$

$$\hat{\beta}_0 = \bar{y} - \hat{\beta}_1\bar{x} = \frac{423}{20} - (.586100044)\left[\frac{540}{20}\right] = 5.325298805$$

$$SSE = SS_{yy} - \hat{\beta}_1 SS_{xy} = 1,670.55 - .586100044(2,648) = 118.5570828$$

$$s^2 = \frac{SSE}{n-2} = \frac{118.5570828}{20-2} = 6.5865046 \qquad s = \sqrt{6.5865046} = 2.5664$$

The fitted line is $\hat{y} = 5.325 + .5861x$

b. It looks like the relationship between x and y might not be linear; it might be cubic. However, it looks like the linear least squares line will provide a useful approximation to the relationship.

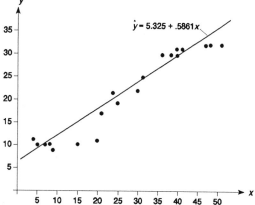

$\hat{y} = 5.325 + .5861x$

c. To determine the usefulness of the model, we test:

H_0: $\beta_1 = 0$
H_a: $\beta_1 \neq 0$

The test statistic is $t = \dfrac{\hat{\beta}_1 - 0}{s_{\hat{\beta}_1}} = \dfrac{.5861 - 0}{\dfrac{2.5664}{\sqrt{4,518}}} = 15.35$

The rejection region requires $\alpha/2 = .05/2 = .025$ in each tail of the t distribution with df $= n - 2 = 20 - 2 = 18$. From Table VI, Appendix B, $t_{.025} = 2.101$. The rejection region is $t > 2.101$ or $t < -2.101$.

Since the observed value of the test statistic falls in the rejection region ($t = 15.35 > 2.101$), H_0 is rejected. There is sufficient evidence to indicate the model is useful at $\alpha = .05$. Therefore, the monthly sales is useful in predicting the number of managers at $\alpha = .05$.

d. For confidence coefficient .90, $\alpha = 1 - .90 = .10$ and $\alpha/2 = .10/2 = .05$. From Table VI, Appendix B, $t_{.05} = 1.734$ with df $= 18$.

For $x_p = 39$, $\bar{x} = \dfrac{\sum x}{n} = \dfrac{540}{20} = 27$, and $\hat{y} = 5.325 + .5861(39) = 28.1829$.

The form of the confidence interval is:

$$\hat{y} \pm t_{\alpha/2}s\sqrt{\dfrac{1}{n} + \dfrac{(x_p - \bar{x})^2}{SS_{xx}}}$$

$$\Rightarrow 28.183 \pm 1.734(2.5664)\sqrt{\dfrac{1}{20} + \dfrac{(39 - 27)^2}{4,518}} \Rightarrow 28.183 \pm 1.273$$

$$\Rightarrow (26.91, 29.456)$$

The form of the prediction interval is:

$$\hat{y} \pm t_{\alpha/2}s\sqrt{1 + \dfrac{1}{n} + \dfrac{(x_p - \bar{x})^2}{SS_{xx}}}$$

$$\Rightarrow 28.183 \pm 1.734(2.5664)\sqrt{1 + \dfrac{1}{20} + \dfrac{(39 - 27)^2}{4,518}} \Rightarrow 28.183 \pm 4.629$$

$$\Rightarrow (23.554, 32.812)$$

e. The 90% prediction interval for the number of managers needed next May is wider than the 90% confidence interval for the mean number of managers needed next May when 39 units are sold.

Interpretation for the 90% confidence interval:

We estimate that this interval from 26.91 to 29.456 encloses the mean number of managers needed when the sales level is 39 units.

Interpretation for the 90% prediction interval:

We predict the number of managers needed when the sales level is 39 units will fall in the interval from 23.554 to 32.812.

10.61 Let us suppose we have taken a sample of size n and fit a linear model to the data. The form of the confidence interval for the mean value of y at a particular x-value, say x_p, is

$$\hat{y} \pm t_{\alpha/2}s\sqrt{1/n + (x_p - \bar{x})^2/SS_{xx}}$$

The width of the confidence interval is:

$$2t_{\alpha/2}s\sqrt{1/n + (x_p - \bar{x})^2/SS_{xx}}$$

Now, for a given confidence coefficient, the following quantities are known and fixed: $t_{\alpha/2}$, s, n, $\bar{x}$, and SS_{xx}; the only quantity which may vary is x_p. As x_p gets farther away from $\bar{x}$, the quantity $(x_p - \bar{x})^2$ gets larger, and the width of the confidence interval is thus increased. Hence, for estimation and prediction purposes, we must exercise caution in making predictions for values of x beyond the region of experimentation.

10.63   a.   From the printout,

$$\hat{\beta}_0(\text{INTERCEP}) = 44.130454$$
$$\hat{\beta}_1(\text{SALARY}) = .236617$$

Thus, $\hat{y} = 44.13 + .2366x$

    b. s = Root MSE = 19.40375

The estimated standard deviation of managerial success index scores is 19.40375. We would expect to be able to predict the managerial success index using the number of interactions with outsiders to within $\pm 2s$ or $\pm 2(19.40375)$ or ± 38.8075.

    c. r^2 = R-SQUARE = .0865

8.65% of the sample variability in the managerial success index is explained by the linear relationship between the success index and the number of interactions with outsiders.

    d.   To determine if the model is useful, we test:

$$H_0: \beta_1 = 0$$
$$H_a: \beta_1 \neq 0$$

The test statistic is $t = 1.269$

The observed significance level is .2216.

For $\alpha = .05$, the observed significance level is greater than α. Therefore, we do not reject H_0. There is insufficient evidence to indicate the model is useful for predicting y at $\alpha = .05$.

10.65   a.

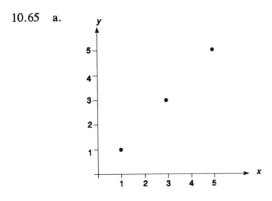

b.   One possible line is $\hat{y} = x$.

x	y	$\hat{y}$	$y - \hat{y}$
1	1	1	0
3	3	3	0
5	5	5	0
			0

For this example, $\sum (y - \hat{y}) = 0$

A second possible line is $\hat{y} = 3$.

x	y	$\hat{y}$	$y - \hat{y}$
1	1	3	-2
3	3	3	0
5	5	3	2
			0

For this example, $\sum (y - \hat{y}) = 0$

c.   Some preliminary calculations are:

$$\sum x_i = 9 \qquad \sum x_i^2 = 35 \qquad \sum x_i y_i = 35$$

$$\sum y_i = 9 \qquad \sum y_i^2 = 35$$

$$SS_{xy} = \sum x_i y_i - \frac{\sum x_i \sum y_i}{n} = 35 - \frac{9(9)}{3} = 8$$

$$SS_{xx} = \sum x_i^2 - \frac{\left(\sum x_i\right)^2}{n} = 35 - \frac{9^2}{3} = 8$$

$$SS_{yy} = \sum y_i^2 - \frac{\left(\sum y_i\right)^2}{n} = 35 - \frac{9^2}{3} = 8$$

$$\hat{\beta}_1 = \frac{SS_{xy}}{SS_{xx}} = \frac{8}{8} = 1$$

$$\hat{\beta}_0 = \bar{y} - \hat{\beta}_1\bar{x} = \frac{9}{3} - 1\left(\frac{9}{3}\right) = 0$$

The least squares line is $\hat{y} = 0 + 1x = x$

d.   For $\hat{y} = x$, SSE $= SS_{yy} - \hat{\beta}_1 SS_{xy} = 8 - 1(8) = 0$
     For $\hat{y} = 3$, SSE $= \sum (y_i - \hat{y}_i)^2 = (1 - 3)^2 + (3 - 3)^2 + (5 + 3)^2 = 8$

The least squares line has the smallest SSE of all possible lines.

10.67 a.   The plot of the data is:

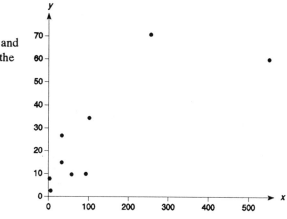

It appears that there is a linear relationship between order size and time. As order size increases, the time tends to increase.

b.   Some preliminary calculations are:

$$\sum x_i = 1149 \qquad \sum x_i^2 = 398,979 \qquad \sum x_i y_i = 58,102$$

$$\sum y_i = 239 \qquad \sum y_i^2 = 11,093$$

$$SS_{xy} = \sum x_i y_i - \frac{\sum x_i \sum y_i}{n} = 58,102 - \frac{1149(239)}{9} = 27,589.66667$$

$$SS_{xx} = \sum x_i^2 - \frac{\left(\sum x_i\right)^2}{n} = 398,979 - \frac{1149^2}{9} = 252,290$$

$$SS_{yy} = \sum y_i^2 - \frac{\left(\sum y_i\right)^2}{n} = 11,093 - \frac{239^2}{9} = 4746.222222$$

$$\hat{\beta}_1 = \frac{SS_{xy}}{SS_{xx}} = \frac{27,589.66667}{252,290} = .109356957 \approx .10936$$

$$\hat{\beta}_0 = \bar{y} - \hat{\beta}_1\bar{x} = \frac{239}{9} - (.109356957)\frac{1149}{9} = 12.59431738 \approx 12.594$$

$$SSE = SS_{yy} - \hat{\beta}_1 SS_{xy} = 4746.222222 - (.109356957)(27,589.66667)$$
$$= 1729.10023$$

$$s^2 = \frac{SSE}{n-2} = \frac{1729.10023}{9-2} = 247.0143186 \qquad s = \sqrt{s^2} = 15.7167$$

The least squares line is $\hat{y} = 12.594 + .10936x$

c. To determine if the mean time to fill an order increases with the size of the order, we test:

$$H_0: \beta_1 = 0$$
$$H_a: \beta_1 > 0$$

The test statistic is $t = \dfrac{\hat{\beta}_1 - 0}{s_{\hat{\beta}_1}} = \dfrac{.1094 - 0}{\dfrac{15.7167}{\sqrt{252,290}}} = 3.50$

The rejection region requires $\alpha = .05$ in the upper tail of the t distribution. From Table VI, Appendix B, $t_{.05} = 1.895$, with df $= n - 2 = 9 - 2 = 7$. The rejection region is $t > 1.895$.

Since the observed value of the test statistic falls in the rejection region ($t = 3.50 > 1.895$), H_0 is rejected. There is sufficient evidence to indicate the mean time to fill an order increases with the size of the order for $\alpha = .05$.

d. For confidence coefficient .95, $\alpha = 1 - .95 = .05$ and $\alpha/2 = .05/2 = .025$. From Table VI, Appendix B, $t_{.025} = 2.365$ with df $= n - 2 = 9 - 2 = 7$.

The confidence interval is:

$$\hat{y} \pm t_{\alpha/2}s\sqrt{\frac{1}{n} + \frac{(x_p - \bar{x})^2}{SS_{xx}}}$$

For $x_p = 150, \hat{y} = 12.594 + .10936(150) = 28.998$, and $\bar{x} = \dfrac{1149}{9} = 127.6667$

$$28.988 \pm 2.365(15.7167)\sqrt{\frac{1}{9} + \frac{(150 - 127.6667)^2}{252,290}} \Rightarrow 28.988 \pm 12.500$$
$$\Rightarrow (16.498, 41.498)$$

10.69 a. Some preliminary calculations are:

$$SS_{xy} = \sum x_i y_i - \frac{\sum x_i \sum y_i}{n} = 1,165,664 - \frac{5,011(3,917)}{16} = -61,091.438$$

$$SS_{xx} = \sum x_i^2 - \frac{\left(\sum x_i\right)^2}{n} = 1,628,915 - \frac{5011^2}{16} = 59,532.437$$

$$SS_{yy} = \sum y_i^2 - \frac{\left(\sum y_i\right)^2}{n} = 1,278,301 - \frac{3917^2}{16} = 319,370.4375$$

$$r = \frac{SS_{xy}}{\sqrt{SS_{xx}SS_{yy}}} = \frac{-61,091.438}{\sqrt{59,532.437(319,370.4375)}} = -.443$$

This implies that the number of passengers carried by air carriers and by railroad are negatively related. Since $r = -.443$ is not close to -1, the relationship is fairly weak.

$$r^2 = (-.443)^2 = .1963$$

19.63% of the sample variability in number of passengers carried by air carriers is explained by the linear relationship between the number of passengers carried by air carriers and by railroad.

b. To determine if x and y are correlated, we test:

H_0: $\rho = 0$
H_a: $\rho \neq 0$

The test statistic is $t = \dfrac{r}{\sqrt{(1 - r^2)/(n - 2)}} = \dfrac{-.443}{\sqrt{(1 - (-.443)^2)/(16 - 2)}} = -1.85$

The rejection region requires $\alpha/2 = .05/2 = .025$ in each tail of the t distribution with df $= n - 2 = 16 - 2 = 14$. From Table VI, Appendix B, $t_{.025} = 2.228$. The rejection region is $t > 2.145$ or $t < -2.145$.

Since the observed value of the test statistic does not fall in the rejection region ($t = -1.85 \not< -2.145$), H_0 is not rejected. There is insufficient evidence to indicate that x and y are correlated at $\alpha = .05$.

10.71 a. Some preliminary calculations are:

$$\sum x_i = 4.7 \qquad \sum x_i^2 = 4.59 \qquad \sum x_i y_i = 156.1$$

$$\sum y_i = 164 \qquad \sum y_i^2 = 5,410$$

$$SS_{xy} = \sum x_i y_i - \frac{\sum x_i \sum y_i}{n} = 156.1 - \frac{4.7(164)}{5} = 1.94$$

$$SS_{xx} = \sum x_i^2 - \frac{\left(\sum x_i\right)^2}{n} = 4.59 - \frac{4.7^2}{5} = .172$$

$$SS_{yy} = \sum y_i^2 - \frac{\left(\sum y_i\right)^2}{n} = 5,410 - \frac{164^2}{5} = 30.8$$

$$\hat{\beta}_1 = \frac{SS_{xy}}{SS_{xx}} = \frac{1.94}{.172} = 11.27906977$$

$$\text{SSE} = \text{SS}_{yy} - \hat{\beta}_1\text{SS}_{xy} = 30.8 - (11.27906977)(1.94) = 8.91860465$$

$$s^2 = \frac{\text{SSE}}{n-2} = \frac{8.91860465}{5-2} = 2.972868217 \qquad s = \sqrt{s^2} = 1.7242$$

$$r = \frac{\text{SS}_{xy}}{\sqrt{\text{SS}_{xx}\text{SS}_{yy}}} = \frac{1.94}{\sqrt{(.172)(30.8)}} = .8429$$

b. To determine if there is a nonzero correlation between sales and advertising expense, we test:

H_0: $\rho = 0$
H_a: $\rho \neq 0$

The test statistic is $t = \dfrac{r}{\sqrt{(1-r^2)/(n-2)}} = \dfrac{.8429}{\sqrt{(1-.8429^2)/(5-2)}} = 2.71$

The rejection region requires $\alpha/2 = .05/2 = .025$ in each tail of the t distribution with df $= n - 2 = 5 - 2 = 3$. From Table VI, Appendix B, $t_{.025} = 3.182$. The rejection region is $t > 3.182$ or $t < -3.182$.

Since the observed value of the test statistic does not fall in the rejection region ($t = 2.71$ $\ngtr 3.182$), H_0 is not rejected. There is insufficient evidence to indicate that there is nonzero correlation between sales and advertising at $\alpha = .05$.

c. We cannot say that additional advertising will cause sales to increase. Even if we would have decided that there was a correlation between sales and advertising expense, we could not say that one caused the other to happen.

10.73 a.

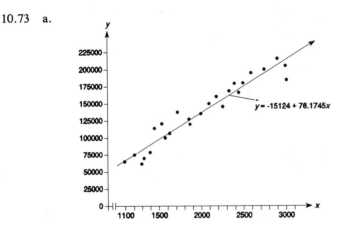

b. From the printout:

$\hat{\beta}_0(\text{INTERCEP}) = -15124$
$\hat{\beta}_1(\text{AREA}) = 76.174547$

The least squares line is $\hat{y} = -15{,}124 + 76.1745x$.

c. r^2 = R-SQUARE = .9185

91.85% of the sample variability in price is explained by the linear relationship between price and area.

d.   To determine if living area contributes information for predicting the price of a home, we test:

$$H_0: \beta_1 = 0$$
$$H_a: \beta_1 \neq 0$$

The test statistic is $t = 15.743$ (from printout)

The p-value is .0001.

Since the p-value is less than $\alpha = .05$, H_0 is rejected. There is sufficient evidence to indicate living area contributes information for predicting the price of a home at $\alpha = .05$.

e.   For confidence coefficient .95, $\alpha = 1 - .95 = .05$ and $\alpha/2 = .05/2 = .025$. From Table VI, Appendix B, $t_{.025} = 2.074$ with df = $n - 2 = 24 - 2 = 22$. The confidence interval is

$$\hat{\beta}_1 \pm t_{\alpha/2}s_{\hat{\beta}_1} \Rightarrow 76.1745 \pm 2.074(4.8385)$$
$$\Rightarrow 76.1745 \pm 10.0350 \Rightarrow (66.1395, 86.2095)$$

Since 0 is not in the confidence interval, it is not a likely value for $\beta_1 \Rightarrow$ reject H_0. This corresponds to the conclusion in d.

f.   The observed significance level is .0001. Since this is less than $\alpha = .05$, H_0 is rejected in part d.

g.   From the 25 observations on the printout, the point estimate for price when $x_p = 2200$ is $\hat{y} = 152,460$. The 95% confidence interval is (146,713, 158,206).

10.75 a.   Some preliminary calculations are:

$$\sum x_i = 78.03 \qquad \sum x_i^2 = 812.4295 \qquad \sum x_i y_i = 2530.517$$

$$\sum y_i = 263.6 \qquad \sum y_i^2 = 15,107$$

$$SS_{xy} = \sum x_i y_i - \frac{\sum x_i \sum y_i}{n} = 2530.517 - \frac{78.03(263.6)}{8} = -40.5715$$

$$SS_{xx} = \sum x_i^2 - \frac{\left(\sum x_i\right)^2}{n} = 812.4295 - \frac{78.03^2}{8} = 51.3443875$$

$$SS_{yy} = \sum y_i^2 - \frac{\left(\sum y_i\right)^2}{n} = 15,107 - \frac{263.6^2}{8} = 6421.38$$

$$\hat{\beta}_1 = \frac{SS_{xy}}{SS_{xx}} = \frac{-40.5715}{51.3443875} = -.790183737$$

$$\hat{\beta}_0 = \bar{y} - \hat{\beta}_1\bar{x} = \frac{263.6}{8} - (-.790183737)\frac{78.03}{8} = 40.65725463$$

$$SSE = SS_{yy} - \hat{\beta}_1SS_{xy} = 6421.38 - (-.790183737)(-40.5715) = 6389.321061$$

$$s^2 = \frac{SSE}{n-2} = \frac{6389.321061}{8-2} = 1064.886844 \quad s = \sqrt{1064.886844} = 32.6326$$

The least squares line is $\hat{y} = 40.657 - .7902x$

b. $\quad r = \frac{SS_{xy}}{\sqrt{SS_{xx}SS_{yy}}} = \frac{-40.5715}{\sqrt{51.3443875(6421.38)}} = .0707$

$r^2 = .0707^2 = .0050$

$s = 32.6326$ (from part a)

Since both r and r^2 are very close to 0, it appears that mean annual prime interest rate and number of business failures are not related. Also, the fairly large value for s indicates that the model would be of little use.

c. To determine whether the prime rate contributes information for the prediction of business failures, we test:

$H_0: \beta_1 = 0$
$H_a: \beta_1 \neq 0$

The test statistic is $t = \dfrac{\hat{\beta}_1 - 0}{s_{\hat{\beta}_1}} = \dfrac{-.7902 - 0}{\dfrac{32.6326}{\sqrt{51.3443875}}} = .17$

The rejection region requires $\alpha/2 = .05/2 = .025$ in each tail of the t distribution with df $= n - 2 = 8 - 2 = 6$. From Table VI, Appendix B, $t_{.025} = 2.447$. The rejection region is $t < -2.447$ or $t > 2.447$.

Since the observed value of the test statistic does not fall in the rejection region ($t = .17$ $\not> 2.447$), H_0 is not rejected. There is insufficient evidence to indicate the price rate contributes information for the prediction of business failure at $\alpha = .05$.

CHAPTER ELEVEN

- -
Multiple Regression

11.1 a. $\hat{\beta}_0$(INTERCEP) = 506.346   The estimated mean value of y when $x_1 = 0$ and $x_2 = 0$ is 506.346.

$\hat{\beta}_1$(X1) = −941.9   The mean value of y is estimated to decrease by 941.9 for each unit increase in x_1, with x_2 held constant.

$\hat{\beta}_2$(X2) = −429.06   The mean value of y is estimated to decrease by 429.06 for each unit increase in x_2, with x_1 held constant.

b. The least squares equation is $\hat{y} = 506.34 - 941.9x_1 - 429.06x_2$

c. SSE = SUM OF SQUARES for error
SSE = 151,015.72376

$$MSE = \frac{151,015.72376}{17} = 8883.27787$$

$$s = \sqrt{\frac{SSE}{n-3}} = \sqrt{MSE} = ROOT\ MSE = 94.25114$$

About 95% of the observations will fall within $\pm 2(94.25114)$ or ± 188.50228 of the fitted regression surface.

d. H_0: $\beta_1 = 0$
H_a: $\beta_1 \neq 0$

The test statistic is $t = -3.424$ (from printout). The p-value = .0032. Since p-value < $\alpha = .05$, we reject H_0 and conclude β_1 is significantly different from zero.

e. A 95% confidence interval for β_2 is

$$\hat{\beta}_2 \pm t_{\alpha/2}s_{\hat{\beta}_2}$$

$s_{\hat{\beta}_2}$ is the standard error for $\hat{\beta}_2$ which is 379.82567.

For confidence coefficient .95, $\alpha = 1 - .95 = .05$ and $\alpha/2 = .05/2 = .025$. From Table VI, Appendix B, with df $= n - 3 = 20 - 3 = 17$, $t_{.025} = 2.110$. The confidence interval is:

$$-429.06 \pm 2.11(379.82567) \Rightarrow -429.06 \pm 801.432 \Rightarrow (-1230.492, 372.372)$$

The p-value associated with $\hat{\beta}_2$ in .2743. Since the p-value is not small and not less than $\alpha = .05$, there is no evidence to reject H_0. There is no evidence to indicate β_2 is different from 0. The confidence interval supports this conclusion since 0 is contained in the interval.

11.3   a.   We are given, $\hat{\beta}_2 = .47$, $s_{\hat{\beta}_2} = .15$, $n = 25$

Test $H_0: E(y) = \beta_0 + \beta_1 x$ or $H_0: \beta_2 = 0$

           $H_a: E(y) = \beta_0 + \beta_1 x_1 + \beta_2 x_2$ $H_a: \beta_2 \neq 0$

The test statistic is $t = \dfrac{\hat{\beta}_2}{s_{\hat{\beta}_2}} = \dfrac{.47}{.15} = 3.13$

The rejection region requires $\alpha/2 = .05/2 = .025$ in each tail of the t distribution with df $= n - (k + 1) = 25 - (2 + 1) = 22$. From Table VI, Appendix B, $t_{.025} = 2.074$. The rejection region is $t < -2.074$ or $t > 2.074$.

Since the observed value of the test statistic falls in the rejection region ($t = 3.13 > 2.074$), H_0 is rejected. There is sufficient evidence to indicate $\beta_2 \neq 0$ at $\alpha = .05$.

b. $H_0: \beta_2 = 0$

     $H_a: \beta_2 > 0$

The test statistic is $t = \dfrac{.47}{.15} = 3.13$

The rejection region requires $\alpha = .05$ in the upper tail of the t distribution with df $= n - (k + 1) = 25 - (2 + 1) = 22$. From Table VI, Appendix B, $t_{.05} = 1.717$. The rejection region is $t > 1.717$.

Since the observed value of the test statistic falls in the rejection region ($t = 3.13 > 1.717$), H_0 is rejected. There is sufficient evidence to indicate $\beta_2 > 0$ at $\alpha = .05$.

11.5   a.   The least squares prediction equation is:

$$\hat{y} = 1.4326 + .01x_1 + .379x_2$$

b.   To determine if the mean food consumption increases with household income, we test:

     $H_0: \beta_1 = 0$

     $H_a: \beta_1 > 0$

The test statistic is $t = 3.15$.

The rejection region requires $\alpha = .01$ in the upper tail of the t distribution with df $= n - (k + 1) = 25 - (2 + 1) = 22$. From Table VI, Appendix B, $t_{.01} = 2.508$. The rejection region is $t > 2.508$.

Since the observed value of the test statistic falls in the rejection region ($t = 3.15 > 2.508$), H_0 is rejected. There is sufficient evidence to indicate the mean food consumption increases with household income.

c. The plot supports the conclusion in part **b** since there is an increasing trend.

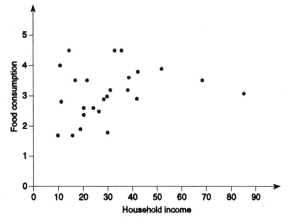

11.7 a. From the output, $\hat{\beta}_0 = 20.09$, $\hat{\beta}_1 = -.6705$ and $\hat{\beta}_2 = .0095$. Therefore, $\hat{y} = 20.09 - .6705x + .0095x^2$.

b. The data do not seem to support the equation, rather, only a linear trend.

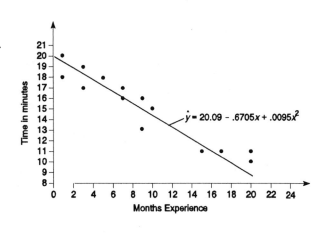

c. Test H_0: $\beta_2 = 0$
   H_a: $\beta_2 \neq 0$

The test statistic is $t = 1.51$.

The rejection region requires $\alpha/2 = .01/2 = .005$ in each tail of the t distribution with df $= n - (k + 1) = 15 - (2 + 1) = 12$. From Table VI, Appendix B, $t_{.005} = 3.055$. The rejection region is $t < -3.055$ or $t > 3.055$.

Since the observed value of the test statistic does not fall in the rejection region ($t = 1.51$ $\not> 3.05$), H_0 is not rejected. There is insufficient evidence to indicate β_2 is not 0 at $\alpha = .01$.

d. Some preliminary calculations are:

$$\sum x = 151 \qquad \sum x^2 = 2295 \qquad \sum xy = 1890$$

$$\sum y = 222 \qquad \sum y^2 = 3456$$

$$SS_{xy} = \sum xy - \frac{\sum x \sum y}{n} = 1890 - \frac{151(222)}{15} = -344.8$$

$$SS_{xx} = \sum x^2 - \frac{(\sum x)^2}{n} = 2295 - \frac{151^2}{15} = 774.933333$$

$$SS_{yy} = \sum y^2 - \frac{(\sum y)^2}{n} = 3456 - \frac{222^2}{15} = 170.4$$

$$\hat{\beta}_1 = \frac{SS_{xy}}{SS_{xx}} = \frac{-344.8}{774.933333} = -.4449415$$

$$\beta_0 = \bar{y} - \hat{\beta}_1 x = \frac{222}{15} - (-.4449415)\left[\frac{151}{15}\right] = 19.2791$$

The fitted "reduced model" is $\hat{y} = 19.2791 - .4449x$.

e. β_1 is the change in mean time to complete the task for each unit increase in months of experience.

$$SSE = SS_{yy} - \hat{\beta}_1 SS_{xy} = 170.4 - (-.4449415)(-334.8) = 16.9841708$$

$$MSE = \frac{SSE}{n-2} = \frac{16.9841708}{15-2} = 1.30647$$

$$s = \sqrt{1.30647} = 1.143$$

For confidence coefficient .90, $\alpha = 1 - .90 = .10$ and $\alpha/2 = .10/2 = .05$. From Table VI, Appendix B, with df $= n - 2 = 15 - 2 = 13$, $t_{.05} = 1.771$. The confidence interval is

$$\hat{\beta}_1 \pm t_{.05}\left(s_{\hat{\beta}_2}\right) \Rightarrow -.4449 \pm 1.771 \sqrt{\frac{MSE}{SS_{xx}}}$$

$$\Rightarrow -.4449 \pm 1.771 \sqrt{\frac{1.30647}{774.9333}} \Rightarrow -.4449 \pm .0727 \Rightarrow (-.5176, -.3722)$$

11.9 H_0: $\beta_2 = 0$
H_a: $\beta_2 > 0$

The test statistic is $t = \dfrac{\hat{\beta}_2}{s_{\hat{\beta}_2}} = \dfrac{.0015}{.000712} = 2.11$

The rejection region requires $\alpha = .05$ in the upper tail of the t distribution with df $= n - (k + 1) = 50 - (2 + 1) = 47$. From Table VI, Appendix B, $t_{.05} \approx 1.684$. The rejection region is $t > 1.684$.

Since the observed value of the test statistic falls in the rejection region ($t = 2.11 > 1.684$), H_0 is rejected. There is sufficient evidence to indicate β_2 is greater than 0 at $\alpha = .05$.

11.11 a. $R^2 = $ R-square $= .8911$

89.11% of the variability in y is explained by the quadratic relationship between y and x.

b. H_0: $\beta_1 = \beta_2 = 0$
    H_a: At least one $\beta_i \neq 0$, for $i = 1, 2$

The test statistic is $F = \dfrac{R^2/k}{(1 - R^2)/[n - (k + 1)]} = \dfrac{.8911/2}{(1 - .8911)/[19 - (2 + 1)]}$

$= 65.462$

The rejection region requires $\alpha = .05$ in the upper tail of the F distribution with df $= \nu_1 = k = 2$ and $\nu_2 = n - (k + 1) = 19 - (2 + 1) = 16$. From Table VIII, Appendix B, $F_{.05} = 3.63$. The rejection region is $F > 3.63$.

Since the observed value of the test statistic falls in the rejection region ($F = 65.462 > 3.63$), H_0 is rejected. There is sufficient evidence to indicate the model is useful in predicting y at $\alpha = .05$.

c. Prob $> F = p$-value $\leq .0001$

The probability of observing a test statistic of 65.462 or anything higher is less than .0001 if H_0 is true. This is very unusual if H_0 is true. There is strong evidence to reject H_0 for $\alpha > .0001$.

d. H_0: $\beta_2 = 0$
    H_a: $\beta_2 \neq 0$

The test statistic is $t = -6.803$ (from printout).

The rejection region requires $\alpha/2 = .05/2 = .025$ in each tail of the t distribution with df $= n - (k + 1) = 19 - (2 + 1) = 16$. From Table VI, Appendix B, $t_{.025} = 2.12$. The rejection region is $t < -2.12$ or $t > 2.12$.

Since the observed value of the test statistic falls in the rejection region ($t = -6.803 < -2.12$), H_0 is rejected. There is sufficient evidence to indicate $\beta_2 \neq 0$ at $\alpha = .05$. Therefore, the squared term is needed in the model.

The p-value $\leq .0001$.

The probability of observing a test statistic of -6.803 or anything more unusual is $.0001$ if H_0 is true. This is very unusual if H_0 is true. There is strong evidence to reject H_0 for $\alpha > .0001$.

11.13 a. Some preliminary calculations are:

$$SSE = \sum (y_i - \hat{y}_i)^2 = 12.37, \text{ df} = n - (k + 1) = 20 - (2 + 1) = 17$$

$$SS(\text{Total}) = \sum (y - \bar{y})^2 = 23.75, \text{ df} = n - 1 = 20 - 1 = 19$$

$$SS(\text{Model}) = SS(\text{Total}) - SSE = 23.75 - 12.37 = 11.38, \text{ df} = k = 2$$

$$MS(\text{Model}) = \frac{SSR}{k} = \frac{11.38}{2} = 5.69$$

$$MS(\text{Error}) = \frac{SSE}{n - (k + 1)} = \frac{12.37}{17} = .72765$$

$$F = \frac{MS(\text{Model})}{MS(\text{Error})} = \frac{5.69}{.72765} = 7.82$$

The analysis of variance table is:

Source	df	SS	MS	F
Model	2	11.38	5.69	7.82
Error	17	12.37	.72765	
Total	19	23.75		

$$R^2 = 1 - \frac{SSE}{SS(\text{Total})} = 1 - \frac{12.37}{23.75} = .4792$$

b. $H_0: \beta_1 = \beta_2 = 0$
$H_a:$ At least one $\beta_i \neq 0, i = 1, 2$

The test statistic is $F = \dfrac{MS(\text{Model})}{MS(\text{Error})} = \dfrac{5.69}{.72765} = 7.82$ or

$$F = \frac{R^2/k}{(1 - R^2)/[n - (k + 1)]} = \frac{.4792/2}{(1 - .4792)/[20 - (2 + 1)]} = 7.82$$

The rejection region requires $\alpha = .05$ in the upper tail of the F distribution with df $= \nu_1 = k = 2$ and $\nu_2 = n - (k + 1) = 17$. From Table VIII, Appendix B, $F_{.05} = 3.59$. The rejection region is $F > 3.59$.

Since the observed value of the test statistic falls in the rejection region ($F = 7.82 > 3.59$), H_0 is rejected. There is sufficient evidence to indicate the model is useful in predicting y at $\alpha = .05$.

11.15 a. The hypothesized model is

$$E(y) = \beta_0 + \beta_1 x_1 + \beta_2 x_2 + \beta_3 x_3 + \beta_4 x_4 + \beta_5 x_5$$

β_0 is the y-intercept.

β_1 is the difference in mean salary between males and females, all other variables held constant.

β_2 is the difference in the mean salary between whites and non-whites, all other variables held constant.

β_3 is the change in mean salary for each additional year of education, all other variables held constant.

β_4 is the change in mean salary for each additional year with the firm, all other variables held constant.

β_5 is the change in mean salary for each additional hour worked per week, all other variables held constant.

b. The least squares equation is:

$$\hat{y} = 15.491 + 12.774x_1 + .713x_2 + 1.519x_3 + .32x_4 + .205x_5$$

$\hat{\beta}_0 = 15.491$ This is the y-intercept.

$\hat{\beta}_1 = 12.774$ The difference in the mean salary between males and females is estimated to be 12.774, all other variables held constant.

$\hat{\beta}_2 = .713$ The difference in the mean salary between whites and non-whites is estimated to be .713, all other variables held constant.

$\hat{\beta}_3 = 1.519$ The mean salary is estimated to increase by 1.519 for each additional year of education, all other variables held constant.

$\hat{\beta}_4 = .320$ The mean salary is estimated to increase by .320 for each additional year with the firm, all other variables held constant.

$\hat{\beta}_5 = .205$ The mean salary is estimated to increase by .205 for each additional hour worked per week, all other variables held constant.

c. $R^2 = .240$ 24% of the variability in the salaries is explained by the model containing the independent variables gender, race, education level, tenure with firm, and number of hours worked per week.

To determine if the model is useful for predicting annual salary, we test:

H_0: $\beta_1 = \beta_2 = \beta_3 = \beta_4 = \beta_5 = 0$
H_a: At least one $\beta_i \neq 0$, for $i = 1, 2, 3, 4, 5$

The test statistic is $F = \dfrac{R^2/k}{(1 - R^2)/[n - (k + 1)]} = \dfrac{.24/5}{(1 - .24)/[191 - (5 + 1)]} = 11.68$

The rejection region requires $\alpha = .05$ in the upper tail of the F distribution with df $= \nu_1 = k = 5$ and $\nu_2 = n - (k + 1) = 191 - (5 + 1) = 185$. From Table VIII, Appendix B, $F_{.05} \approx 2.21$. The rejection region is $F > 2.21$.

Since the observed value of the test statistic falls in the rejection ($F = 11.68 > 2.21$), H_0 is rejected. There is sufficient evidence to indicate the model is useful for predicting annual salary at $\alpha = .05$.

d. To determine whether the gender variable indicates the male managers are paid more than female managers, after adjusting for and holding the other variables constant, we test:

H_0: $\beta_1 = 0$
H_a: $\beta_1 > 0$

The p-value is $< .05/2 < .025$. Since the p-value is less than $\alpha = .05$, there is evidence to reject H_0. There is sufficient evidence to indicate that the male managers are paid more than female managers, after adjusting for and holding the other variables constant at $\alpha = .05$

e. Many factors other than gender affect salaries. Some of those factors have been included in this study such as level of education, years with the firm, number of hours worked per week, etc.

f. If gender and tenure with the firm interact, then the effect of tenure on annual salary depends on the gender. In other words, the relationship between salary and tenure depends on which gender is being considered.

11.17 a. The least squares equation is:

$$\hat{y} = 131.924 + 2.726x_1 + .0472x_2 - 2.587x_3$$

b. To test the usefulness of the model, we test:

H_0: $\beta_1 = \beta_2 = \beta_3 = 0$
H_a: At least one $\beta_i \neq 0$, for $i = 1, 2, 3$

The test statistic is $F = \dfrac{\text{MSR}}{\text{MSE}} = \dfrac{1719.4379}{96.242886} = 17.87$

The rejection region requires $\alpha = .01$ in the upper tail of the F distribution with $\nu_1 = k = 3$ and $\nu_2 = n - (k + 1) = 20 - (3 + 1) = 16$. From Table X, Appendix B, $F_{.01} = 5.29$. The rejection region is $F > 5.29$.

Since the observed value of the test statistic falls in the rejection region ($F = 17.87 > 5.29$), H_0 is rejected. There is sufficient evidence to indicate a relationship exists between hours of labor and at least one of the independent variables at $\alpha = .01$.

c. H_0: $\beta_2 = 0$
    H_a: $\beta_2 \neq 0$

   The test statistic is $t = .51$. The p-value $= .6199$. We reject H_0 if p-value $< \alpha$. Since .6199 $>$ .05, do not reject H_0. There is insufficient evidence to indicate a relationship exists between hours of labor and percentage of units shipped by truck, all other variables held constant, at $\alpha = .05$.

d. R^2 is printed as R-SQUARE. $R^2 = .7701$. We conclude that 77% of the sample variation of the labor hours is explained by the regression model, including the independent variables pounds shipped, percentage of units shipped by truck, and weight.

e. If the average number of pounds per shipment increases from 20 to 21, the estimated change in mean number of hours of labor is -2.587. Thus, it will cost $\$7.50(2.587) = \$19.4025$ less, if the variables x_1 and x_2 are constant.

f. Since $s = $ ROOT MSE $= 9.81$, we can estimate approximately with $\pm 2s$ precision or $\pm 2(9.81)$ or ± 19.62 hours.

11.19 a. The value $R^2 = .87$. 87% of the sample variation in attitude is explained by the regression model containing gender, years of experience, years squared, and the interaction between gender and years of experience.

b. H_0: $\beta_1 = \beta_2 = \beta_3 = \beta_4 = 0$
    H_a: At least one $\beta_i \neq 0$, for $i = 1, 2, \ldots, 4$

   The test statistic $F = \dfrac{R^2/k}{\dfrac{1 - R^2}{(n - (k + 1))}}$ where $k = 4$, $n = 40$

   $$= \frac{.87/4}{(1 - .87)/(40 - (4 + 1))} = \frac{.2175}{.0037} = 58.56$$

   The rejection region requires $\alpha = .05$ in the upper tail of the F distribution with $\nu_1 = k = 4$ and $\nu_1 = n - (k + 1) = 40 - (4 + 1) = 35$. From Table X, Appendix B, $F_{.05} \approx 2.69$. The rejection region is $F > 2.69$.

   Since the observed value of the test statistic falls in the rejection region ($F = 58.56 > 2.69$), H_0 is rejected. There is sufficient evidence to indicate the model is useful in predicting attitude at $\alpha = .05$. Attitude is related to at least one of the independent variables.

c. To determine if the interaction between sex and years of experience is useful in the prediction model, we test:

   H_0: $\beta_4 = 0$
   H_a: $\beta_4 \neq 0$

   The test statistic is $t = \dfrac{\hat{\beta}_4}{s_{\hat{\beta}_4}} = \dfrac{-1}{.02} = -50$

The rejection region requires $\alpha/2 = .05/2 = .025$ in each tail of the t distribution with df $= n - (k + 1) = 40 - (4 + 1) = 35$. From Table VI, Appendix B, $t_{.025} \approx 2.042$. The rejection region is $t < -2.042$ or $t > 2.042$.

Since the observed value of the test statistic falls in the rejection region ($t = -50 < -2.042$), H_0 is rejected. There is sufficient evidence to indicate the interaction between sex and years of experience is useful in the prediction model at $\alpha = .05$.

d. For $x_1 = 0$, $\hat{y} = 50 + 5(0) + 6x_2 - .2x_2^2 - 0(x_2)$

$$= 50 + 6x_2 - .2x_2^2$$

For $x_1 = 1$, $\hat{y} = 50 + 5(1) + 6x_2 - .2x_2^2 - x_2$

$$= 55 + 5x_2 - .2x_2^2$$

$x_1 = 0$		$x_1 = 1$	
x_2	$\hat{y}$	x_2	$\hat{y}$
0	50	0	55
2	61.2	2	64.2
4	70.8	4	71.8
6	78.8	6	77.8
8	85.2	8	82.2
10	90	10	85

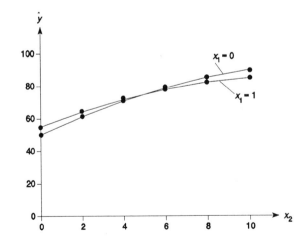

11.21 The SAS output is:

```
DEP VARIABLE: TIME
                              ANALYSIS OF VARIANCE

                           SUM OF          MEAN
            SOURCE    DF    SQUARES        SQUARE      F VALUE      PROB>F

            MODEL     2    251.60167     125.80083     54.807      0.0001
            ERROR     7    16.06733367   2.29533338
            C TOTAL   9    267.66900

                 ROOT MSE    1.515036    R-SQUARE     0.9400
                 DEP MEAN        8.19    ADJ R-SQ     0.9228
                 C.V.       18.49861

                              PARAMETER ESTIMATES

                         PARAMETER      STANDARD     T FOR HO:
            VARIABLE  DF   ESTIMATE       ERROR      PARAMETER=0    PROB > |T|

            INTERCEP   1   2.40524442   1.24836624      1.927       0.0954
            MICRO      1   1.42614858   0.14062909     10.141       0.0001
            EXPER      1  -0.36629552   0.13168833     -2.782       0.0272
```

a. Fitting the model to the data, the least squares prediction equation is

$$\hat{y} = 2.41 + 1.43x_1 - .366x_2$$

b. To test if the model is useful, we test:

H_0: $\beta_1 = \beta_2 = 0$
H_a: At least one $\beta_i \neq 0$, $i = 1, 2$

The test statistic is $F = \dfrac{\text{MS(Model)}}{\text{MS(Error)}} = 54.807$

To see if this is significant, that is, if we can reject H_0, we compare the p-value to our α level. Since $\alpha = .1$ and the p-value $\leq .0001 < .1$, we reject H_0. We conclude that at least one variable is significant in predicting maintenance time at $\alpha = .1$.

c. R^2, which is printed next to R-SQUARE, is .9400. This tells us we can explain approximately 94% of the sample variation in maintenance time with this model.

d. The SAS output is:

```
DEP VARIABLE: TIME
                              ANALYSIS OF VARIANCE

                           SUM OF          MEAN
            SOURCE    DF    SQUARES        SQUARE      F VALUE      PROB>F

            MODEL     3    264.05135     88.01711754   145.980     0.0001
            ERROR     6    3.61764738    0.60294123
            C TOTAL   9    267.66900

                 ROOT MSE    0.7764929   R-SQUARE     0.9865
                 DEP MEAN        8.19    ADJ R-SQ     0.9797
                 C.V.        9.480988
```

VARIABLE	DF	PARAMETER ESTIMATE	STANDARD ERROR	T FOR HO: PARAMETER=0	PROB > |T|
INTERCEP	1	-0.34875688	0.88129883	-0.396	0.7060
MICRO	1	2.06539551	0.15806738	13.067	0.0001
EXPER	1	0.02152454	0.10880943	0.198	0.8497
INTER	1	-0.09187342	0.02021846	-4.544	0.0039

The least squares prediction equation is:

$$\hat{y} = -.349 + 2.065x_1 + .0215x_2 - .0919x_1x_2$$

e. R^2, which is printed next to R-SQUARE, is .9865. This tells us we can explain approximately 98.65% of the sample variation in maintenance time with this model.

f. In comparing R^2, we must realize that additional variables added to the existing model will always increase R^2. We need to determine if the increase is significant.

g. H_0: $\beta_3 = 0$
H_a: $\beta_3 \neq 0$

The test statistic is $t = -4.544$.

Since the p-value is .0039 < .05, reject H_0. We conclude the interaction is significant in predicting maintenance time at $\alpha = .05$.

h. No. We must test to make sure each term in the model contributes information. Also, there may be other independent variables that have not been considered.

11.23 a. The regression equation is $\hat{y} = 506.35 - 941.9x_1 - 429.1x_2$ (from the MINITAB printout).

b. $s = 94.25$

In predicting y, we would expect 95% of the observations to be $\pm 2s$ or $\pm 2(94.25)$ or ± 188.5 from the mean y value.

c. The test statistic is $F = \dfrac{\text{Mean Square (Regression)}}{\text{Mean Square (Error)}} = \dfrac{64,165}{8,883} = 7.22$

To determine if the model contributes information to predict y, we test:

H_0: $\beta_1 = \beta_2 = 0$
H_a: At least one $\beta_i \neq 0$, $i = 1, 2$

The test statistic is $F = 7.22$.

The rejection region requires $\alpha = .05$ in the upper tail of the F distribution with $\nu_1 = k = 2$ and $\nu_2 = n - (k + 1) = 20 - (2 + 1) = 17$. From Table VIII, Appendix B, $F_{.05} = 3.59$. The rejection region is $F > 3.59$.

Since the observed value of the test statistic falls in the rejection region ($F = 7.22 > 3.59$), H_0 is rejected. There is sufficient evidence to indicate the model is useful in predicting y at $\alpha = .05$.

d. The standard error of $\hat{\beta}_2$ is the standard deviation of the coefficient of x_2 which is 379.8. For confidence coefficient .90, $\alpha = 1 - .90 = .10$ and $\alpha/2 = .10/2 = .05$. From Table VI, Appendix B, with df $= n - (k + 1) = 20 - (2 + 1) = 17$, $t_{.05} = 1.74$. The confidence interval is:

$$\hat{\beta}_2 \pm t_{\alpha/2} s_{\hat{\beta}_2} \Rightarrow -429.1 \pm 1.74(379.8) \Rightarrow -429.1 \pm 660.85$$
$$\Rightarrow (-1089.95, 231.75)$$

e. R^2 is labeled R-squared on the printout and is .459 (45.9%).

45.9% of the sample variation in the y's can be explained by the model.

11.25 In the residual plot for the straight-line model, there is a definite mound shape to the plot. This implies there is a need for a quadratic term in the model.

In the residual plot for the quadratic model, there is no longer a mound shape evident. The quadratic term seemed to help. From Exercise 11.11, the standard deviation for the quadratic model is $s = .43$. Two standard deviations from the mean is $\pm 2(.43) \Rightarrow \pm .86$. There are no points more than 2 standard deviations from the mean for the quadratic model.

In Exercise 11.11, part d, there was sufficient evidence to indicate the squared term was needed in the model at $\alpha = .05$. This agrees with our conclusion from analyzing the residual plots.

11.27 a. For the model fit with 25 observations:

$$\hat{y}_1 = 1.43 + .00999x_1 + .379x_2$$

For the model fit with 26 observations:

$$\hat{y}_2 = 1.16 + .0187x_1 + .406x_2$$

$\hat{\beta}_0$ decreased for the new model while $\hat{\beta}_1$ and $\hat{\beta}_2$ increased.

For the model fit with 25 observations:

$\hat{\beta}_0 = 1.43$	This is the y-intercept. It has no other meaning.
$\hat{\beta}_1 = .009999$	For each additional one-thousand dollars of income, we estimate the mean food consumption will increase by $9.99, for a fixed household size.
$\beta_2 = .379$	For each additional person in the household, we estimate the mean food consumption will increase by $379.00 for a fixed household income.

For the model fit with 26 observations:

$\hat{\beta}_0 = 1.16$ This is the y-intercept. It has no other meaning.

$\hat{\beta}_1 = .0187$ For each additional one-thousand dollars of income, we estimate the mean food consumption will increase by $18.70, for a fixed household size.

$\hat{\beta}_2 = .406$ For each additional person in the household, we estimate the mean food consumption will increase by $406.00, for a fixed household income.

b.   For the first model ($n = 25$):

$s = .2769$

In predicting the 1990 food consumption expenditure, we would expect 95% of the observations to lie within $\pm 2s \Rightarrow \pm 2(.2769) \Rightarrow \pm .5538$ thousand dollars of the mean food consumption expenditure.

For the second model ($n = 26$):

$s = .5581$

In predicting the 1990 food consumption expenditure, we would expect 95% of the observations to lie within $\pm 2s \Rightarrow \pm 2(.5881) \Rightarrow \pm 1.1162$ thousand dollars of the mean food consumption expenditure.

c.   For the first model ($n = 25$):

H_0: $\beta_1 = \beta_2 = 0$
H_a: At least one $\beta_i \neq 0$, $i = 1, 2$

The test statistic is $F_1 = \dfrac{MSR}{MSE} = \dfrac{7.7311}{.0767} = 100.80$

The rejection region requires $\alpha = .05$ in the upper tail of the F distribution with $\nu_1 = k = 2$ and $\nu_2 = n - (k + 1) = 25 - (2 + 1) = 22$. From Table VIII, Appendix B, $F_{.05} = 3.44$. The rejection region is $F_1 > 3.44$.

Since the observed value of the test statistic falls in the rejection region ($F = 100.80 > 3.44$), H_0 is rejected. There is sufficient evidence to indicate the model is useful for predicting the 1990 food consumption expenditure at $\alpha = .05$.

For the second model ($n = 26$):

H_0: $\beta_1 = \beta_2 = 0$
H_a: At least one $\beta_i \neq 0$, $i = 1, 2$

The test statistic is $F_2 = \dfrac{MSR}{MSE} = \dfrac{10.538}{.311} = 33.88$

The rejection region requires $\alpha = .05$ in the upper tail of the F distribution with $\nu_1 = k = 2$ and $\nu_2 = n - (k + 1) = 26 - (2 + 1) = 23$. From Table VIII, Appendix B, $F_{.05} = 3.42$. The rejection region is $F_2 > 3.42$.

Since the observed value of the test statistic falls in the rejection region ($F_2 = 34.88 > 3.42$), H_0 is rejected. There is sufficient evidence to indicate the model is useful for predicting the 1990 food consumption expenditure at $\alpha = .05$.

d.   For the first model ($n = 25$):

The standard error for $\hat{\beta}_2$ is .02772 (from printout).

For confidence coefficient .95, $\alpha = 1 - .95 = .05$ and $\alpha/2 = .05/2 = .025$. From Table VI, Appendix B, with df $= 22$, $t_{.025} = 2.074$. The confidence interval is:

$$\hat{\beta}_2 \pm t_{.025}s_{\hat{\beta}_2} \Rightarrow .379 \pm 2.074(.02772) \Rightarrow .379 \pm .0575 \Rightarrow (.3215, .4365)$$

For the second model ($n = 26$):

The standard error for $\hat{\beta}_2$ is .05551 (from printout).

For confidence coefficient .95, $\alpha = 1 - .95 = .05$ and $\alpha/2 = .05/2 = .025$. From Table VI, Appendix B, with df $= 23$, $t_{.025} = 2.069$. The confidence interval is:

$$\hat{\beta}_2 \pm t_{.025}s_{\hat{\beta}_2} \Rightarrow .406 \pm 2.069(.05551) \Rightarrow .406 \pm .1149 \Rightarrow (.2911, .5209)$$

e.   With the 26th observation, the standard deviation for the model increases to more than double the old amount. This brings the test statistic down and increases the width of the confidence interval for β_2. The observation appears to have a large amount of influence.

11.29   The stem-and-leaf display for the first model's residuals, ($n = 25$), appears to be somewhat normal. However when the 26th observation is included, the stem-and-leaf display becomes skewed to the right. The 26th observation has destroyed the normality of the residuals.

11.31   For the straight-line model, from Exercise 11.30 we know $s = 1.14301$. Two standard deviations from the mean is $\pm 2(1.14301) \Rightarrow \pm 2.28602$. Three standard deviations from the mean is $\pm 3(1.14301) \Rightarrow \pm 3.42903$. There may be one observation more than 2 standard deviations from the mean, but none beyond 3 standard deviations. Thus, there is no evidence to indicate outliers are present. Also, there is no mound or bowl shape to the plot. This implies the quadratic term is not necessary.

For the quadratic model, from Exercise 11.30 we know $s = 1.09089$. Two standard deviations from the mean is $\pm 2(1.09089) \Rightarrow -2.18178$ to 2.18178. There are no observations more than 2 observations from the mean. Thus, there is no evidence to indicate outliers are present. The plot is very similar to that for the straight-line model, again implying the quadratic term is not necessary. This agrees with the conclusion in Exercise 11.30d.

11.33 **a.** $R^2 = .45$. 45% of the variability in the suicide rates is explained by the model containing the independent variables unemployment rate, percentage of females in the labor force, divorce rate, log of GNP, and annual percent change in GNP.

To determine if the model is useful for predicting the suicide rate, we test:

H_0: $\beta_1 = \beta_2 = \beta_3 = \beta_4 = \beta_5 = 0$
H_a: At least one $\beta_i \neq 0$, $i = 1, 2, 3, 4, 5$

The test statistic is $F = \dfrac{R^2/k}{(1 - R^2)/[n - (k + 1)]} = \dfrac{.45/5}{(1 - .45)/[45 - (5 + 1)]} = 6.38$

The rejection region requires $\alpha = .05$ in the upper tail of the F distribution with $\nu_1 = k = 5$ and $\nu_2 = n - (k + 1) = 45 - (5 + 1) = 39$. From Table VIII, Appendix B, $F_{.05} \approx 2.45$. The rejection region is $F > 2.45$.

Since the observed value of the test statistic falls in the rejection region ($F = 6.38 > 2.45$), H_0 is rejected. There is sufficient evidence to indicate the model is useful for predicting suicide rate at $\alpha = .05$.

b. $\hat{\beta}_0 = .002$ — This is the y-intercept. It has no other meaning in this problem.

$\hat{\beta}_1 = .0204$ — For each additional unit increase in unemployment rate, the mean suicide rate is estimated to increase by .0204, all other variables held constant. The p-value is .002. Since the p-value is so small, there is evidence to indicate β_1 is not 0, adjusted for all the other variables being in the model.

$\hat{\beta}_2 = -.0231$ — For each additional unit increase in percentage of females in the labor force, the mean suicide rate is estimated to decrease by .0231, all other variables held constant. The p-value is .02. Since the p-value is so small, there is evidence to indicate β_2 is not 0, adjusted for all the other variables being in the model.

$\hat{\beta}_3 = .0765$ — For each additional unit increase in divorce rate, the mean suicide rate is estimated to increase by .0765, all other variables held constant. The p-value is $> .10$. Since the p-value is so large, there is no evidence to indicate β_3 is not 0, adjusted for all the other variables being in the model.

$\hat{\beta}_4 = .2760$ — For each additional unit increase in the log of GNP, the mean suicide rate is estimated to increase by .2760, all other variables held constant. The p-value is $> .10$. Since the p-value is so large, there is no evidence to indicate β_4 is not 0, adjusted for all the other variables being in the model.

$\hat{\beta}_5 = .0018$ — For each additional unit increase in annual percent change in GNP, the mean suicide rate is estimated to increase by .0018, all other variables held constant. The p-value is $> .10$. Since the p-value is so large, there is no evidence to indicate β_5 is not 0, adjusted for all the other variables being in the model.

From the *p*-values, there is evidence that only the independent variables unemployment rate and percentage of females in the labor force should be in the model. Further investigation of the other variables is warranted.

c. To determine if unemployment rate is a useful predictor of the suicide rate, we test:

$$H_0: \beta_1 = 0$$
$$H_a: \beta_1 \neq 0$$

The *p*-value for the test is .002. Since the *p*-value is so small, there is evidence to indicate β_1 is not 0, adjusted for all the other variables being in the model. There is sufficient evidence to indicate that the unemployment rate is a useful predictor of suicide rate.

d. We still need to check several terms to see if the model can be improved. Curvature means that the relationship between suicide and some of the independent variables may not be linear but curved. The three independent variables that do not appear to be related to the suicide rate may be related, but not linearly.

Also, there may be significant interaction among the independent variables. This implies that the effect of one variable on the suicide rate may depend on the level of a second independent variable. Again, the interaction terms need to be checked.

Finally, there may be a multicollinearity problem among the independent variables. If the independent variables are highly correlated with each other, then having more than one in the model at the same time can lead to wrong conclusions.

11.35　a. The least squares prediction equation is

$$\hat{y} = .601 + .595x_1 - 3.725x_2 - 16.232x_3 + .235x_1x_2 + .308x_1x_3$$

b. $R^2 = .928$ which tells us we can explain about 92.8% of the variation in test scores with the model

$$E(y) = \beta_0 + \beta_1x_1 + \beta_2x_2 + \beta_3x_3 + \beta_4x_1x_2 + \beta_5x_1x_3$$

To see whether this model is useful for predicting achievement test scores, we test:

$$H_0: \beta_1 = \beta_2 = \beta_3 = \beta_4 = \beta_5 = 0$$
$$H_a: \text{At least one } \beta_i \neq 0, \text{ for } i = 1, 2, \dots, 5$$

The test statistic is $F = \dfrac{MSR}{MSE} = 139.42$

The *p*-value is $\leq .0001$. Since $.0001 < \alpha = .05$, reject H_0. We conclude the model is useful in predicting achievement test scores at $\alpha = .05$.

c. For $x_2 = 0$ and $x_3 = 0$ (SES low)

$$\hat{y} = .601 + .595x_1$$

For $x_2 = 1$ and $x_3 = 0$ (SES medium)

$$\hat{y} = .601 + .595x_1 - 3.725(1) + .235(1)x_1$$
$$= -3.124 + .83x_1$$

For $x_2 = 0$ and $x_3 = 1$ (SES high)

$$\hat{y} = .601 + .595x_1 - 16.232(1) + .308(1)x_1$$
$$= -15.631 + .903x_1$$

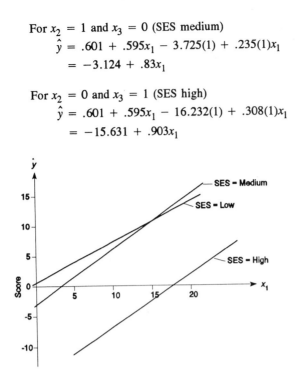

11.37 If you use your prediction equation to predict y when $x_1 = 30$, $x_2 = .6$, and $x_3 = 1{,}300$, all the independent values are inside the observed ranges of the variables used to develop the model. However, $x_1 = 60$, $x_2 = .4$, and $x_3 = 900$ are not in the observed ranges of the variables you used to develop your model. Since the model is not necessarily valid outside the sampled ranges, the results can be misleading.

11.39 a. The least squares prediction equation is

$$\hat{y} = .0562 + .273x_1 + .0006x_2$$

b. To determine if the model contributes information for predicting the number of positions filled, we test:

H_0: $\beta_1 = \beta_2 = 0$
H_a: At least one $\beta_i \neq 0$, $i = 1, 2$

The test statistic is $F = \dfrac{\text{MSR}}{\text{MSE}} = \dfrac{291.59}{1.77} = 164.74$

The rejection region requires $\alpha = .05$ in the upper tail of the F distribution with $\nu_1 = k = 2$ and $\nu_2 = n - (k + 1) = 10 - (2 + 1) = 7$. From Table VIII, Appendix B, $F_{.05} = 4.74$. The rejection region is $F > 4.74$.

Since the observed value of the test statistic falls in the rejection region ($F = 164.74 > 4.74$), H_0 is rejected. There is sufficient evidence to indicate the model is useful for predicting the number of positions filled at $\alpha = .05$.

c. H_0: $\beta_2 = 0$
   H_a: $\beta_2 \neq 0$

   The test statistic is $t = 4.34$ (from printout).

   The rejection region requires $\alpha/2 = .05/2 = .025$ in the each tail of the t distribution with df $= n - (k + 1) = 10 - (2 + 1) = 7$. From Table VI, Appendix B, $t_{.05} = 2.365$. The rejection region is $t < -2.365$ or $t > 2.365$.

   Since the observed value of the test statistic falls in the rejection region ($t = 4.34 > 2.365$), H_0 is rejected. There is sufficient evidence to indicate the recruiting budget contributes information for predicting the number of positions that are filled, adjusted for number of positions open at $\alpha = .05$.

d. If $x_1 = 30$ and $x_2 = 10,000$,

   $$\hat{y} = .0562 + .273(30) + .0006(10,000)$$
   $$= 14.25$$

e. Since the dependent variable is discrete (number of positions filled), it is unlikely that ϵ is normally distributed.

11.41 To determine if the model is useful for predicting y, we test:

   H_0: $\beta_1 = \beta_2 = 0$
   H_a: At least one $\beta_i \neq 0$, for $i = 1, 2$

   The test statistic is $F = \dfrac{R^2/k}{(1 - R^2)/[n - (k + 1)]} = \dfrac{.24/2}{(1 - .24)/[100 - (2 + 1)]} = 15.32$

   The rejection region requires $\alpha = .05$ in the upper tail of the F distribution with $\nu_1 = k = 2$ and $\nu_2 = n - (k + 1) = 100 - (2 + 1) = 97$. From Table X, Appendix B, $F_{.05} \approx 3.10$. The rejection region is $F > 3.10$.

   Since the observed value of the test statistic falls in the rejection region ($F = 15.32 > 3.10$), H_0 is rejected. There is sufficient evidence to indicate the model is useful in predicting kilowatt usage during the winter months at $\alpha = .05$.

11.43 a. $\hat{y} = 12.3 + .25(6) - .0033(6^2) = 13.6812$

   b. We are 95% confident the actual time lost after 6 hours of work prior to an accident is between 1.35 and 26.01.

   Yes. This interval is so wide that it is virtually worthless.

11.45 a. $\hat{\beta}_1 = -.01$. It is estimated that the change in the mean proportion of customers who buy its product for each additional dollar in price is $-.01$, holding years of experience constant.

   $\hat{\beta}_2 = .10$. It is estimated that the change in the mean proportion of customers who buy its product for each additional year of experience of the salesperson is .10, holding price constant.

Multiple Regression

b. To determine if the model is useful in the prediction of the proportion of customers who buy its product, we test:

H_0: $\beta_1 = \beta_2 = 0$
H_a: At least one $\beta_i \neq 0$, for $i = 1, 2$

The test statistic is $F = \dfrac{R^2/k}{(1 - R^2)/[n - (k + 1)]} = \dfrac{.86/2}{(1 - .86)/[20 - (2 + 1)]} = 52.21$

The rejection region requires $\alpha = .05$ in the upper tail of the F distribution with $\nu_1 = k = 2$ and $\nu_2 = n - (k + 1) = 20 - (2 + 1) = 17$. From Table VIII, Appendix B, $F_{.05} = 3.59$. The rejection region is $F > 3.59$.

Since the observed value of the test statistic falls in the rejection region ($F = 52.21 > 3.59$), H_0 is rejected. There is sufficient evidence to indicate the model is useful in predicting the proportion of customers who buy its product at $\alpha = .05$.

c. To determine if as the price increases the mean proportion of buyers will decrease, we test:

H_0: $\beta_1 = 0$
H_a: $\beta_1 < 0$

The test statistic is $t = \dfrac{\hat{\beta}_1 - 0}{s_{\hat{\beta}_1}} = \dfrac{-.01 - 0}{.003} = -3.33$

The rejection region requires $\alpha = .05$ in the lower tail of the t distribution with df $= n - (k + 1) = 20 - (2 + 1) = 17$. From Table VI, Appendix B, $t_{.05} = 1.74$. The rejection region is $t < -1.74$.

Since the observed value of the test statistic falls in the rejection region ($t = -3.33 < -1.74$), H_0 is rejected. There is sufficient evidence to indicate that the mean proportion of buyers will decrease as the price increases at $\alpha = .05$.

d. To determine if as the experience of the salesperson increases the mean proportion of buyers increases, we test:

H_0: $\beta_2 = 0$
H_a: $\beta_2 > 0$

The test statistic is $t = \dfrac{\hat{\beta}_2}{s_{\hat{\beta}_2}} = \dfrac{.10}{.025} = 4$

The rejection region requires $\alpha = .05$ in the upper tail of the t distribution with df $= n - (k + 1) = 17$. From Table VI, Appendix B, $t_{.05} = 1.74$. The rejection region is $t > 1.74$.

Since the observed value of the test statistic falls in the rejection region ($t = 4 > 1.74$), H_0 is rejected. There is sufficient evidence to indicate the mean proportion of buyers will increase as the experience of the salesperson increases at $\alpha = .05$.

11.47 a. $\hat{\beta}_0 = -105$. This is the y-intercept. It has no other meaningful interpretation.

$\hat{\beta}_1 = 25$. It is estimated that the difference in mean attendance between weekend and non-weekend days is 25 people, all other variables held constant.

$\hat{\beta}_2 = 100$. It is estimated that the difference in mean attendance between sunny and overcast days is 100 people, all other variables held constant.

$\hat{\beta}_3 = 10$. It is estimated that the change in mean attendance for each additional degree is 10 people, all other variables held constant.

b. To determine if the model is useful in the prediction of daily attendance, we test:

H_0: $\beta_1 = \beta_2 = \beta_3 = 0$
H_a: At least one $\beta_i \neq 0$, for $i = 1, 2, 3$

The test statistic is $F = \dfrac{R^2/k}{(1 - R^2)/(n - (k + 1))}$

$$F = \frac{.65/3}{(1 - .65).(30 - (3 + 1))} = \frac{.21667}{.01346} = 16.10$$

The rejection region requires $\alpha = .05$ in the upper tail of the F distribution with $\nu_1 = k = 3$ and $\nu_2 = n - (k + 1) = 30 - (3 + 1) = 26$. From Table VIII, Appendix B, $F_{.05} = 2.98$. The rejection region is $F > 2.98$.

Since the observed value of the test statistic falls in the rejection region ($F = 16.10 > 2.98$), H_0 is rejected. There is sufficient evidence to indicate the model is useful in the prediction of daily attendance.

c. To determine if the mean attendance rises on weekends, we test:

H_0: $\beta_1 = 0$
H_a: $\beta_1 > 0$

The test statistic is $t = \dfrac{\hat{\beta}_1 - 0}{s_{\hat{\beta}_1}} = \dfrac{25 - 0}{10} = 2.5$

The rejection region requires $\alpha = .10$ in the upper tail of the t distribution with df $= n - (k + 1) = 30 - (3 + 1) = 26$. From Table VI, Appendix B, $t_{.10} = 1.315$. The rejection region is $t > 1.315$.

Since the observed value of the test statistic falls in the rejection region ($t = 2.5 > 1.315$), H_0 is rejected. There is sufficient evidence to indicate the mean attendance increases on weekends at $\alpha = .10$.

d. For $x_1 = 0$, $x_2 = 1$, and $x_3 = 95$,

$$\hat{y} = -105 + 25(0) + 100(1) + 10(95) = 945$$

e. We are 90% confident the actual number of people attending the park will be between 645 and 1245 when the predicted temperature is 95° and the day is a sunny weekday.

11.49 Because of the interaction term, the true coefficient for x_1 is $(-700 + 15x_3)$. As long as the temperature, x_3, is more than $47°$, $-700 + 15x_3$ will be a positive term.

11.51 a. To convert the data, we use

$$y_2 = \frac{y_1}{PPI} \times 100, \; x_4 = \frac{x_2}{PPI} \times 100, \; x_5 = \frac{x_3}{PPI} \times 100$$

The new data are:

Year	y_2	x_4	x_5
1967	50.00	30.00	25.00
1968	117.07	29.27	25.37
1969	131.46	30.99	26.29
1970	122.28	30.80	27.17
1971	142.98	28.95	27.19
1972	195.63	30.23	28.55
1973	178.92	29.70	27.47
1974	159.28	28.11	26.23
1975	163.52	28.59	27.44
1976	180.33	28.96	29.51
1977	200.31	29.87	29.87
1978	203.06	28.67	29.14
1979	188.88	30.14	30.56
1980	175.60	29.76	30.13
1981	170.76	30.67	31.70
1982	170.40	30.74	30.74
1983	161.66	30.35	29.69
1984	162.75	30.29	30.29

b. The SAS printout is:

DEP VARIABLE: Y2

ANALYSIS OF VARIANCE

SOURCE	DF	SUM OF SQUARES	MEAN SQUARE	F VALUE	PROB>F
MODEL	3	15118.16609	5039.38870	8.609	0.0017
ERROR	14	8194.78934	585.34210		
C TOTAL	17	23312.95543			

ROOT MSE	24.19384	R-SQUARE	0.6485
DEP MEAN	159.71611	ADJ R-SQ	0.5732
C.V.	15.14803		

PARAMETER ESTIMATES

VARIABLE	DF	PARAMETER ESTIMATE	STANDARD ERROR	T FOR H0: PARAMETER=0	PROB > |T|
INTERCEP	1	238.005635	208.63040485	1.141	0.2731
PEOPLE	1	-0.441832	1.45797635	-0.303	0.7663
X4	1	-19.379015	9.29112882	-2.086	0.0558
X5	1	17.929540	8.86812439	2.022	0.0627

The fitted model is $\hat{y}_2 = 238.006 - .4418x_1 - 19.379x_4 + 17.9295x_5$

The mean sales revenue (in constant dollars) is estimated to decrease by .4418 (thousands of dollars) for each additional sales person, with mean price and mean competitor's price held constant.

The mean sales revenue (in constant dollars) is estimated to decrease by 19.379 (thousands of dollars) for each additional one constant dollar increase in mean price, with sales people and mean competitor's price held constant.

The mean sales revenue (in constant dollars) is estimated to increase by 17.9295 (thousands of dollars) for each additional one constant dollar increase in the competitor's mean price, mean price and sales people held constant.

c. From the printout $R^2 = .6485$. About 65% of the sample variability in sales revenue (in constant dollars) is explained by the fitted model containing sales people, mean price (constant dollars), and mean competitor's price (constant dollar).

d. To investigate the usefulness of the model, we test:

H_0: $\beta_1 = \beta_2 = \beta_3 = 0$
H_a: At least one $\beta_i \neq 0$, $i = 1, 2, 3$

e. The test statistic is $F = \dfrac{MSR}{MSE} = 8.609$

The p-value is .0017 which is less than $\alpha = .05$. Thus, we reject H_0. There is sufficient evidence to indicate the model is useful for predicting sales revenue at $\alpha = .05$.

The p-value is .0017.

f. H_0: $\beta_2 = 0$
H_a: $\beta_2 < 0$

The test statistic is $t = \dfrac{\hat{\beta}_2 - 0}{s_{\hat{\beta}_2}} = -2.086$

The rejection region requires $\alpha = .05$ in the lower tail of the t distribution with df $= n - (k + 1) = 18 - (3 + 1) = 14$. From Table VI, Appendix B, $t_{.05} = 1.761$. The rejection region is $t < -1.761$.

Since the observed value of the test statistic falls in the rejection region ($t = -2.086 < -1.761$), H_0 is rejected. There is sufficient evidence to indicate that the mean sales price decreases as the mean price (constant dollars) increases, with the other variables held constant, at $\alpha = .05$.

g. For confidence coefficient .95, $\alpha = 1 - .95 = .05$ and $\alpha/2 = .05/2 = .025$. From Table VI, Appendix B, with df $= 14$, $t_{.025} = 2.145$. The confidence interval is:

$$\hat{\beta}_1 \pm t_{.025}s_{\hat{\beta}_1} = -.442 \pm 2.145(1.45798)$$
$$\Rightarrow -.442 \pm 3.127 \Rightarrow (-3.569, 2.685)$$

We are 95% confident that the value of β_1 will fall between -3.569 and 2.685. Because the interval contains 0, there is insufficient evidence to conclude sales revenue and the number of sales people are related.

h. $x_4 = \dfrac{x_2}{\text{PPI}} \times 100 = \dfrac{90}{315} \times 100 = 28.57$

$x_5 = \dfrac{x_3}{\text{PPI}} \times 100 = \dfrac{92}{315} \times 100 = 29.21$

Thus, $\hat{y} = 238.006 - .442(35) - 19.379(28.57) + 17.930(29.21)$
$= 192.613$

i. For $x_1 > 38$, $x_4 > \$31$, and $x_5 > \$32$, all values are outside the observed ranges. We have no idea what the relationship between y and the independent variables looks like outside the observed range.

11.53 **Residual Plot for ϵ vs. x_1**

There is a possible mound shape to the plot. We may want to try adding a quadratic term for verbal score $\left(x_1^2\right)$ to the model. From Exercise 11.52, $s = .40228$. Two standard deviations from the mean is $\pm 2(.40228) \Rightarrow -.80456$ to $.80456$. Three standard deviations from the mean is $\pm 3(.40228) \Rightarrow -1.20684$ to 1.20684. There are two points more than 2 standard deviations from the mean, but none more than 3 standard deviations from the mean. Thus, there is no evidence to indicate outliers are present.

Residual Plot for ϵ vs. x_2

There appears to be a very slight mound or bowl shape to the plot. We may want to try adding the quadratic term for mathematics score $\left(x_2^2\right)$ to the model. Again there are two data points more than 2 standard deviations from the mean, but none more than 3 standard deviations from the mean.

11.55 The second-order model is preferable as a predictor of G.P.A. The first-order model's residual plots were mound-shaped indicating that quadratic terms were needed. The second-order model residual plots no longer look mound-shaped. There are also no observations more than three standard deviations from the mean $(\pm 3(.18714) \Rightarrow \pm .56142)$.

CHAPTER TWELVE

....

Introduction to Model Building

12.1 a. Weight would be quantitative. It could take on values greater than 0.

  b. Color would be qualitative. Values color could take on are blue, red, green, etc.

  c. Age would be quantitative. Age could take on values greater than 0.

  d. Hardness would be quantitative. It could take on values greater than 0.

  e. Package design would be qualitative. Values it could take on include design A, design B, etc.

12.3 The assumption that the distribution of ϵ is normal prohibits the use of a qualitative variable as a dependent variable.

12.5 Since the lines graphed are both parabolas, the order of the polynomial in both cases is second order. In the graph **a**, β_0 is 4 (value of y when $x = 0$) and β_2 is negative because the parabola opens downward. In graph **b**, β_0 is 8 (value of y when $x = 0$) and β_2 is positive because the parabola opens upward.

12.7 $E(y) = \beta_0 + \beta_1 x + \beta_2 x^2$

12.9 a. Because the company suspects that new and old copiers require more service calls than those in the middle, the model would be

$$E(y) = \beta_0 + \beta_1 x + \beta_2 x^2$$

  where y = number of service calls
  x = age of copier

  b. $\beta_0 > 0$ and $\beta_2 > 0$. Since the parabola will open upwards (number of service calls larger for young and old machines), both β_0 and β_2 will be greater than 0.

12.11 The model would be $E(y) = \beta_0 + \beta_1 x + \beta_2 x^2$. Since the value of y is expected to increase and then decrease as x gets larger, β_2 will be negative. A sketch of the model would be:

12.13   a.   A model that would allow the assembly time to decrease and then increase would be

$$E(y) = \beta_0 + \beta_1 x + \beta_2 x^2$$

where y = assembly time
      x = time since lunch
      β_0 = y-intercept
      β_1 = shifts parabola to right or left
      β_2 = rate of curvature

   b.   The sketch of the model is:

12.15   a.   The model for a straight line is $E(y) = \beta_0 + \beta_1 x$.

Some preliminary calculations are:

$$\sum x_i = 20{,}066.7 \quad \sum x_i^2 = 16{,}797{,}815.35 \quad \sum x_i y_i = 187{,}148.25$$

$$\sum y_i = 224.5 \qquad \sum y_i^2 = 2116.75$$

$$SS_{xy} = \sum x_i y_i - \frac{\sum x_i \sum y_i}{n} = 187{,}148.25 - \frac{(20{,}066.7)(224.5)}{24} = -559.0063$$

$$SS_{xx} = \sum x_i^2 - \frac{\left(\sum x_i\right)^2}{n} = 16{,}797{,}815.35 - \frac{20.066.7^2}{24} = 19{,}796.65$$

$$SS_{yy} = \sum y_i^2 - \frac{\left(\sum y_i\right)^2}{n} = 2116.75 - \frac{224.5^2}{24} = 16.739583$$

$$\hat{\beta}_1 = \frac{SS_{xy}}{SS_{xx}} = \frac{-559.0063}{19{,}796.65} = -.028237418 \approx -.0282$$

$$\hat{\beta}_0 = \bar{y} - \hat{\beta}_1 \bar{x} = \frac{224.5}{24} - (-.028237418)\frac{20{,}066.7}{24} = 32.96382563 \approx 32.964$$

The fitted line is $\hat{y} = 32.964 - .0282x$

$$SSR = \frac{SS_{xy}^2}{SS_{xx}} = \frac{-559.0063^2}{19{,}796.65} = 15.78489509$$

$$SS(Total) = SS_{yy} = 16.739583$$

$$R^2 = \frac{SSR}{SS(Total)} = \frac{15.78489509}{16.739583} = .9430$$

b. Based on the results of **a**, it appears that as the rate of growth in money supply increases, the prime interest rate decreases (because $\hat{\beta}_1$ is negative).

c. The scatter plot, using SAS, is:

From the scattergram and the least squares line, there is no indication that a second-order model would better explain the variation in interest rates.

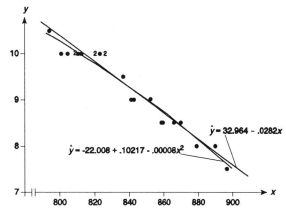

$$\hat{y} = 32.964 - .0282x$$

$$\hat{y} = -22.008 + .10217 - .00008x^2$$

d. The SAS computer printout for fitting the model to the data is:

```
Model: MODEL1
Dependent Variable: Y
```

Analysis of Variance

Source	DF	Sum of Squares	Mean Square	F Value	Prob>F
Model	2	15.87098	7.93549	191.854	0.0001
Error	21	0.86861	0.04136		
C Total	23	16.73958			

Root MSE	0.20338	R-square	0.9481	
Dep Mean	9.35417	Adj R-sq	0.9432	
C.V.	2.17418			

Parameter Estimates

Variable	DF	Parameter Estimate	Standard Error	T for H0: Parameter=0	Prob > |T|
INTERCEP	1	-22.008134	38.12441073	-0.577	0.5699
X	1	0.102169	0.09040592	1.130	0.2712
XSQ	1	-0.000077242	0.00005354	-1.443	0.1639

The fitted model is $\hat{y} = -22.008 + .102x - .00008x^2$

$R^2 = R\text{-square} = .9481$

f. To determine if $\beta_2 \neq 0$, we test:

H_0: $\beta_2 = 0$
H_a: $\beta_2 \neq 0$

The test statistic is $t = -1.443$ (From printout.)

The *p*-value $= .1639$. Since the *p*-value $> \alpha = .05$, H_0 is not rejected. There is insufficient evidence to indicate $\beta_2 \neq 0$ at $\alpha = .05$.

Based on the results above and the graph of the second-order model, there is no indication that the second-order term is necessary. The first-order model is sufficient to represent the relationship between money supply and prime interest rate.

12.17 a. $E(y) = \beta_0 + \beta_1 x_1 + \beta_2 x_2$

b. $E(y) = \beta_0 + \beta_1 x_1 + \beta_2 x_2 + \beta_3 x_1 x_2$

c. $E(y) = \beta_0 + \beta_1 x_1 + \beta_2 x_2 + \beta_3 x_1 x_2 + \beta_4 x_1^2 + \beta_5 x_2^2$

12.19 a. The order of the model is second order, because it has the term $x_1 x_2$ in it.

b. The response surface is a twisted plane.

c. The contour line when $x_1 = 0$ is $E(y) = 4 - 0 + 2x_2 + 0(x_2) = 4 + 2x_2$

The contour line when $x_1 = 1$ is $E(y) = 4 - 1 + 2x_2 + x_2 = 3 + 3x_2$

The contour line when $x_1 = 2$ is $E(y) = 4 - 2 + 2x_2 + 2x_2 = 2 + 4x_2$

The plots are:

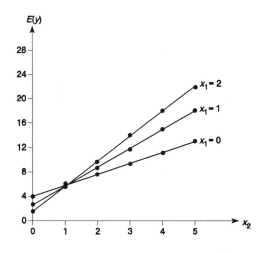

d. The interactive term $x_1 x_2$ allows the contour lines to be non-parallel.

e. When $x_1 = 0$, the contour line is $E(y) = 4 + 2x_2$. For each unit change in x_2, $E(y)$ increases by 2 units.

When $x_1 = 1$, the contour line is $E(y) = 3 + 3x_2$. For each unit change in x_2, $E(y)$ increases by 3 units.

When $x_1 = 2$, the contour line is $E(y) = 2 + 4x_2$. For each unit change in x_2, $E(y)$ increases by 4 units.

f. When $x_1 = 2$ and $x_2 = 4$, $E(y) = 18$. When $x_1 = 0$ and $x_2 = 5$, $E(y) = 14$. Thus, the change in $E(y)$ is from 18 to 14 or -4.

12.21 a. The prediction equation is $\hat{y} = \hat{\beta}_0 + \hat{\beta}_1 x_1 + \hat{\beta}_2 x_2 + \hat{\beta}_3 x_1 x_2$ or
$\hat{y} = -2.550 + 3.82x_1 + 2.63x_2 - 1.29x_1x_2$

b. The response surface is a twisted plane.

c. For $x_2 = 1$,

$$\hat{y} = -2.55 + 3.82x_1 + 2.63(1) - 1.29x_1(1)$$
$$= .08 + 2.53x_1$$

For $x_2 = 3$,

$$\hat{y} = -2.55 + 3.82x_1 + 2.63(3) - 1.29x_1(3)$$
$$= 5.34 - .05x_1$$

For $x_2 = 5$,

$$\hat{y} = -2.55 + 3.82x_1 + 2.63(5) - 1.29x_1(5)$$
$$= 10.6 - 2.63x_1$$

d. If x_1 and x_2 interact, the effect of x_1 on y is not the same at each level of x_2. When $x_2 = 1$, as x_1 increases from 0 to 4, $\hat{y}$ increases from .08 to 10.2. When $x_2 = 5$, as x_1 increases from 0 to 4, $\hat{y}$ decreases from 10.6 to .08.

e. H_0: $\beta_3 = 0$
H_a: $\beta_3 \neq 0$

f. The test statistic is $t = \dfrac{\hat{\beta}_3 - 0}{s(\hat{\beta}_3)} = \dfrac{-1.285 - 0}{.1594} = -8.06$

The rejection region requires $\alpha/2 = .01/2 = .005$ in each tail of the t distribution. From Table VI, Appendix B, $t_{.005} = 3.106$ with df $= n - 4 = 15 - 4 = 11$. The rejection region is $t > 3.106$ or $t < -3.106$.

Since the observed value of the test statistic falls in the rejection region ($t = -8.06 < -3.106$), H_0 is rejected. There is sufficient evidence to indicate that x_1 and x_2 interact at $\alpha = .01$.

12.23 a. Both independent variables are quantitative.

b. $E(y) = \beta_0 + \beta_1 x_1 + \beta_2 x_2$

c. $E(y) = \beta_0 + \beta_1 x_1 + \beta_2 x_2 + \beta_3 x_1 x_2$

If x_1 and x_2 interact, the plot of the contour lines may look like:

Each line corresponds to a different value of x_1.

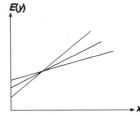

d. $E(y) = \beta_0 + \beta_1 x_1 + \beta_2 x_2 + \beta_3 x_1 x_2 + \beta_4 x_1^2 + \beta_5 x_2^2$

12.25 a. The prediction equation is $\hat{y} = 149 + .472 x_1 - .0993 x_2 - .0005 x_1^2 + .000015 x_2^2$

b. Since the signs of x_1^2 and x_2^2 differ, the response surface will be saddle-shaped.

c. To test to see if the model is useful for predicting quarterly sales, we test:

H_0: $\beta_1 = \beta_2 = \beta_3 = \beta_4 = 0$
H_a: At least one $\beta_i \neq 0$, $i = 1, 2, 3, 4$

The test statistic is $F = \dfrac{MSR}{MSE} = \dfrac{1500.52}{47.40} = 31.66$

The rejection region requires $\alpha = .01$ in the upper tail of the F distribution. From Table X, Appendix B, $F = 5.67$ with numerator df $= k = 4$ and denominator df $= n - (k + 1) = 16 - (4 + 1) = 11$. The rejection region is $F > 5.67$.

Since the observed value of the test statistic falls in the rejection region ($F = 31.66 > 5.67$), H_0 is rejected. There is sufficient evidence to indicate the model is useful for predicting quarterly sales at $\alpha = .01$.

d. It appears the variation in air conditioner sales could be explained by a less complex model. The t-values associated with $\hat{\beta}_2$, $\hat{\beta}_3$, and $\hat{\beta}_4$ are all fairly small. This would indicate that the terms x_2, x_1^2, and x_2^2 may not be necessary.

12.27 a. H_a: At least one of the β's (β_3, β_4, or β_5) is not 0.

b. To compute the test statistic, F, the following calculations are necessary. First the complete model with all terms is fit, and the sum of squares for error is computed, SSE_c. Then, the reduced model without the terms associated with β_3, β_4 and β_5 is fit, and the sum of squares for error is computed, SSE_r. The test statistic, then, is:

$$F = \frac{(SSE_r - SSE_c)/(k - g)}{SSE_c/(n - (k + 1))}$$

where n = total sample size, $k + 1$ = number of β parameters in the complete model, and $k - g$ = number of β parameters in H_0.

c.   The numerator degrees of freedom is $k - g = 3$ and the denominator degrees of freedom is $n - (k + 1) = 30 - (5 + 1) = 24$.

12.29   By adding variables to the model, SSE will either decrease or stay the same. Thus, $\text{SSE}_c \leq \text{SSE}_r$. Therefore, the numerator of the F, $(\text{SSE}_c - \text{SSE}_r)/(k - g)$, is always positive. The only time H_0 will be rejected is when the difference between SSE_r and SSE_c is large. Therefore, the F is only a one-tailed, upper-tailed test.

12.31   a.   To determine whether the complete model contributes information for the prediction of y, we test:

H_0: $\beta_1 = \beta_2 = \beta_3 = \beta_4 = \beta_5 = 0$
H_a: At least one of the β_i's is not 0, $i = 1, 2, 3, 4, 5$

b.   To determine whether a second-order model contributes more information than a first-order model for the prediction of y, we test:

H_0: $\beta_3 = \beta_4 = \beta_5 = 0$
H_a: At least one of the parameters, β_3, β_4, or β_5, is not 0

c.   The test statistic is $F = \dfrac{\text{MSR}}{\text{MSE}} = \dfrac{982.31}{53.84} = 18.24$

The rejection region requires $\alpha = .05$ in the upper tail of the F distribution with numerator df $= k = 5$ and denominator df $= n - (k + 1) = 40 - (5 + 1) = 34$. From Table VIII, Appendix B, $F_{.05} \approx 2.53$. The rejection region is $F > 2.53$.

Since the observed value of the test statistic falls in the rejection region ($F = 18.24 > 2.53$), H_0 is rejected. There is sufficient evidence to indicate that the complete model contributes information for the prediction of y at $\alpha = .05$.

d.   The test statistic is $F = \dfrac{(\text{SSE}_r - \text{SSE}_c)/(k - g)}{\text{SSE}_c/(n - (k + 1))}$

$$= \frac{(3197.16 - 1830.44)/(5 - 2)}{1830.44/(40 - (5 + 1))} = \frac{455.5733}{53.8365} = 8.4622$$

The rejection region requires $\alpha = .05$ in the upper tail of the F distribution with numerator df $= k - g = 3$ and denominator df $= n - (k + 1) = 40 - (5 + 1) = 34$. From Table VIII, Appendix B, $F_{.05} \approx 2.92$. The rejection region is $F > 2.92$.

Since the observed value of the test statistic falls in the rejection region ($F = 8.4622 > 2.92$), H_0 is rejected. There is sufficient evidence to indicate the second-order model contributes more information than a first-order model for the prediction of y at $\alpha = .05$.

12.33 To determine if the second-order model contributes more information for the prediction of y than a first-order model, we test:

H_0: $\beta_3 = \beta_4 = \beta_5 = 0$
H_a: At least one of the parameters, β_3, β_4, or β_5, is not 0

The test statistic is $F = \dfrac{(SSE_r - SSE_c)/(k - g)}{SSE_c/(n - (k + 1))}$

$$= \frac{(2094.4 - 159.94)/(5 - 2)}{159.94/(12 - (5 + 1))} = \frac{644.82}{26.6567} = 24.19$$

The rejection region requires $\alpha = .05$ in the upper tail of the F distribution with numerator df $= k - g = 5 - 2 = 3$ and denominator df $= n - (k + 1) = 12 - (5 + 1) = 6$. From Table VIII, Appendix B, $F_{.05} = 4.76$. The rejection region is $F > 4.76$.

Since the observed value of the test statistic falls in the rejection region ($F = 24.19 > 4.76$), H_0 is rejected. There is sufficient evidence to conclude that a second-order model contributes more information for the prediction of y than a first-order model at $\alpha = .05$.

12.35 a. The complete second-order model is

$$E(y) = \beta_0 + \beta_1 x_1 + \beta_2 x_2 + \beta_3 x_3 + \beta_4 x_1 x_2 + \beta_5 x_1 x_3 + \beta_6 x_2 x_3 + \beta_7 x_1^2 + \beta_8 x_2^2$$
$$+ \beta_9 x_3^2$$

b. The prediction equation is

$$\hat{y} = 655.8 - 57.33 x_1 - 3.39 x_2 - 28.27 x_3 + .224 x_1 x_2 + 2.20 x_1 x_3 + .089 x_2 x_3$$
$$+ .453 x_1^2 + .004 x_2^2 + .208 x_3^2$$

c. To determine if the model is useful for predicting y, we test:

H_0: $\beta_1 = \beta_2 = \cdots = \beta_9 = 0$
H_a: At least one $\beta_i \neq 0$, $i = 1, 2, \ldots, 9$

The test statistic is $F = \dfrac{MSR}{MSE} = \dfrac{671.4895}{65.4795} = 10.25$

The rejection region requires $\alpha = .05$ in the upper tail of the F distribution with numerator df $= k = 9$ and denominator df $= n - (k + 1) = 20 - (9 + 1) = 10$. From Table VIII, Appendix B, $F_{.05} = 3.02$. The rejection region is $F > 3.02$.

Since the observed value of the test statistic falls in the rejection region ($F = 10.25 > 3.02$), H_0 is rejected. There is sufficient evidence to indicate the model is useful for predicting y at $\alpha = .05$.

d. To determine if the second-order terms are useful, we test:

H_0: $\beta_4 = \beta_5 = \cdots = \beta_9 = 0$
H_a: At least one $\beta_i \neq 0$, $i = 4, 5, \ldots, 9$

The test statistic is $F = \dfrac{(\text{SSE}_r - \text{SSE}_c)/(k - g)}{\text{SSE}_c[n - (k + 1)]}$

$$= \dfrac{(1539.8862 - 654.7947)/(9 - 3)}{654.7947/[20 - (9 + 1)]} = \dfrac{147.51525}{65.47947} = 2.25$$

where SSE_r is found from Exercise 11.17.

The rejection region requires $\alpha = .05$ in the upper tail of the F distribution with numerator df $= k - g = 9 - 3 = 6$ and denominator df $= n - (k + 1) = 20 - (9 + 1) = 10$. From Table VIII, Appendix B, $F_{.05} = 3.22$. The rejection region is $F > 3.22$.

Since the observed value of the test statistic does not fall in the rejection region ($F = 2.25 \not> 3.22$), H_0 is not rejected. There is insufficient evidence to indicate the second-order terms are useful at $\alpha = .05$.

12.37   Let $x_1 = \begin{cases} 1 \text{ if qualitative variable assumes 2nd level} \\ 0 \text{ otherwise} \end{cases}$

The model is $E(y) = \beta_0 + \beta_1 x_1$

β_0 = mean value of y when the qualitative variable assumes the first level
β_1 = difference in the mean value of y between levels 2 and 1 of the qualitative variable

12.39   a.   Level 1 implies $x_1 = x_2 = x_3 = 0$. $\hat{y} = 10.2$

b.   Level 2 implies $x_1 = 1$ and $x_2 = x_3 = 0$. $\hat{y} = 10.2 - 4(1) = 6.2$

c.   Level 3 implies $x_2 = 1$ and $x_1 = x_3 = 0$. $\hat{y} = 10.2 + 12(1) = 22.2$

d.   Level 4 implies $x_3 = 1$ and $x_1 = x_2 = 0$. $\hat{y} = 10.2 + 2(1) = 12.2$

e.   The hypotheses are:

H_0: $\beta_1 = \beta_2 = \beta_3 = 0$
H_a: At least one $\beta_i \neq 0$, $i = 1, 2, 3$

12.41   a.   A confidence interval for the difference of two population means could be used. Since both sample sizes are over 30, the large sample confidence interval is used (with independent samples).

b.   Let $x_1 = \begin{cases} 1 \text{ if public college} \\ 0 \text{ otherwise} \end{cases}$

The model is $E(y) = \beta_0 + \beta_1 x_1$

c. β_1 is the difference between the two population means. A point estimate for β_1 is $\hat{\beta}_1$. A confidence interval for β_1 could be used to estimate the difference in the two population means.

12.43   a.   To determine if there is a difference in mean monthly sales among the three incentive plans, we test:

H_0: $\beta_1 = \beta_2 = 0$
H_a: At least one $\beta_i \neq 0$, $i = 1, 2$

The test statistic is $F = \dfrac{MSR}{MSE} = \dfrac{201.8}{42} = 4.80$

The rejection region requires $\alpha = .05$ in the upper tail of the F distribution with numerator df $= k = 2$ and denominator df $= n - (k + 1) = 15 - (2 + 1) = 12$. From Table VIII, Appendix B, $F_{.05} = 3.89$. The rejection region is $F > 3.89$.

Since the observed value of the test statistic falls in the rejection region ($F = 4.80 > 3.89$), H_0 is rejected. There is sufficient evidence to conclude that there is a difference in mean monthly sales among the three incentive plans at $\alpha = .05$.

b.   The least squares prediction equation is

$$\hat{y} = 20.0 - 8.60x_1 + 3.80x_2$$

$x_1 = 1$ if salesperson is paid a straight salary and $x_2 = 0$.

Thus, an estimate of the mean sales for those on a straight salary is

$$\hat{y} = 20.0 - 8.60(1) + 3.80(0) = 11.4$$

Thus, the mean sales is estimated to be $11,400.

c.   For those on commission only, $x_1 = 0$ and $x_2 = 0$. The estimate of the mean sales for those on commission only is

$$\hat{y} = 20.0 - 8.60(0) + 3.80(0) = 20.0$$

Thus, the mean sales is estimated to be $20,000.

12.45   a.   Brand of beer is qualitative.

b. $E(y) = \beta_0 + \beta_1 x_1 + \beta_2 x_2$

where $x_1 = \begin{cases} 1 \text{ if Brand } B_2 \\ 0 \text{ if not} \end{cases}$ $x_2 = \begin{cases} 1 \text{ if Brand } B_3 \\ 0 \text{ if not} \end{cases}$

c. $\beta_0 = $ mean sales for brand B_1

$\beta_1 = $ difference in mean sales between brand B_2 and brand B_1

$\beta_2 = $ difference in mean sales between brand B_3 and brand B_1

d.   From part c, $\beta_0 = \mu_{B_1}$, $\beta_1 = \mu_{B_2} - \mu_{B_1}$, and $\beta_2 = \mu_{B_3} - \mu_{B_1}$

Thus, $\mu_{B_2} = \beta_1 + \beta_0$

12.47 a. The model is $E(y) = \beta_0 + \beta_1 x_1 + \beta_2 x_2 + \beta_3 x_3 + \beta_4 x_4$

$$\text{where} \quad x_1 = \begin{cases} 1 \text{ if pea A} \\ 0 \text{ if not} \end{cases} \quad x_3 = \begin{cases} 1 \text{ if pea C} \\ 0 \text{ if not} \end{cases}$$

$$x_2 = \begin{cases} 1 \text{ if pea B} \\ 0 \text{ if not} \end{cases} \quad x_4 = \begin{cases} 1 \text{ if pea D} \\ 0 \text{ if not} \end{cases}$$

β_0 = mean yield for variety E
β_1 = difference in mean yield between peas A and E.
β_2 = difference in mean yield between peas B and E.
β_3 = difference in mean yield between peas C and E.
β_4 = difference in mean yield between peas D and E.

b. The least squares model is $\hat{y} = 20.35 + 4.75x_1 + 8.6x_2 + 11.3x_3 + 2.1x_4$

c. The hypotheses are:

$$H_0: \beta_1 = \beta_2 = \beta_3 = \beta_4 = 0$$
$$H_a: \text{At least one } \beta_i \neq 0, \; i = 1, 2, 3, 4$$

The null hypothesis says all β_i's are 0 or that the mean yields of varieties A, B, C, and D are not different from the mean yield of variety E.

The alternative hypothesis says at least one $\beta_i \neq 0$ or that the mean yield of at least one of the varieties A, B, C, or D differs from the mean yield of variety E.

d. The test statistic is $F = \dfrac{\text{MSR}}{\text{MSE}} = \dfrac{85.51}{3.568} = 23.966$

The rejection region requires $\alpha = .05$ in the upper tail of the F distribution with numerator df $= k = 4$ and denominator df $= n - (k + 1) = 20 - (4 + 1) = 15$. From Table VIII, Appendix B, $F_{.05} = 3.06$. The rejection region is $F > 3.06$.

Since the observed value of the test statistic falls in the rejection region ($F = 23.966 > 3.06$), H_0 is rejected. There is sufficient evidence to indicate a difference in mean yields among the 5 varieties at $\alpha = .05$.

e. β_4 = difference in mean yield between peas D and E.

For confidence coefficient .95, $\alpha = 1 - .95 = .05$ and $\alpha/2 = .05/2 = .025$. From Table VI, Appendix B, $t_{.025} = 2.131$ with df $= n - (k + 1) = 20 - (4 + 1) = 15$. The 95% confidence interval is

$$\hat{\beta}_4 \pm t_{.025} s_{\hat{\beta}_4} \Rightarrow 2.10 \pm 2.131(1.3357) \Rightarrow 2.10 \pm 2.846 \Rightarrow (-.746, 4.946)$$

12.49 a. $E(y) = \beta_0 + \beta_1 x_1$

b. $E(y) = \beta_0 + \beta_1 x_1 + \beta_2 x_2 + \beta_3 x_3$

where $x_2 = \begin{cases} 1 \text{ if qualitative variable at level 2} \\ 0 \text{ if not} \end{cases}$

$x_3 = \begin{cases} 1 \text{ if qualitative variable at level 3} \\ 0 \text{ if not} \end{cases}$

c. $E(y) = \beta_0 + \beta_1 x_1 + \beta_2 x_2 + \beta_3 x_3 + \beta_4 x_1 x_2 + \beta_5 x_1 x_3$

d. The response lines will be parallel if β_4 and β_5 are 0.

e. The model will have only one response line if β_2, β_3, β_4, and β_5 are all 0.

12.51 a. The type of juice extractor is qualitative.
    The size of the orange is quantitative.

b. The model is $E(y) = \beta_0 + \beta_1 x_1 + \beta_2 x_2$

where $x_1 =$ diameter of orange

$x_2 = \begin{cases} 1 \text{ if Brand B} \\ 0 \text{ if not} \end{cases}$

c. To allow the lines to differ, the interaction term is added:

$E(y) = \beta_0 + \beta_1 x_1 + \beta_2 x_2 + \beta_3 x_1 x_2$

d. For part **b**:

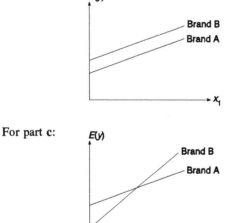

For part **c**:

e. To determine whether the model in part **c** provides more information for predicting yield than does the model in part **b**, we test:

H_0: $\beta_3 = 0$
H_a: $\beta_3 \neq 0$

f.   The test statistic would be $F = \dfrac{(SSE_r - SSE_c)/(k - g)}{SSE_c/(n - (k + 1))}$

To compute SSE_r: The model in part **b** is fit and SSE_r is the sum of squares for error.

To compute SSE_c: The model in part **c** is fit and SSE_c is the sum of squares for error.

$k - g$ = number of parameters in $H_0 = 1$
$n - (k + 1)$ = degrees of freedom for error in the complete model

12.53 a.   The model is $E(y) = \beta_0 + \beta_1 x_1 + \beta_2 x_2 + \beta_3 x_3$

where x_1 = number of apartment units

$x_2 = \begin{cases} 1 \text{ if Condition E} \\ 0 \text{ if not} \end{cases}$ $x_3 = \begin{cases} 1 \text{ if Condition G} \\ 0 \text{ if not} \end{cases}$

b.   It does appear that the model in
a is appropriate. The lines
appear to be approximately
parallel.

c.   The SAS printout is:

DEPENDENT VARIABLE: Y

SOURCE	DF	SUM OF SQUARES	MEAN SQUARE	F VALUE
MODEL	3	986169504463.134	328723168154.378	78.71
ERROR	21	87700442851.106	4176211564.338	PR > F
CORRECTED TOTAL	24	1073869947314.240		0.0001

R-SQUARE	C.V.	ROOT MSE	Y MEAN
0.918332	22.2400	64623.615	290573.52000000

PARAMETER	ESTIMATE	T FOR H0: PARAMETER = 0	PR > ¦T¦	STD ERROR OF ESTIMATE
INTERCEPT	36387.63972373	1.19	0.2454	30450.08697797
X1	15616.92891959	14.66	0.0001	1065.54040134
X2	152487.42810722	3.89	0.0008	39157.32396921
X3	49441.14619353	1.45	0.1619	34099.03698648

The least squares prediction equation is:

$$\hat{y} = 36{,}388 + 15{,}617x_1 + 152{,}487x_2 + 49{,}441x_3$$

The least squares prediction equation for condition E ($x_2 = 1$, $x_3 = 0$) is:

$$\hat{y} = 36{,}388 + 15{,}617x_1 + 152{,}487(1) = 188{,}875 + 15{,}617x_1$$

For condition G ($x_2 = 0$, $x_3 = 1$):

$$\hat{y} = 36{,}388 + 15{,}617x_1 + 49{,}441(1) = 85{,}829 + 15{,}617x_1$$

For condition F ($x_2 = x_3 = 0$):

$$\hat{y} = 36{,}388 + 15{,}617x_1$$

e. To determine if the relationship between the mean sale price and number of units differs depending on the physical condition of the apartments, we test:

H_0: $\beta_2 = \beta_3 = 0$
H_a: At least one $\beta_i \neq 0$

To calculate the test statistic, we must first fit the reduced model, $E(y) = \beta_0 + \beta_1 x_1$.

The SAS printout is:

DEPENDENT VARIABLE: Y

SOURCE	DF	SUM OF SQUARES	MEAN SQUARE	F VALUE
MODEL	1	915775843108.351	915775843108.351	133.23
ERROR	23	158094104205.889	6873656704.604	PR > F
CORRECTED TOTAL	24	1073869947314.240		0.0001

R-SQUARE	C.V.	ROOT MSE	Y MEAN
0.852781	28.5324	82907.519	290573.52000000

PARAMETER	ESTIMATE	T FOR H0: PARAMETER = 0	PR > ¦IT¦	STD ERROR OF ESTIMATE
INTERCEPT	101786.14688790	4.37	0.0002	23290.75036701
X1	15525.27739409	11.54	0.0001	1345.05082746

The test statistic is $F = \dfrac{(SSE_r - SSE_c)/(k - g)}{SSE_c/[n - (k + 1)]}$

$$= \frac{(158{,}094{,}104{,}205.889 - 87{,}700{,}442{,}851.106)/(3 - 1)}{87{,}700{,}442{,}851.106/[25 - (3 + 1)]}$$

$$= \frac{35{,}196{,}830{,}680}{4{,}176{,}211{,}564} = 8.43$$

The rejection region requires $\alpha = .05$ in the upper tail of the F distribution with numerator df $= k - g = 3 - 1 = 2$ and denominator df $= n - (k + 1) = 25 - (3 + 1) = 21$. From Table VIII, Appendix B, $F_{.05} = 3.47$. The rejection region is $F > 3.47$.

Since the observed value of the test statistic falls in the rejection region ($F = 8.43 > 3.47$), H_0 is rejected. There is sufficient evidence to indicate the relationship between mean sale price and number of units differ depending on the physical condition of the apartments at $\alpha = .05$.

12.55 To determine whether the mean work-hours lost differ for the 3 training programs, we test:

H_0: $\beta_2 = \beta_3 = 0$
H_a: At least one $\beta_i \neq 0$, $i = 2, 3$

The test statistic is $F = \dfrac{(SSE_r - SSE_c)/(k - g)}{SSE_c/(n - (k + 1))}$

$= \dfrac{(3113.14 - 1527.27)/(3 - 1)}{1527.27/(9 - (3 + 1))} = \dfrac{792.935}{305.454} = 2.60$

The rejection region requires $\alpha = .05$ in the upper tail of the F distribution with numerator df $= k - g = 3 - 1 = 2$ and denominator df $= n - (k + 1) = 9 - (3 + 1) = 5$. From Table VIII, Appendix B, $F_{.05} = 5.79$. The rejection region is $F > 5.79$.

Since the observed value of the test statistic does not fall in the rejection region ($F = 2.60 \not> 5.79$), H_0 is not rejected. There is insufficient evidence to indicate the mean work-hours lost differ for the 3 training programs at $\alpha = .05$.

12.57 a. Remember that y is the per capita consumption, obtained by dividing total consumption (column 1) by population (column 2). Then the scattergram of the data is:

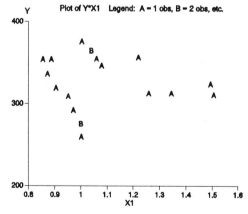

b. The model would be $E(y) = \beta_0 + \beta_1 x_1 + \beta_2 x_2 + \beta_3 x_1 x_2$

where x_1 = relative price of a gallon of gasoline

$x_2 = \begin{cases} 1 \text{ if after 1973} \\ 0 \text{ otherwise} \end{cases}$

c. Using SAS, the output from fitting the model is:

DEPENDENT VARIABLE: Y

ANALYSIS OF VARIANCE

SOURCE	DF	SUM OF SQUARES	MEAN SQUARE	F VALUE	PROB>F
MODEL	3	19335.15329	6445.05110	36.538	0.0001
ERROR	15	2645.87850	176.39190		
C TOTAL	18	21981.03179			

ROOT MSE	13.28126	R-SQUARE	0.8796	
DEP MEAN	326.59692	ADJ R-SQ	0.8556	
C.V.	4.06656			

PARAMETER ESTIMATE

VARIABLE	DF	PARAMETER ESTIMATE	STANDARD ERROR	T FOR HO: PARAMETER=0	PROB > ¦T¦
INTERCEP	1	884.841568	76.33041486	11.592	0.0001
X1	1	-612.577652	81.07545937	-7.556	0.0001
X2	1	-409.361031	81.37043612	-5.031	0.0001
X1X2	1	502.201751	84.30593412	5.957	0.0001

The fitted model is:

$$\hat{y} = 884.8416 - 612.5777x_1 - 409.3610x_2 + 502.2018x_1x_2$$

d. Using SAS, the plots of the fitted regression lines are:

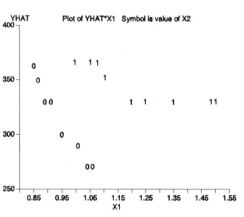

where the 1's correspond to the post-OPEC period, and the 0's correspond to the pre-OPEC period. The slope of the pre-OPEC line is much steeper than the slope of the post-OPEC line. It appears that OPEC had an effect on the consumption of gasoline.

e. To determine if the relationship between demand for motor fuel and the relative price of gasoline changed after 1973, we test:

H_0: $\beta_2 = \beta_3 = 0$
H_a: At least one of the β_i's is not 0, $i = 2, 3$

To calculate the test statistic, we must first fit the reduced model, $E(y) = \beta_0 + \beta_1x_1$

Using SAS, the output from fitting the model is:

DEPENDENT VARIABLE: Y

ANALYSIS OF VARIANCE

SOURCE	DF	SUM OF SQUARES	MEAN SQUARE	F VALUE	PROB>F
MODEL	1	172.38330	172.38330	0.134	0.7185
ERROR	17	21808.64849	1282.86168		
C TOTAL	18	21981.03179			

ROOT MSE	35.81706	R-SQUARE	0.0078	
DEP MEAN	326.59692	ADJ R-SQ	-0.0505	
C.V.	10.96675			

PARAMETER ESTIMATE

VARIABLE	DF	PARAMETER ESTIMATE	STANDARD ERROR	T FOR HO: PARAMETER=0	PROB > |T|
INTERCEP	1	343.634284	47.19849286	7.281	0.0001
X1	1	-15.775341	43.03492828	-0.367	0.7185

The test statistic is $F = \dfrac{(SSE_r - SSE_c)/(k - g)}{SSE_c/[n - (k + 1)]}$

$$= \frac{(21808.65 - 2645.88)/(3 - 1)}{2645.88/[19 - (3 + 1)]} = \frac{9581.385}{176.39} = 54.32$$

The rejection region requires $\alpha = .05$ in the upper tail of the F distribution with numerator df $= k - g = 3 - 1 = 2$ and denominator df $= n - (k + 1) = 19 - (3 + 1) = 15$. From Table VIII, Appendix B, $F_{.05} = 3.68$. The rejection region is $F > 3.68$.

Since the observed value of the test statistic falls in the rejection region ($F = 54.32 > 3.68$), H_0 is rejected. There is sufficient evidence to indicate the relationship between demand for motor fuel and the relative price of gasoline changed after 1973 at $\alpha = .05$.

f. To determine if the slopes differ, we test:

H_0: $\beta_3 = 0$
H_a: $\beta_3 \neq 0$

The test statistic is $t = 5.957$ (from the printout of fitting the complete model).

The rejection region requires $\alpha/2 = .05/2 = .025$ in each tail of the t distribution with df $= n - (k + 1) = 19 - (3 + 1) = 15$. From Table VI, Appendix B, $t_{.025} = 2.131$. The rejection region is $t > 2.131$ or $t < -2.131$.

Since the observed value of the test statistic falls in the rejection region ($t = 5.957 > 2.131$), H_0 is rejected. There is sufficient evidence to indicate the slopes differ at $\alpha = .05$.

12.59 a. Using the model $E(y) = \beta_0 + \beta_1 x_1 + \beta_2 x_1^2 + \beta_3 x_2 + \beta_4 x_3 + \beta_5 x_1 x_2 + \beta_6 x_1 x_3$
$$+ \beta_7 x_1^2 x_2 + \beta_8 x_1^2 x_3$$

where x_1 = quantitative variable

$$x_2 = \begin{cases} 1 \text{ if level 2 of qualitative variable} \\ 0 \text{ if not} \end{cases}$$

$$x_3 = \begin{cases} 1 \text{ if level 3 of qualitative variable} \\ 0 \text{ if not} \end{cases}$$

The response curves will have the same slope but different y-intercepts if $\beta_5 = \beta_6 = \beta_7 = \beta_8 = 0$ and β_3 and β_4 are not 0.

b. In order for the response curves to be parallel lines, they must be straight lines. They will be parallel if $\beta_2 = \beta_5 = \beta_6 = \beta_7 = \beta_8 = 0$ and β_3 and β_4 are not 0.

c. The response curves will be identical if $\beta_3 = \beta_4 = \beta_5 = \beta_6 = \beta_7 = \beta_8 = 0$.

12.61 a. $\hat{y} = 48.8 - 3.36x_1 + .0749x_1^2 - 2.36x_2 - 7.6x_3 + 3.71x_1 x_2 + 2.66x_1 x_3 - .0183x_1^2 x_2$
$$- .0372x_1^2 x_3$$

b. When $x_2 = 0$ and $x_3 = 0$, $\hat{y} = 48.8 - 3.36x_1 + .0749x_1^2$

When $x_2 = 1$ and $x_3 = 0$,

$$\hat{y} = 48.8 - 3.36x_1 + .0749x_1^2 - 2.36(1) + 3.71x_1(1) - .0183x_1^2(1)$$
$$= 46.44 + .35x_1 + .0566x_1^2$$

When $x_2 = 0$ and $x_3 = 1$,

$$\hat{y} = 48.8 - 3.36x_1 + .0749x_1^2 - 7.6(1) + 2.66x_1(1) - .0372x_1^2(1)$$
$$= 41.2 - .7x_1 + .0377x_1^2$$

c.

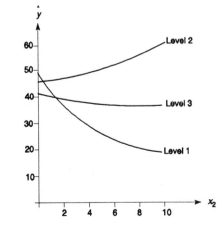

d. To determine whether the second-order response curves differ for the three levels, we test:

H_0: $\beta_3 = \beta_4 = \beta_5 = \beta_6 = \beta_7 = \beta_8 = 0$
H_a: At least one of the parameters, $\beta_3 = \beta_4 = \beta_5 = \beta_6 = \beta_7 = \beta_8$, is not 0

e. The test statistic is $F = \dfrac{(SSE_r - SSE_c)/(k - g)}{SSE_c/[n - (k + 1)]}$

$$= \frac{(14986 - 162.9)/(8 - 2)}{162.9/(25 - (8 + 1))} = \frac{2470.5167}{10.18125} = 242.65$$

The rejection region requires $\alpha = .05$ in the upper tail of the F distribution with numerator df $= k - g = 8 - 2 = 6$ and denominator df $= n - (k + 1) = 25 - (8 + 1) = 16$. From Table VIII, Appendix B, $F_{.05} = 2.74$. The rejection region is $F > 2.74$.

Since the observed value of the test statistic falls in the rejection region ($F = 242.65 > 2.74$), H_0 is rejected. There is sufficient evidence to indicate the second-order response curves differ for the three levels at $\alpha = .05$.

12.63 a. To determine whether the rate of increase of mean salary with experience is different for males and females, we test:

H_0: $\beta_4 = \beta_5 = 0$
H_a: At least one of the parameters, β_4 and β_5, is not 0

b. To determine whether there are differences in mean salaries that are attributable to sex, we test:

H_0: $\beta_3 = \beta_4 = \beta_5 = 0$
H_a: At least one of the parameters, β_3, β_4, and β_5, is not 0

12.65 To determine whether the speed limit contributes information for the prediction of the annual number of highway deaths, we test:

H_0: $\beta_3 = \beta_4 = \beta_5 = 0$
H_a: At least one of the parameters, β_3, β_4 and β_5, is not 0

The test statistic is $F = \dfrac{(SSE_r - SSE_c)/(k - g)}{SSE_c/(n - (k + 1))}$

$$= \frac{(456.45996 - 177.08741)/(5 - 2)}{177.08741/(36 - (5 + 1))} = \frac{93.124183}{5.902914} = 15.78$$

The rejection region requires $\alpha = .05$ in the upper tail of the F distribution with numerator df $= k - g = 5 - 2 = 3$ and denominator df $= n - (k + 1) = 36 - (5 + 1) = 30$. From Table VIII, Appendix B, $F_{.05} = 2.92$. The rejection region is $F > 2.92$.

Since the observed value of the test statistic falls in the rejection region ($F = 15.78 > 2.92$), H_0 is rejected. There is sufficient evidence to indicate that the speed limit contributes information for the prediction of the annual number of highway deaths at $\alpha = .05$.

12.67 a.   Let $x_2 = \begin{cases} 1 \text{ if bear market} \\ 0 \text{ otherwise} \end{cases}$ and x_1 = monthly rate of return

The model would be $E(y) = \beta_0 + \beta_1 x_1 + \beta_2 x_2 + \beta_3 x_1 x_2$

b.   From the model in a, the slopes of the regression lines for bull and bear markets may not be the same, and the y-intercepts may not be the same.

c.   To determine if the mutual fund's beta coefficient is different during bull and bear markets, we test:

H_0: $\beta_3 = 0$
H_a: $\beta_3 \neq 0$

d.   The hypotheses are:

H_0: $\beta_2 = \beta_3 = 0$
H_a: At least one $\beta_i \neq 0$, $i = 2, 3$

12.69 a.   The quantitative variables are:

x_1 = number of people in household,
x_2 = miles to bank,
x_3 = number of cars in household,
x_4 = years of education,
x_5 = miles to work,
x_6 = total family income, and
x_{11} = number of shopping trips

The qualitative variables are:

$x_7, x_8, x_9,$ and x_{10}

b.   The variable, kind of work, can assume 5 levels. The number of dummy variables associated with a qualitative variable is equal to the number of levels of the qualitative variable minus 1. The levels are not employed, white collar, blue collar, farm-related, and other.

c. $R^2 = .1836$. This implies that 18.36% of the sample variation in the number of trips to the bank is explained by the model containing the 11 variables.

d.   To test the usefulness of the model, we test:

H_0: $\beta_1 = \beta_2 = \cdots = \beta_{11} = 0$
H_a: At least one of the β_i parameters $\neq 0$, $i = 1, 2, \ldots, 11$

The test statistic is $F = \dfrac{R^2/k}{(1 - R^2)/[n - (k + 1)]}$

$$= \dfrac{.1836/11}{(1 - .1836)/[597 - (11 + 1)]} = \dfrac{.01669}{.0014} = 11.96$$

The rejection region requires $\alpha = .01$ in the upper tail of the F distribution with numerator df $= k = 11$ and denominator df $= n - (k + 1) = 597 - (11 + 1) = 585$. From Table X, Appendix B, $F_{.01} \approx 2.25$. The rejection region is F > 2.25.

Since the observed value of the test statistic falls in the rejection region ($F = 11.96 >$ 2.25), H_0 is rejected. There is sufficient evidence to indicate the model is useful in predicting the number of trips to the bank at $\alpha = .01$.

e. The hypotheses would be:

H_0: $\beta_7 = \beta_8 = \beta_9 = \beta_{10} = 0$
H_a: At least one of the β_i parameters $\neq 0$, $i = 7, 8, 9, 10$

12.71 a. To determine whether the quadratic terms are useful, we test:

H_0: $\beta_2 = \beta_5 = 0$
H_a: At least one β_i parameter $\neq 0$, $i = 2, 5$

b. To determine whether there is a difference in mean delivery time by rail and truck, we test:

H_0: $\beta_3 = \beta_4 = \beta_5 = 0$
H_a: At least one β_i parameter $\neq 0$, $i = 3, 4, 5$

12.73 a. Using SAS to fit the model $E(y) = \beta_0 + \beta_1 x_1 + \beta_2 x_2 + \beta_3 x_3$, we get:

DEPENDENT VARIABLE: Y

SOURCE	DF	SUM OF SQUARES	MEAN SQUARE	F VALUE
MODEL	3	13.34333333	4.44777778	63.09
ERROR	20	1.41000000	0.07050000	PR > F
CORRECTED TOTAL	23	14.75333333		0.0001

R-SQUARE	C.V.	ROOT MSE	Y MEAN
0.904428	2.4065	0.26551836	11.03333333

PARAMETER		ESTIMATE	T FOR HO: PARAMETER = 0	PR > ¦IT¦	STD ERROR OF ESTIMATE
INTERCEPT		10.23333333	94.41	0.0001	0.10839742
X	VH	0.50000000	3.26	0.0039	0.15329710
	H	2.01666667	13.16	0.0001	0.15329710
	M	0.68333333	4.46	0.0002	0.15329710

The fitted model is $\hat{y} = 10.2 + .5x_1 + 2.02x_2 + .683x_3$

$$x_1 = \begin{cases} 1 & \text{if VH} \\ 0 & \text{otherwise} \end{cases}$$

$$x_2 = \begin{cases} 1 & \text{if H} \\ 0 & \text{otherwise} \end{cases}$$

$$x_3 = \begin{cases} 1 & \text{if M} \\ 0 & \text{otherwise} \end{cases}$$

To determine if the firm's expected market share differs for different levels of advertising exposure, we test:

H_0: $\beta_1 = \beta_2 = \beta_3 = 0$
H_a: At least one $\beta_i \neq 0$, $i = 1, 2, 3$

The test statistic is $F = 63.09$.

The rejection region requires $\alpha = .05$ in the upper tail of the F distribution with numerator df $= k = 3$ and denominator df $= n - (k + 1) = 24 - (3 + 1) = 20$. From Table VIII, Appendix B, $F_{.05} = 3.10$. The rejection region is $F > 3.10$.

Since the observed value of the test statistic falls in the rejection region ($F = 63.09 > 3.10$), H_0 is rejected. There is sufficient evidence to indicate the firm's expected market share differs for different levels of advertising exposure at $\alpha = .05$.

12.75 The model relating durability rating to bake time (second-order) and type of paint is:

$$E(y) = \beta_0 + \beta_1 x_1 + \beta_2 x_1^2 + \beta_3 x_2 + \beta_4 x_3 + \beta_5 x_1 x_2 + \beta_6 x_1 x_3 + \beta_7 x_1^2 x_2 + \beta_8 x_1^2 x_3$$

where

$$x_1 = \text{bake time} \quad x_2 = \begin{cases} 1 & \text{if paint B} \\ 0 & \text{otherwise} \end{cases} \quad x_3 = \begin{cases} 1 & \text{if paint C} \\ 0 & \text{otherwise} \end{cases}$$

12.77 When fitting models, we are trying to obtain a "general" relationship between the dependent and independent variables that can be used to estimate values of the dependent variable at different levels of the independent variables. By fitting a fourth-order model with 5 data points, we would have 0 degrees of freedom for error; thus, we could run no tests. Also, if a 6th observation is obtained, the model could change dramatically to fit all 6 variables instead of 5.

12.81 a. To determine if the second-order terms are important, we test:

H_0: $\beta_3 = \beta_5 = 0$
H_a: At least one $\beta_i \neq 0$, $i = 3, 5$

The test statistic is $F = \dfrac{(\text{SSE}_r - \text{SSE}_c)/(k - g)}{\text{SSE}_c/[n - (k + 1)]}$

$$= \frac{(8.548 - 6.133)/(5 - 3)}{6.133/[25 - (5 + 1)]} = \frac{1.2075}{.3228} = 3.74$$

The rejection region requires $\alpha = .05$ in the upper tail of the F distribution with numerator df $= k - g = 5 - 3 = 2$ and denominator df $= n - (k + 1) = 25 - (5 + 1) = 19$. From Table VIII, Appendix B, $F_{.05} = 3.52$. The rejection region is $F > 3.52$.

Since the observed value of the test statistic falls in the rejection region ($F = 3.74 > 3.52$), H_0 is rejected. There is sufficient evidence to indicate the second-order terms are important for predicting the mean cost at $\alpha = .05$.

b. To determine if the main effects model is useful, we test

H_0: $\beta_1 = \beta_2 = 0$
H_a: At least one $\beta_i \neq 0$, $i = 1, 2$

The test statistic is: $F = \dfrac{R^2/k}{(1 - R^2)/[n - (k + 1)]}$

$$= \frac{.950/2}{(1 - .950)/[25 - (2 + 1)]} = \frac{.475}{.0023} = 209$$

The rejection region requires $\alpha = .05$ in the upper tail of the F distribution with numerator df $= k = 2$ and denominator df $= n - (k + 1) = 25 - (2 + 1) = 22$. From Table VIII, Appendix B, $F_{.05} = 3.44$. The rejection region is $F > 3.44$.

Since the observed value of the test statistic falls in the rejection region ($F = 209 > 3.44$), H_0 is rejected. There is sufficient evidence to indicate the main effects model is useful for predicting costs at $\alpha = .05$.

CHAPTER THIRTEEN

....

Methods for Quality Improvement

13.1   A control chart is a time series plot of individual measurements or means of a quality variable to which a centerline and two other horizontal lines called control limits have been added. The center line represents the mean of the process when the process is in a state of statistical control. The upper control limit and the lower control limit are positioned so that when the process is in control the probability of an individual measurement or mean falling outside the limits is very small. A control chart is used to determine if a process is in control (only common causes of variation present) or not (both common and special causes of variation present). This information helps us to determine when to take action to find and remove special causes of variation and when to leave the process alone.

13.3   When a control chart is first constructed, it is not known whether the process is in control or not. If the process is found not to be in control, then the centerline and control limits should not be used to monitor the process in the future.

13.5   Even if all the points of an $\bar{x}$-chart fall within the control limits, the process may be out of control. Nonrandom patterns may exist among the plotted points that are within the control limits, but are very unlikely if the process is in control. Examples include six points in a row steadily increasing or decreasing and fourteen points in a row alternating up and down.

13.7   Rule 1:   One point beyond Zone A: No points are beyond Zone A.

       Rule 2:   Nine points in a row in Zone C or beyond: No sequence of 9 points are in Zone C (on one side of the centerline) or beyond.

       Rule 3:   Six points in a row steadily increasing or decreasing: No sequence of 6 points steadily increase or decrease.

       Rule 4:   Fourteen points in a row alternating up and down: This pattern does not exist.

       Rule 5:   Two out of three points in Zone A or beyond: There are no groups of three consecutive points that have two or more in Zone A or beyond.

       Rule 6:   Four out of five points in a row in Zone B or beyond: Points 18 through 21 are all in Zone B or beyond. This indicates the process is out of control.

    Thus, Rule 6 indicates this process is out of control.

13.9   Using Table XVII, Appendix B:

    a.   With $n = 3$, $A_2 = 1.023$

    b.   With $n = 10$, $A_2 = 0.308$

    c.   With $n = 22$, $A_2 = 0.167$

13.11   a.   For each sample, we compute $\bar{x} = \dfrac{\sum x}{n}$ and R = range = largest measurement − smallest measurement. The results are listed in the table:

Sample No.	$\bar{x}$	R	Sample No.	$\bar{x}$	R
1	20.225	1.8	11	21.225	3.2
2	19.750	2.8	12	20.475	0.9
3	20.425	3.8	13	19.650	2.6
4	19.725	2.5	14	19.075	4.0
5	20.550	3.7	15	19.400	2.2
6	19.900	5.0	16	20.700	4.3
7	21.325	5.5	17	19.850	3.6
8	19.625	3.5	18	20.200	2.5
9	19.350	2.5	19	20.425	2.2
10	20.550	4.1	20	19.900	5.5

b. $\bar{\bar{x}} = \dfrac{\bar{x}_1 + \bar{x}_2 + \cdots + \bar{x}_{20}}{n} = \dfrac{402.325}{20} = 20.11625$

$\bar{R} = \dfrac{R_1 + R_2 + \cdots + R_{20}}{n} = \dfrac{66.2}{20} = 3.31$

c.   Centerline = $\bar{\bar{x}} = 20.116$

From Table XVII, Appendix B, with $n = 4$, $A_2 = .729$.

Upper control limit = $\bar{\bar{x}} + A_2\bar{R} = 20.116 + .729(3.31) = 22.529$
Lower control limit = $\bar{\bar{x}} - A_2\bar{R} = 20.116 - .729(3.31) = 17.703$

d.   Upper A−B Boundary = $\bar{\bar{x}} + \dfrac{2}{3}(A_2\bar{R}) = 20.116 + \dfrac{2}{3}(.729)(3.31) = 21.725$

Lower A−B Boundary = $\bar{\bar{x}} - \dfrac{2}{3}(A_2\bar{R}) = 20.116 - \dfrac{2}{3}(.729)(3.31) = 18.507$

Upper B−C Boundary = $\bar{\bar{x}} + \dfrac{1}{3}(A_2\bar{R}) = 20.116 + \dfrac{1}{3}(.729)(3.31) = 20.920$

Lower B−C Boundary = $\bar{\bar{x}} - \dfrac{1}{3}(A_2\bar{R}) = 20.116 - \dfrac{1}{3}(.729)(3.31) = 19.312$

e. The $\bar{x}$-chart is:

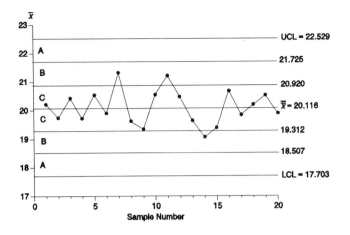

Rule 1: One point beyond Zone A: No points are beyond Zone A.

Rule 2: Nine points in a row in Zone C or beyond: No sequence of 9 points are in Zone C (on one side of the centerline) or beyond.

Rule 3: Six points in a row steadily increasing or decreasing: No sequence of 6 points steadily increase or decrease.

Rule 4: Fourteen points in a row alternating up and down: This pattern does not exist.

Rule 5: Two out of three points in Zone A or beyond: There are no groups of three consecutive points that have two or more in Zone A or beyond.

Rule 6: Four out of five points in a row in Zone B or beyond: No sequence of 5 points has 4 or more in Zone B or beyond.

The process appears to be in control.

13.13 a. For each sample, we compute $\bar{x} = \dfrac{\sum x}{n}$ and R = range = largest measurement − smallest measurement. The results are listed in the table:

Sample No.	$\bar{x}$	R	Sample No.	$\bar{x}$	R
1	24.048	.22	11	24.008	.21
2	23.958	.12	12	24.016	.16
3	23.990	.32	13	24.058	.16
4	24.022	.16	14	23.960	.15
5	23.946	.26	15	24.032	.12
6	24.012	.25	16	24.030	.18
7	23.974	.17	17	23.972	.19
8	23.984	.12	18	23.998	.21
9	23.948	.28	19	24.018	.13
10	23.996	.07	20	23.972	.15

$$\overline{\overline{x}} = \frac{\overline{x}_1 + \overline{x}_2 + \cdots + \overline{x}_{20}}{n} = \frac{479.942}{20} = 23.9971$$

$$\overline{R} = \frac{R_1 + R_2 + \cdots + R_{20}}{n} = \frac{3.63}{20} = .1815$$

Centerline $= \overline{\overline{x}} = 23.9971$

From Table XVII, Appendix B, with $n = 5$, $A_2 = .577$.

Upper control limit $= \overline{\overline{x}} + A_2\overline{R} = 23.9971 + .577(.1815) = 24.102$
Lower control limit $= \overline{\overline{x}} - A_2\overline{R} = 23.9971 - .577(.1815) = 23.892$

Upper A−B Boundary $= \overline{\overline{x}} + \frac{2}{3}(A_2\overline{R}) = 23.9971 + \frac{2}{3}(.577)(.1815) = 24.067$

Lower A−B Boundary $= \overline{\overline{x}} - \frac{2}{3}(A_2\overline{R}) = 23.9971 - \frac{2}{3}(.577)(.1815) = 23.927$

Upper B−C Boundary $= \overline{\overline{x}} + \frac{1}{3}(A_2\overline{R}) = 23.9971 + \frac{1}{3}(.577)(.1815) = 24.032$

Lower B−C Boundary $= \overline{\overline{x}} - \frac{1}{3}(A_2\overline{R}) = 23.9971 - \frac{1}{3}(.577)(.1815) = 23.962$

e.   The $\overline{x}$-chart is:

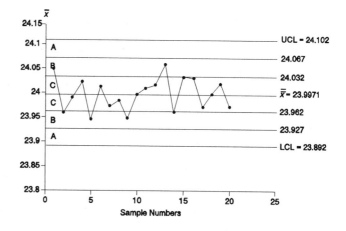

b.   To determine if the process is in or out of control, we check the 6 rules:

Rule 1:   One point beyond Zone A: No points are beyond Zone A.
Rule 2:   Nine points in a row in Zone C or beyond: No sequence of 9 points are in Zone C (on one side of the centerline) or beyond.
Rule 3:   Six points in a row steadily increasing or decreasing: No sequence of 6 points steadily increase or decrease.
Rule 4:   Fourteen points in a row alternating up and down: This pattern does not exist.

Rule 5: Two out of three points in Zone A or beyond: There are no groups of three consecutive points that have two or more in Zone A or beyond.

Rule 6: Four out of five points in a row in Zone B or beyond: No sequence of 5 points has 4 or more in Zone B or beyond.

The process appears to be in control.

c. Since the process is in control, these limits should be used to monitor future process output.

d. The rational subgrouping strategy used by K-Company will facilitate the identification of process variation caused by differences in the two shifts. All observations within one sample are from the same shift. The shift change is at 3:00 P.M. The samples are selected at 8:00 A.M., 11:00 A.M., 2:00 P.M., 5:00 P.M., and 8:00 P.M. No samples will contain observations from both shifts.

13.15 a. The mechanical process within the freezer that produces and monitors the freezer temperature.

b. The $\bar{x}$-chart monitors the process mean, or in this case, the mean temperature in the freezer.

c. For each sample, we compute $\bar{x} = \dfrac{\sum x}{n}$ and R = range = largest measurement − smallest measurement. The results are listed in the table:

Sample No.	$\bar{x}$	R	Sample No.	$\bar{x}$	R
1	5.070	.51	11	5.010	.55
2	4.860	1.63	12	4.942	.63
3	5.044	.88	13	4.776	1.25
4	5.060	1.41	14	5.108	.81
5	4.946	1.25	15	4.848	.92
6	5.364	1.09	16	5.054	.51
7	5.330	.97	17	4.726	1.42
8	4.802	.73	18	4.934	.88
9	5.442	1.21	19	5.160	1.08
10	5.306	.85	20	5.064	.79

$$\bar{\bar{x}} = \frac{\bar{x}_1 + \bar{x}_2 + \cdots + \bar{x}_{20}}{n} = \frac{100.846}{20} = 5.0423$$

$$\bar{R} = \frac{R_1 + R_2 + \cdots + R_{20}}{n} = \frac{19.37}{20} = .9685$$

Centerline = $\bar{\bar{x}} = 5.042$

From Table XVII, Appendix B, with $n = 5$, $A_2 = .577$.

Upper control limit $= \bar{\bar{x}} + A_2\bar{R} = 5.042 + .577(.9685) = 5.601$

Lower control limit $= \bar{\bar{x}} - A_2\bar{R} = 5.042 - .577(.9685) = 4.483$

Upper A−B Boundary $= \bar{\bar{x}} + \dfrac{2}{3}(A_2\bar{R}) = 5.042 + \dfrac{2}{3}(.577)(.9685) = 5.415$

Lower A−B Boundary $= \bar{\bar{x}} - \dfrac{2}{3}(A_2\bar{R}) = 5.042 - \dfrac{2}{3}(.577)(.9685) = 4.669$

Upper B−C Boundary $= \bar{\bar{x}} + \dfrac{1}{3}(A_2\bar{R}) = 5.042 + \dfrac{1}{3}(.577)(.9685) = 5.228$

Lower B−C Boundary $= \bar{\bar{x}} - \dfrac{1}{3}(A_2\bar{R}) = 5.042 - \dfrac{1}{3}(.577)(.9685) = 4.856$

The $\bar{x}$-chart is:

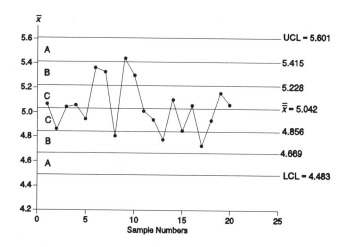

d. To determine if the process is in or out of control, we check the 6 rules:

Rule 1: One point beyond Zone A: No points are beyond Zone A.

Rule 2: Nine points in a row in Zone C or beyond: No sequence of 9 points are in Zone C (on one side of the centerline) or beyond.

Rule 3: Six points in a row steadily increasing or decreasing: No sequence of 6 points steadily increase or decrease.

Rule 4: Fourteen points in a row alternating up and down: This pattern does not exist.

Rule 5: Two out of three points in Zone A or beyond: There are no groups of three consecutive points that have two or more in Zone A or beyond.

Rule 6: Four out of five points in a row in Zone B or beyond: Points 6, 7, 9, and 10 are in Zone B or beyond. This indicates the process is out of control.

Rule 6 indicates that the process is out of control.

e.   Since the process is out of control, the control limits should not be used to monitor future temperatures. Control limits must be determined during a period when the process is under control.

13.17   The control limits of the $\bar{x}$-chart are a function of and reflect the variation in the process. If the variation were unstable (i.e., out of control), the control limits would not be constant. Under these circumstances, the fixed control limits of the $\bar{x}$-chart would have little meaning. We use the R-chart to determine whether the variation of the process is stable. If it is, the $\bar{x}$-chart is meaningful. Thus, we interpret the R-chart prior to the $\bar{x}$-chart.

13.19   a.   From Exercise 13.10,

$$\bar{R} = \frac{R_1 + R_2 + \cdots + R_{25}}{n} = \frac{198.7}{25} = 7.948$$

Centerline $= \bar{R} = 7.948$

From Table XVII, Appendix B, with $n = 5$, $D_4 = 2.114$ and $D_3 = 0$.

Upper control limit $= \bar{R}D_4 = 7.948(2.114) = 16.802$

Since $D_3 = 0$, the lower control limit is negative and is not included on the chart.

b.   From Table XVII, Appendix B, with $n = 5$, $d_2 = 2.326$, and $d_3 = .864$.

Upper A−B Boundary $= \bar{R} + 2d_3 \dfrac{\bar{R}}{d_2} = 7.948 + 2(.864)\dfrac{7.948}{2.326} = 13.853$

Lower A−B Boundary $= \bar{R} - 2d_3 \dfrac{\bar{R}}{d_2} = 7.948 - 2(.864)\dfrac{7.948}{2.326} = 2.043$

Upper B−C Boundary $= \bar{R} + d_3 \dfrac{\bar{R}}{d_2} = 7.948 + (.864)\dfrac{7.948}{2.326} = 10.900$

Lower B−C Boundary $= \bar{R} - d_3 \dfrac{\bar{R}}{d_2} = 7.948 - (.864)\dfrac{7.948}{2.326} = 4.996$

c.   The R-chart is:

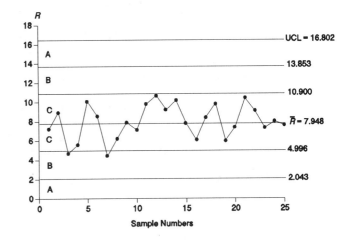

To determine if the process is in or out of control, we check the 6 rules:

Rule 1: One point beyond Zone A: No points are beyond Zone A.

Rule 2: Nine points in a row in Zone C or beyond: No sequence of 9 points are in Zone C (on one side of the centerline) or beyond.

Rule 3: Six points in a row steadily increasing or decreasing: No sequence of 6 points steadily increase or decrease.

Rule 4: Fourteen points in a row alternating up and down: This pattern does not exist.

Rule 5: Two out of three points in Zone A or beyond: There are no groups of three consecutive points that have two or more in Zone A or beyond.

Rule 6: Four out of five points in a row in Zone B or beyond: No sequence of 5 points has 4 or more in Zone B or beyond.

The process appears to be in control.

13.21 First, we construct an R-chart.

$$\bar{R} = \frac{R_1 + R_2 + \cdots + R_{20}}{n} = \frac{80.6}{20} = 4.03$$

Centerline $= \bar{R} = 4.03$

From Table XVII, Appendix B, with $n = 7$, $D_4 = 1.924$ and $D_3 = .076$.

Upper control limit $= \bar{R}D_4 = 4.03(1.924) = 7.754$
Lower control limit $= \bar{R}D_3 = 4.03(0.076) = 0.306$

From Table XVII, Appendix B, with $n = 7$, $d_2 = 2.704$ and $d_3 = .833$.

Upper A$-$B Boundary $= \bar{R} + 2d_3 \dfrac{\bar{R}}{d_2} = 4.03 + 2(.833)\dfrac{4.03}{2.704} = 6.513$

Lower A$-$B Boundary $= \bar{R} - 2d_3 \dfrac{\bar{R}}{d_2} = 4.03 - 2(.833)\dfrac{4.03}{2.704} = 1.547$

Upper B$-$C Boundary $= \bar{R} + d_3 \dfrac{\bar{R}}{d_2} = 4.03 + (.833)\dfrac{4.03}{2.704} = 5.271$

Lower B$-$C Boundary $= \bar{R} - d_3 \dfrac{\bar{R}}{d_2} = 4.03 - (.833)\dfrac{4.03}{2.704} = 2.789$

The *R*-chart is:

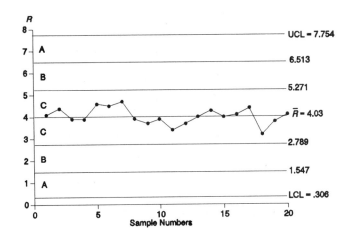

To determine if the process is in or out of control, we check the 6 rules:

Rule 1: One point beyond Zone A: No points are beyond Zone A.

Rule 2: Nine points in a row in Zone C or beyond: No sequence of 9 points are in Zone C (on one side of the centerline) or beyond.

Rule 3: Six points in a row steadily increasing or decreasing: No sequence of 6 points steadily increase or decrease.

Rule 4: Fourteen points in a row alternating up and down: This pattern does not exist.

Rule 5: Two out of three points in Zone A or beyond: There are no groups of three consecutive points that have two or more in Zone A or beyond.

Rule 6: Four out of five points in a row in Zone B or beyond: No sequence of 5 points has 4 or more in Zone B or beyond.

The process appears to be in control. Since the process variation is in control, it is appropriate to construct the $\bar{x}$-chart.

To construct an $\bar{x}$-chart, we first calculate the following:

$$\bar{\bar{x}} = \frac{\bar{x}_1 + \bar{x}_2 + \cdots + \bar{x}_{20}}{n} = \frac{434.56}{20} = 21.728$$

$$\bar{R} = \frac{R_1 + R_2 + \cdots + R_{20}}{n} = \frac{80.6}{20} = 4.03$$

Centerline = $\bar{\bar{x}}$ = 21.728

From Table XVII, Appendix B, with $n = 7$, $A_2 = .419$.

Upper control limit = $\bar{\bar{x}} + A_2\bar{R} = 21.728 + .419(4.03) = 23.417$
Lower control limit = $\bar{\bar{x}} - A_2\bar{R} = 21.728 - .419(4.03) = 20.039$

$$\text{Upper A}-\text{B Boundary} = \bar{\bar{x}} + \frac{2}{3}(A_2\bar{R}) = 21.728 + \frac{2}{3}(.419)(4.03) = 22.854$$

$$\text{Lower A}-\text{B Boundary} = \bar{\bar{x}} - \frac{2}{3}(A_2\bar{R}) = 21.728 - \frac{2}{3}(.419)(4.03) = 20.602$$

$$\text{Upper B}-\text{C Boundary} = \bar{\bar{x}} + \frac{1}{3}(A_2\bar{R}) = 21.728 + \frac{1}{3}(.419)(4.03) = 22.291$$

$$\text{Lower B}-\text{C Boundary} = \bar{\bar{x}} - \frac{1}{3}(A_2\bar{R}) = 21.728 - \frac{1}{3}(.419)(4.03) = 21.165$$

The $\bar{x}$-chart is:

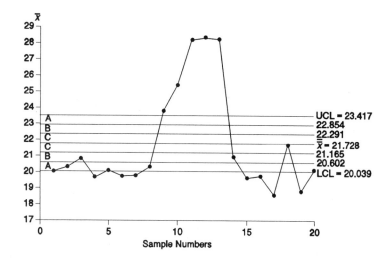

To determine if the process is in or out of control, we check the 6 rules:

Rule 1: One point beyond Zone A: There are 12 points beyond Zone A. This indicates the process is out of control.

Rule 2: Nine points in a row in Zone C or beyond: No sequence of 9 points are in Zone C (on one side of the centerline) or beyond.

Rule 3: Six points in a row steadily increasing or decreasing: Points 6 through 12 steadily increase. This indicates the process is out of control.

Rule 4: Fourteen points in a row alternating up and down: This pattern does not exist.

Rule 5: Two out of three points in Zone A or beyond: There are several groups of three consecutive points that have two or more in Zone A or beyond. This indicates the process is out of control.

Rule 6: Four out of five points in a row in Zone B or beyond: Several sequences of 5 points have 4 or more in Zone B or beyond. This indicates the process is out of control.

Rules 1, 3, 5, and 6 indicate that the process is out of control.

13.23   a.   Yes. Because all five observations in each sample were selected from the same dispenser, the rational subgrouping will enable the company to detect variation in fill caused by differences in the carbon dioxide dispensers.

b. For each sample, we compute the range = R = largest measurement − smallest measurement. The results are listed in the table:

Sample No.	R	Sample No.	R
1	.05	13	.05
2	.06	14	.04
3	.06	15	.05
4	.05	16	.05
5	.07	17	.06
6	.07	18	.06
7	.09	19	.05
8	.08	20	.08
9	.08	21	.08
10	.11	22	.12
11	.14	23	.12
12	.14	24	.15

$$\bar{R} = \frac{R_1 + R_2 + \cdots + R_{24}}{n} = \frac{1.91}{24} = .0796$$

Centerline = $\bar{R}$ = .0796

From Table XVII, Appendix B, with $n = 5$, $D_4 = 2.114$, and $D_3 = 0$.

Upper control limit = $\bar{R}D_4$ = .0796(2.114) = .168

Since $D_3 = 0$, the lower control limit is negative and is not included on the chart.

From Table XVII, Appendix B, with $n = 5$, $d_2 = 2.326$, and $d_3 = .864$.

Upper A−B Boundary = $\bar{R} + 2d_3 \dfrac{\bar{R}}{d_2}$ = $.0796 + 2(.864)\dfrac{.0796}{2.326}$ = .139

Lower A−B Boundary = $\bar{R} - 2d_3 \dfrac{\bar{R}}{d_2}$ = $.0796 - 2(.864)\dfrac{.0796}{2.326}$ = .020

Upper B−C Boundary = $\bar{R} + d_3 \dfrac{\bar{R}}{d_2}$ = $.0796 + (.864)\dfrac{.0796}{2.326}$ = .109

Lower B−C Boundary = $\bar{R} - d_3 \dfrac{\bar{R}}{d_2}$ = $.0796 - (.864)\dfrac{.0796}{2.326}$ = .050

The R-chart is:

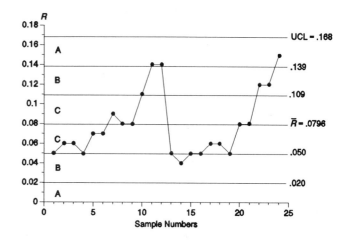

c.   To determine if the process is in or out of control, we check the 6 rules:

Rule 1:   One point beyond Zone A: No points are beyond Zone A.

Rule 2:   Nine points in a row in Zone C or beyond: No sequence of 9 points are in Zone C (on one side of the centerline) or beyond.

Rule 3:   Six points in a row steadily increasing or decreasing: No sequence of 6 points steadily increase or decrease.

Rule 4:   Fourteen points in a row alternating up and down: This pattern does not exist.

Rule 5:   Two out of three points in Zone A or beyond: Points 11 and 12 are in Zone A or beyond. This indicates the process is out of control.

Rule 6:   Four out of five points in a row in Zone B or beyond: No sequence of 5 points has 4 or more in Zone B or beyond.

Rule 5 indicates that the process is out of control. The process is unstable.

d.   Since the process variation is out of control, the R-chart should not be used to monitor future process output.

e.   The $\bar{x}$-chart should not be constructed. The control limits of the $\bar{x}$-chart depend on the variation of the process. (In particular, they are constructed using $\bar{R}$.) If the variation of the process is out of control, the control limits of the $\bar{x}$-chart are meaningless.

13.25   a.   From Exercise 13.14, we get the following data:

Sample No.	R	Sample No.	R
1	.20	14	.06
2	.19	15	.15
3	.17	16	.09
4	.22	17	.20
5	.29	18	.20
6	.24	19	.13
7	.13	20	.25
8	.21	21	.09
9	.27	22	.19
10	.24	23	.09
11	.22	24	.17
12	.16	25	.20
13	.31		

$$\bar{R} = \frac{R_1 + R_2 + \cdots + R_{25}}{n} = \frac{4.67}{25} = .1868$$

Centerline $= \bar{R} = .1868$

From Table XVII, Appendix B, with $n = 4$, $D_4 = 2.282$, and $D_3 = 0$.

Upper control limit $= \bar{R}D_4 = .1868(2.282) = .426$

Since $D_3 = 0$, the lower control limit is negative and is not included on the chart.

From Table XVII, Appendix B, with $n = 4$, $d_2 = 2.059$, and $d_3 = .880$.

$$\text{Upper A$-$B Boundary} = \bar{R} + 2d_3\frac{\bar{R}}{d_2} = .1868 + 2(.880)\frac{.1868}{2.059} = .346$$

$$\text{Lower A$-$B Boundary} = \bar{R} - 2d_3\frac{\bar{R}}{d_2} = .1868 - 2(.880)\frac{.1868}{2.059} = .027$$

$$\text{Upper B$-$C Boundary} = \bar{R} + d_3\frac{\bar{R}}{d_2} = .1868 + (.880)\frac{.1868}{2.059} = .267$$

$$\text{Lower B$-$C Boundary} = \bar{R} - d_3\frac{\bar{R}}{d_2} = .1868 - (.880)\frac{.1868}{2.059} = .107$$

The *R*-chart is:

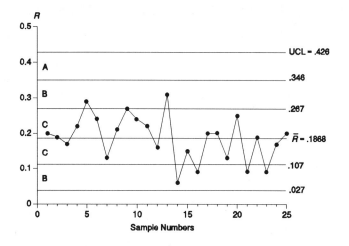

b. To determine if the process is in or out of control, we check the 6 rules:

Rule 1: One point beyond Zone A: No points are beyond Zone A.

Rule 2: Nine points in a row in Zone C or beyond: No sequence of 9 points are in Zone C (on one side of the centerline) or beyond.

Rule 3: Six points in a row steadily increasing or decreasing: No sequence of 6 points steadily increase or decrease.

Rule 4: Fourteen points in a row alternating up and down: This pattern does not exist.

Rule 5: Two out of three points in Zone A or beyond: No group of three consecutive points have two or more in Zone A or beyond.

Rule 6: Four out of five points in a row in Zone B or beyond: No sequence of 5 points has 4 or more in Zone B or beyond.

The process appears to be in control. There do not appear to be any special causes of variation during the time the data were collected.

c. Since the process appears to be in control, it is appropriate to use these limits to monitor future process output.

13.27 The sample size is determined as follows:

$$n > \frac{9(1 - p_0)}{p_0} = \frac{9(1 - .08)}{.08} = 103.5 \approx 104$$

13.29 a. We must first calculate $\bar{p}$. To do this, it is necessary to find the total number of defectives in all the samples. To find the number of defectives per sample, we multiple the proportion by the sample size, 150. The number of defectives per sample are shown in the table:

Sample No.	p	No. Defectives	Sample No.	p	No. Defectives
1	.03	4.5	11	.07	10.5
2	.05	7.5	12	.04	6.0
3	.10	15.0	13	.06	9.0
4	.02	3.0	14	.05	7.5
5	.08	12.0	15	.07	10.5
6	.09	13.5	16	.06	9.0
7	.08	12.0	17	.07	10.5
8	.05	7.5	18	.02	3.0
9	.07	10.5	19	.05	7.5
10	.06	9.0	20	.03	4.5

Note: There cannot be a fraction of a defective. The proportions presented in the exercise have been rounded off. I have used the fractions to minimize the roundoff error.

To get the total number of defectives, sum the number of defectives for all 20 samples. The sum is 172.5. To get the total number of units sampled, multiply the sample size by the number of samples:

$$150(20) = 3000$$

$$\bar{p} = \frac{\text{Total defective in all samples}}{\text{Total units sampled}} = \frac{172.5}{3000} = .0575$$

Centerline $= \bar{p} = .0575$

$$\text{Upper control limit} = \bar{p} + 3\sqrt{\frac{\bar{p}(1-\bar{p})}{n}} = .0575 + 3\sqrt{\frac{.0575(.9425)}{150}} = .1145$$

$$\text{Lower control limit} = \bar{p} - 3\sqrt{\frac{\bar{p}(1-\bar{p})}{n}} = .0575 - 3\sqrt{\frac{.0575(.9425)}{150}} = .0005$$

b.

$$\text{Upper A}-\text{B boundary} = \bar{p} + 2\sqrt{\frac{\bar{p}(1-\bar{p})}{n}} = .0575 + 2\sqrt{\frac{.0575(.9425)}{150}} = .0955$$

$$\text{Lower A}-\text{B boundary} = \bar{p} - 2\sqrt{\frac{\bar{p}(1-\bar{p})}{n}} = .0575 - 2\sqrt{\frac{.0575(.9425)}{150}} = .0195$$

$$\text{Upper B}-\text{C boundary} = \bar{p} + \sqrt{\frac{\bar{p}(1-\bar{p})}{n}} = .0575 + \sqrt{\frac{.0575(.9425)}{150}} = .0765$$

$$\text{Lower B}-\text{C boundary} = \bar{p} - \sqrt{\frac{\bar{p}(1-\bar{p})}{n}} = .0575 - \sqrt{\frac{.0575(.9425)}{150}} = .0385$$

c.   The *p*-chart is:

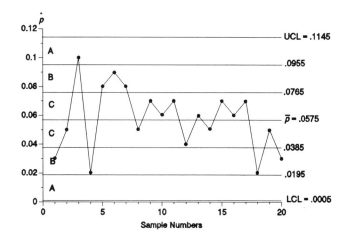

Sample Numbers

d.   To determine if the process is in or out of control, we check the 6 rules:

Rule 1:   One point beyond Zone A: No points are beyond Zone A.

Rule 2:   Nine points in a row in Zone C or beyond: No sequence of 9 points are in Zone C (on one side of the centerline) or beyond.

Rule 3:   Six points in a row steadily increasing or decreasing: No sequence of 6 points steadily increase or decrease.

Rule 4:   Fourteen points in a row alternating up and down: Points 7 through 20 alternate up and down. This indicates the process is out of control.

Rule 5:   Two out of three points in Zone A or beyond: No group of three consecutive points have two or more in Zone A or beyond.

Rule 6:   Four out of five points in a row in Zone B or beyond: Points 3, 5, 6, and 7 are in Zone B or beyond. This indicates the process is out of control.

Rules 4 and 6 indicate that the process is out of control.

e.   Since the process is out of control, the centerline and control limits should not be used to monitor future process output. The centerline and control limits are intended to represent the behavior of the process when it is under control.

13.31   a.   Yes. The minimum sample size necessary so the lower control limit is not negative is:

$$n > \frac{9(1 - p_0)}{p_0}$$

From the data, $p_0 \approx .01$

Thus, $n > \dfrac{9(1 - .01)}{.01} = 891$. Our sample size was 1000.

b. $\quad$ Upper control limit $= \bar{p} + 3\sqrt{\dfrac{\bar{p}(1-\bar{p})}{n}} = .01047 + 3\sqrt{\dfrac{.01047(.98953)}{1000}} = .02013$

$\quad$ Lower control limit $= \bar{p} - 3\sqrt{\dfrac{\bar{p}(1-\bar{p})}{n}} = .01047 - 3\sqrt{\dfrac{.01047(.98953)}{1000}} = .00081$

c. $\quad$ To determine if special causes are present, we must complete the p-chart.

$\quad$ Upper A$-$B boundary $= \bar{p} + 2\sqrt{\dfrac{\bar{p}(1-\bar{p})}{n}} = .01047 + 2\sqrt{\dfrac{.01047(.98953)}{1000}} = .01691$

$\quad$ Lower A$-$B boundary $= \bar{p} - 2\sqrt{\dfrac{\bar{p}(1-\bar{p})}{n}} = .01047 - 2\sqrt{\dfrac{.01047(.98953)}{1000}} = .00403$

$\quad$ Upper B$-$C boundary $= \bar{p} + \sqrt{\dfrac{\bar{p}(1-\bar{p})}{n}} = .01047 + \sqrt{\dfrac{.01047(.98953)}{1000}} = .01369$

$\quad$ Lower B$-$C boundary $= \bar{p} - \sqrt{\dfrac{\bar{p}(1-\bar{p})}{n}} = .01047 - \sqrt{\dfrac{.01047(.98953)}{1000}} = .00725$

The p-chart is:

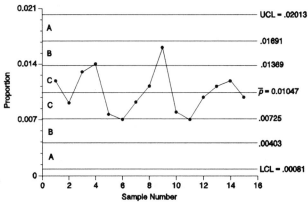

To determine if the process is in control, we check the 6 rules.

Rule 1: $\quad$ One point beyond Zone A: No points are beyond Zone A.

Rule 2: $\quad$ Nine points in a row in Zone C or beyond: There are not 9 points in a row in Zone C (on one side of the centerline) or beyond.

Rule 3: $\quad$ Six points in a row steadily increasing or decreasing: No sequence of 6 points steadily increase or decrease.

Rule 4: $\quad$ Fourteen points in a row alternating up and down: This pattern does not exist.

Rule 5: $\quad$ Two out of three points in Zone A or beyond: There are no group of three consecutive points that have two or more in Zone A or beyond.

Rule 6: $\quad$ Four out of five points in a row in Zone B or beyond: No sequence of 5 points has 4 or more in Zone B or beyond.

It appears that the process is in control.

d. The rational subgrouping strategy says that samples should be chosen so that it gives the maximum chance for the measurements in each sample to be similar and so that it gives the maximum chance for the samples to differ. By selecting 1000 consecutive disks each time, this gives the maximum chance for the measurements in the sample to be similar. By selecting the samples every other day, there is a relatively large chance that the samples differ.

13.33 a. To compute the proportion of defectives in each sample, divide the number of defectives by the number in the sample, 250:

$$\hat{p} = \frac{\text{No. of defectives}}{\text{No. in sample}}$$

The sample proportions are listed in the table:

Sample No.	$\hat{p}$	Sample No.	$\hat{p}$
1	.044	13	.048
2	.036	14	.016
3	.032	15	.036
4	.056	16	.040
5	.064	17	.052
6	.048	18	.076
7	.028	19	.072
8	.068	20	.040
9	.056	21	.064
10	.012	22	.076
11	.044	23	.032
12	.024	24	.028

To get the total number of defectives, sum the number of defectives for all 24 samples. The sum is 273. To get the total number of units sampled, multiply the sample size by the number of samples: $250(24) = 6000$.

$$\bar{p} = \frac{\text{Total defective in all samples}}{\text{Total units sampled}} = \frac{273}{6000} = .0455$$

Centerline $= \bar{p} = .0455$

$$\text{Upper control limit} = \bar{p} + 3\sqrt{\frac{\bar{p}(1-\bar{p})}{n}} = .0455 + 3\sqrt{\frac{.0455(.9545)}{250}} = .085$$

$$\text{Lower control limit} = \bar{p} - 3\sqrt{\frac{\bar{p}(1-\bar{p})}{n}} = .0455 - 3\sqrt{\frac{.0455(.9545)}{250}} = .006$$

$$\text{Upper A}-\text{B boundary} = \bar{p} + 2\sqrt{\frac{\bar{p}(1-\bar{p})}{n}} = .0455 + 2\sqrt{\frac{.0455(.9545)}{250}} = .072$$

$$\text{Lower A}-\text{B boundary} = \bar{p} - 2\sqrt{\frac{\bar{p}(1-\bar{p})}{n}} = .0455 - 2\sqrt{\frac{.0455(.9545)}{250}} = .019$$

$$\text{Upper B}-\text{C boundary} = \bar{p} + \sqrt{\frac{\bar{p}(1-\bar{p})}{n}} = .0455 + \sqrt{\frac{.0455(.9545)}{250}} = .059$$

$$\text{Lower B}-\text{C boundary} = \bar{p} - \sqrt{\frac{\bar{p}(1-\bar{p})}{n}} = .0455 - \sqrt{\frac{.0455(.9545)}{250}} = .032$$

c. The p-chart is:

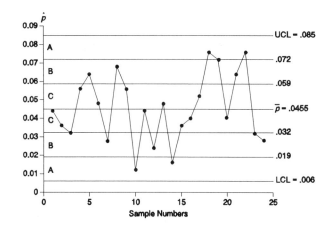

b. To determine if the process is in or out of control, we check the 6 rules.

Rule 1: One point beyond Zone A: No points are beyond Zone A.

Rule 2: Nine points in a row in Zone C or beyond: No sequence of 9 points are in Zone C (on one side of the centerline) or beyond.

Rule 3: Six points in a row steadily increasing or decreasing: No sequence of 6 points steadily increase or decrease.

Rule 4: Fourteen points in a row alternating up and down: This pattern does not exist.

Rule 5: Two out of three points in Zone A or beyond: Points 18 and 19 are in Zone A or beyond. This indicates the process is out of control.

Rule 6: Four out of five points in a row in Zone B or beyond: Points 18, 19, 21, and 22 are in Zone B or beyond. This indicates the process is out of control.

Rules 5 and 6 indicate that the process is out of control.

c. Since the process is out of control, the control limits should not be used to monitor future process output.

13.35 The quality of a good or service is indicated by the extent to which it satisfies the needs and preferences of its users. Its eight dimensions are: performance, features, reliability, conformance, durability, serviceability, aesthetics, and other perceptions that influence judgments of quality.

13.37 A process is a series of actions or operations that transform inputs to outputs. A process produces output over time. Organizational process: Manufacturing a product. Personnel Process: Balancing a checkbook.

13.39 The five major sources of process variation are: people, machines, materials, methods, and environment.

13.41 The use of a hierarchical systems model to facilitate and guide the description, exploration, analysis, and/or understanding of organizations and their problems is called systems thinking. It has many benefits, but one of the most important is that it focuses attention on the processes through which things get done rather than on final outcomes. Thus, it facilitates the prevention of problems rather than after-the-fact correction of problems.

13.45 The two properties of a system that make it an "indivisible whole" are:
1. The performance of a system is dependent on the performance of every one of its processes.
2. The way any individual process affects system performance depends on the performance of at least one other process.

13.49 If a process is in control and remains in control, its future will be like its past. It is predictable in that its output will stay within certain limits. If a process is out of control, there is no way of knowing what the future pattern of output from the process may look like.

13.51 The upper control limit and the lower control limit are positioned so that when the process is in control, the probability of an individual value falling outside the control limits is very small. Most practitioners position the control limits a distance of three standard deviations from the centerline. If the process is in control and follows a normal distribution, the probability of an individual measurement falling outside the control limits is .0026.

13.53 The probability of observing a value of $\overline{x}$ more than 3 standard deviations from its mean is:

$$P(\overline{x} > \mu + 3\sigma_{\overline{x}}) + P(\overline{x} < \mu - 3\sigma_{\overline{x}}) = P(z > 3) + P(z < 3)$$
$$= .5000 - .4987 + .5000 - .4987 = .0026$$

If we want to find the number of standard deviations from the mean the control limits should be set so the probability of the chart falsely indicating the presence of a special cause of variation is .10, we must find the z score such that:

$$P(z > z_0) + P(z < -z_0) = .1000 \text{ or } P(z > z_0) = .0500$$

Using Table IV, Appendix B, $z_0 = 1.645$. Thus the control limits should be set 1.645 standard deviations above and below the mean.

13.55 a. The centerline $= \bar{x} = \dfrac{\sum x}{n} = \dfrac{150.58}{20} = 7.529$

The time series plot is:

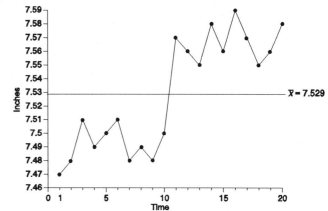

b. The variation pattern that best describes the pattern in this time series is the level shift. Points 1 through 10 all have fairly low values, while points 11 through 20 all have fairly high values.

13.57 a. Yes. The minimum sample size necessary so the lower control limit is not negative is:

$$n > \frac{9(1 - p_0)}{p_0}$$

From the data, $p_0 \approx .06$

Thus, $n > \dfrac{9(1 - .06)}{.06} = 141$. Our sample size was 200.

b. To compute the proportion of defectives in each sample, divide the number of defectives by the number in the sample, 200:

$$\bar{p} = \frac{\text{No. of defectives}}{\text{No. in sample}}$$

The sample proportions are listed in the table:

Sample No.	p	Sample No.	p
1	.02	12	.10
2	.03	13	.10
3	.055	14	.085
4	.06	15	.065
5	.025	16	.05
6	.05	17	.055
7	.04	18	.035
8	.08	10	.03
9	.085	20	.04
10	.10	21	.045
11	.14		

To get the total number of defectives, sum the number of defectives for all 21 samples. The sum is 258. To get the total number of units sampled, multiply the sample size by the number of samples: $200(21) = 4200$.

$$\bar{p} = \frac{\text{No. of defectives}}{\text{No. in sample}} = \frac{258}{4200} = .0614$$

Centerline $= \bar{p} = .0614$

$$\text{Upper control limit} = \bar{p} + 3\sqrt{\frac{\bar{p}(1 - \bar{p})}{n}} = .0614 + 3\sqrt{\frac{.0614(.9386)}{200}} = .1123$$

$$\text{Lower control limit} = \bar{p} - 3\sqrt{\frac{\bar{p}(1 - \bar{p})}{n}} = .0614 - 3\sqrt{\frac{.0614(.9386)}{200}} = .0105$$

$$\text{Upper A$-$B boundary} = \bar{p} + 2\sqrt{\frac{\bar{p}(1 - \bar{p})}{n}} = .0614 + 2\sqrt{\frac{.0614(.9386)}{200}} = .0953$$

$$\text{Lower A$-$B boundary} = \bar{p} - 2\sqrt{\frac{\bar{p}(1 - \bar{p})}{n}} = .0614 - 2\sqrt{\frac{.0614(.9386)}{200}} = .0275$$

$$\text{Upper B$-$C boundary} = \bar{p} + \sqrt{\frac{\bar{p}(1 - \bar{p})}{n}} = .0614 + \sqrt{\frac{.0614(.9386)}{200}} = .0784$$

$$\text{Lower B$-$C boundary} = \bar{p} - \sqrt{\frac{\bar{p}(1 - \bar{p})}{n}} = .0614 - \sqrt{\frac{.0614(.9386)}{200}} = .0444$$

The *p*-chart is:

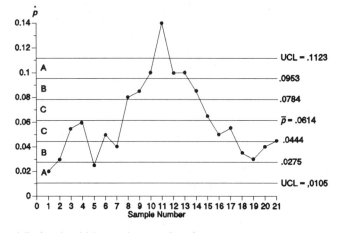

c. To determine if the control limits should be used to monitor future process output, we need to check the 6 rules.

Rule 1: One point beyond Zone A: The 11th point is beyond Zone A. This indicates the process is out of control.

Rule 2: Nine points in a row in Zone C or beyond: There are not 9 points in a row in Zone C (on one side of the centerline) or beyond.

Rule 3: Six points in a row steadily increasing or decreasing: No sequence of 6 points steadily increase or decrease.

Rule 4: Fourteen points in a row alternating up and down: This pattern does not exist.

Rule 5: Two out of three points in Zone A or beyond: Points 10 through 13 are in Zone A or beyond. This indicates the process is out of control.

Rule 6: Four out of five points in a row in Zone B or beyond: Points 8 through 14 are in Zone B or beyond. This indicates the process is out of control.

Rules 1, 5, and 6 indicate the process is out of control. These control limits should not be used to monitor future process output.

13.59 a. In order for the $\bar{x}$-chart to be meaningful, we must assume the variation in the process is constant (i.e., stable).

For each sample, we compute $\bar{x} = \dfrac{\sum x}{n}$ and R = range = largest measurement − smallest measurement. The results are listed in the table:

Sample No.	$\bar{x}$	R	Sample No.	$\bar{x}$	R
1	32.325	11.6	13	31.050	13.3
2	30.825	12.4	14	34.400	9.6
3	30.450	7.8	15	31.350	7.3
4	34.525	10.2	16	28.150	8.6
5	31.725	9.1	17	30.950	7.6
6	33.850	10.4	18	32.225	5.6
7	32.100	10.1	19	29.050	10.0
8	28.250	6.8	20	31.400	8.7
9	32.375	8.7	21	30.350	8.9
10	30.125	6.3	22	34.175	10.5
11	32.200	7.1	23	33.275	13.0
12	29.150	9.3	24	30.950	8.9

$$\bar{\bar{x}} = \frac{\bar{x}_1 + \bar{x}_2 + \cdots + \bar{x}_{24}}{n} = \frac{755.225}{24} = 31.4677$$

$$\bar{R} = \frac{R_1 + R_2 + \cdots + R_{24}}{n} = \frac{221.8}{24} = 9.242$$

Centerline = $\bar{\bar{x}}$ = 31.468

From Table XVII, Appendix B, with $n = 4$, $A_2 = .729$.

Upper control limit = $\bar{\bar{x}} + A_2\bar{R} = 31.468 + .729(9.242) = 38.205$
Lower control limit = $\bar{\bar{x}} - A_2\bar{R} = 31.468 - .729(9.242) = 24.731$

Upper A−B Boundary = $\bar{\bar{x}} + \dfrac{2}{3}(A_2\bar{R}) = 31.468 + \dfrac{2}{3}(.729)(9.242) = 35.960$

Lower A−B Boundary = $\bar{\bar{x}} - \dfrac{2}{3}(A_2\bar{R}) = 31.468 - \dfrac{2}{3}(.729)(9.242) = 26.976$

Upper B−C Boundary = $\bar{\bar{x}} + \dfrac{1}{3}(A_2\bar{R}) = 31.468 + \dfrac{1}{3}(.729)(9.242) = 33.714$

Lower B−C Boundary = $\bar{\bar{x}} - \dfrac{1}{3}(A_2\bar{R}) = 31.468 - \dfrac{1}{3}(.729)(9.242) = 29.222$

The $\bar{x}$-chart is:

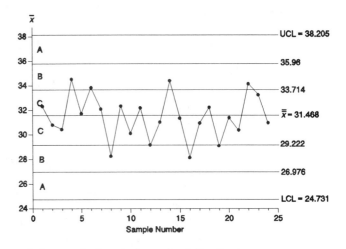

b. To determine if the process is in or out of control, we check the 6 rules.

   Rule 1:   One point beyond Zone A: No points are beyond Zone A.
   Rule 2:   Nine points in a row in Zone C or beyond: No sequence of 9 points are in Zone C (on one side of the centerline) or beyond.
   Rule 3:   Six points in a row steadily increasing or decreasing: No sequence of 6 points steadily increase or decrease.
   Rule 4:   Fourteen points in a row alternating up and down: This pattern does not exist.
   Rule 5:   Two out of three points in Zone A or beyond: There are no groups of three consecutive points that have two or more in Zone A or beyond.
   Rule 6:   Four out of five points in a row in Zone B or beyond: No sequence of 5 points has 4 or more in Zone B or beyond.

   The process appears to be in control. There are no indications that special causes of variation are affecting the process.

c. Since the process appears to be in control, these limits should be used to monitor future process output.

13.61 a. The sample size is determined by the following:

$$n > \frac{9(1 - p_0)}{p_0} = \frac{9(1 - .06)}{.06} = 141$$

   The minimum sample size is 141. Since the sample size of 150 was used, it is large enough.

b. To compute the proportion of defectives in each sample, divide the number of defectives by the number in the sample, 150:

$$\hat{p} = \frac{\text{No. of defectives}}{\text{No. in sample}}$$

The sample proportions are listed in the table:

Sample No.	$\hat{p}$	Sample No.	$\hat{p}$
1	.060	11	.047
2	.073	12	.040
3	.080	13	.080
4	.053	14	.067
5	.067	15	.073
6	.040	16	.047
7	.087	17	.040
8	.060	18	.080
9	.073	19	.093
10	.033	20	.067

To get the total number of defectives, sum the number of defectives for all 20 samples. The sum is 189. To get the total number of units sampled, multiply the sample size by the number of samples: $150(20) = 3000$

$$\bar{p} = \frac{\text{Total defectives in all samples}}{\text{Total units sampled}} = \frac{189}{3000} = .063$$

Centerline $= \bar{p} = .063$

$$\text{Upper control limit} = \bar{p} + 3\sqrt{\frac{\bar{p}(1 - \bar{p})}{n}} = .063 + 3\sqrt{\frac{.063(.937)}{150}} = .123$$

$$\text{Lower control limit} = \bar{p} - 3\sqrt{\frac{\bar{p}(1 - \bar{p})}{n}} = .063 - 3\sqrt{\frac{.063(.937)}{150}} = .003$$

$$\text{Upper A}-\text{B boundary} = \bar{p} + 2\sqrt{\frac{\bar{p}(1 - \bar{p})}{n}} = .063 + 2\sqrt{\frac{.063(.937)}{150}} = .103$$

$$\text{Lower A}-\text{B boundary} = \bar{p} - 2\sqrt{\frac{\bar{p}(1 - \bar{p})}{n}} = .063 - 2\sqrt{\frac{.063(.937)}{150}} = .023$$

$$\text{Upper B}-\text{C boundary} = \bar{p} + \sqrt{\frac{\bar{p}(1 - \bar{p})}{n}} = .063 + \sqrt{\frac{.063(.937)}{150}} = .083$$

$$\text{Lower B}-\text{C boundary} = \bar{p} - \sqrt{\frac{\bar{p}(1 - \bar{p})}{n}} = .063 - \sqrt{\frac{.063(.937)}{150}} = .043$$

c.   The *p*-chart is:

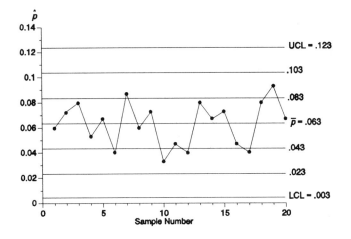

To determine if the process is in or out of control, we check the 6 rules.

Rule 1:   One point beyond Zone A: No points are beyond Zone A.
Rule 2:   Nine points in a row in Zone C or beyond: No sequence of 9 points are in Zone C (on one side of the centerline) or beyond.
Rule 3:   Six points in a row steadily increasing or decreasing: No sequence of 6 points steadily increase or decrease.
Rule 4:   Fourteen points in a row alternating up and down: Points 2 through 16 alternate up and down. This indicates the process is out of control.
Rule 5:   Two out of three points in Zone A or beyond: No group of three consecutive points have two or more in Zone A or beyond.
Rule 6:   Four out of five points in a row in Zone B or beyond: No sequence of 5 points has 4 or more in Zone B or beyond.

Rules 4 indicates the process is out of control. Special causes of variation appear to be present.

e.   Since the process is out of control, the control limits should not be used to monitor future process output. It would not be appropriate to evaluate whether the process is in control using control limits determined during a period when the process was out of control.

CHAPTER FOURTEEN

. . . .

Time Series: Index Numbers and Descriptive Analyses

14.1   To calculate a simple index number, first obtain the prices or quantities over a time period and select a base year. For each time period, the index number is the number at that time period divided by the values at the base period multiplied by 100.

14.3   To compute the simple index, divide each U.S. Beer Production value by the 1977 value, 170.5, and then multiply by 100.

Year	Simple Index	Year	Simple Index
1970	$(133.1/170.5) \times 100 = 78.06$	1980	$(194.1/170.5) \times 100 = 113.84$
1971	$(137.4/170.5) \times 100 = 80.59$	1981	$(193.7/170.5) \times 100 = 113.61$
1972	$(141.3/170.5) \times 100 = 82.87$	1982	$(196.2/170.5) \times 100 = 115.07$
1973	$(148.6/170.5) \times 100 = 87.16$	1983	$(195.4/170.5) \times 100 = 114.60$
1974	$(156.2/170.5) \times 100 = 91.61$	1984	$(192.2/170.5) \times 100 = 112.73$
1975	$(160.6/170.5) \times 100 = 94.19$	1985	$(194.3/170.5) \times 100 = 113.96$
1976	$(163.7/170.5) \times 100 = 96.01$	1986	$(194.4/170.5) \times 100 = 114.02$
1977	$(170.5/170.5) \times 100 = 100.00$	1987	$(195.9/170.5) \times 100 = 114.90$
1978	$(179.1/170.5) \times 100 = 105.04$	1988	$(197.4/170.5) \times 100 = 115.78$
1979	$(184.2/107.5) \times 100 = 108.04$	1989	$(197.8/170.5) \times 100 = 116.01$

14.5   Using 1980 as the base period, the simple index is found by dividing each entry by 194.1 and then multiplying by 100.

Year	Simple Index	Year	Simple Index
1970	$(133.1/194.1) \times 100 = 68.57$	1980	$(194.1/194.1) \times 100 = 100.00$
1971	$(137.4/194.1) \times 100 = 70.79$	1981	$(193.7/194.1) \times 100 = 99.79$
1972	$(141.3/194.1) \times 100 = 72.80$	1982	$(196.2/194.1) \times 100 = 101.08$
1973	$(148.6/194.1) \times 100 = 76.56$	1983	$(195.4/194.1) \times 100 = 100.67$
1974	$(156.2/194.1) \times 100 = 80.47$	1984	$(192.2/194.1) \times 100 = 99.02$
1975	$(160.6/194.1) \times 100 = 82.74$	1985	$(194.3/194.1) \times 100 = 100.10$
1976	$(163.7/194.1) \times 100 = 84.34$	1986	$(194.4/194.1) \times 100 = 100.15$
1977	$(170.5/194.1) \times 100 = 87.84$	1987	$(195.9/194.1) \times 100 = 100.93$
1978	$(179.1/194.1) \times 100 = 92.27$	1988	$(197.4/194.1) \times 100 = 101.70$
1979	$(184.2/194.1) \times 100 = 94.90$	1989	$(197.8/194.1) \times 100 = 101.91$

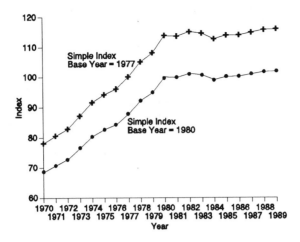

14.7 a. To compute the simple index, divide each price by the 1982 price, 222.2, and multiply by 100. The simple index values are:

Year	Simple Index
1982	$(222.2/222.2) \times 100 = 100.00$
1983	$(232.3/222.2) \times 100 = 104.55$
1984	$(239.9/222.2) \times 100 = 107.97$
1985	$(225.7/222.2) \times 100 = 101.58$
1986	$(174.8/222.2) \times 100 = 78.67$
1987	$(150.2/222.2) \times 100 = 67.60$
1988	$(152.4/222.2) \times 100 = 68.59$
1989	$(152.7/222.2) \times 100 = 68.72$
1990	$(155.4/222.2) \times 100 = 69.94$

The plot of the simple index is:

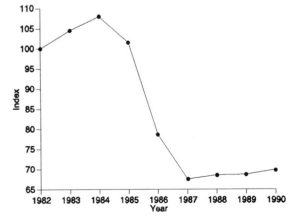

b.　The natural gas prices were fairly stable in the late 1980's, and were approximately 31% less than what they were in 1982.

c.　This is a price index because the values are the prices of natural gas, not the quantities.

14.9　a.　The simple composite index is calculated as follows:

First, sum the observations for all the series of interest at each time period. Select the base time period. Divide each sum by the sum in the base time period and multiply by 100.

b.　To calculate a weighted composite index, we follow the following steps:

First, multiply the observations in each time series by its appropriate weight. Then sum the weighted observations across all times series for each time period. Select the base time period. Divide each weighted sum by the weighted sum in the base time period and multiply by 100.

c.　The steps necessary to compute a Laspeyres Index are:

1.　Collect data for each of k price series.
2.　Select a base time period and collect purchase quantity information for each of the k series at the base time period.
3.　Using the purchase quantity values at the base period as weights, multiply each value in the kth series by its corresponding weight.
4.　Sum the products for each time period.
5.　Divide each sum by the sum corresponding to the base period and multiply by 100.

d.　The steps necessary to compute a Paasche index are:

1.　Collect data for each of k price series.
2.　Select a base period.
3.　Collect purchase quantity information for each series at each time period.
4.　For each time period, multiply the value in each price series by its corresponding purchase quantity for that time period. Sum the products for each time period.
5.　To find the value of the Paasche index at a particular time period, multiply the purchase quantity values (weights) for that time period by the corresponding price values of the base time period. Sum the results for the base period. The Paasche Index is then found by dividing the sum found in (4) by the sum found in (5).

14.11　A Laspeyres index uses the purchase quantity at the base period as the weights for all other time periods. A Paasche index uses the purchase quantity at each time period as the weight for that time period. The weights at the specified time period are also used with the base period to find the index.

14.13 To find the simple composite index for 1970, first sum the three values for each year, divide by 646.5 (the sum for the base year, 1970), and multiply by 100. To update the index by using 1980 as the base year, take the values of the 1970 index and divide by 270.39, (the value of the 1970 index for the base year 1980), and multiply by 100. The two indexes are:

Changing the base year from 1970 to 1980 flattens out the graph. Also, the spread of the values for the 1980 index is much smaller than the spread of the values for the 1970 index.

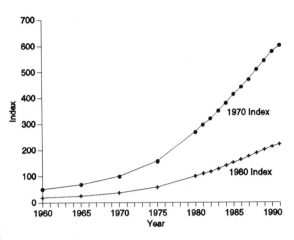

Year	Total	Index (1970)	Index (1980)	Year	Total	Index (1970)	Index (1980)
1960	332.5	51.43	19.02	1984	2460.3	380.56	140.74
1965	444.6	68.77	25.43	1985	2667.4	412.59	152.59
1970	646.5	100.00	36.98	1986	2850.6	440.93	163.07
1975	1024.8	158.52	58.62	1987	3052.2	472.11	174.60
1980	1748.1	270.39	100.00	1988	3296.1	509.84	188.56
1981	1926.2	297.94	110.19	1989	3517.9	544.15	201.24
1982	2059.2	318.52	117.80	1990	3742.6	578.90	214.10
1983	2257.5	349.19	129.14	1991	3889.0	601.55	222.47

14.15 To find Laspeyres index, we multiply the durable goods by 10.9, the nondurable goods by 140.2, and the services by 42.6. The index is found by dividing the weighted sum at each time period by the weighted sum of 1970 and then multiplying by 100. Laspeyres index and the simple composite indexes are:

Year	Simple Composite Index	Laspeyres Index	Year	Simple Composite Index	Laspeyres Index
1960	51.43	54.13	1984	380.56	351.25
1965	68.77	69.60	1985	412.59	375.14
1970	100.00	100.00	1986	440.93	394.36
1975	158.52	156.17	1987	472.11	421.47
1980	270.39	262.33	1988	509.84	451.63
1981	297.94	287.82	1989	544.15	482.60
1982	318.52	303.74	1990	578.90	514.39
1983	349.19	326.52	1991	601.55	534.36

The simple composite index starts out slightly below the Laspeyres index. Both indexes are the same in 1970, and then the simple composite index is above Laspeyres index after 1970.

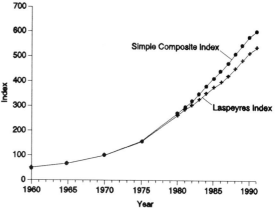

14.17 a. The following steps are used to compute the Paasche index.

1. First, multiply the price × production for copper, pig iron, and lead for each month. The numerator of the index is the sum of these 3 quantities at each month.

2. Next, multiply the production values of copper by 1361.6, the production of pig iron by 213, and the production of lead by 530. The denominator of the index is the sum of these 3 quantities at each month.

3. The values of the Paasche index is the ratio of these two values times 100.

The Laspeyres index is computed by multiplying the price of copper by 100.7 for each month, multiplying the price of pig iron by 4,311 for each month, and multiplying the price of lead by 46.1 for each month. The sum of these quantities for each month is then multiplied by 100 and divided by 1,079,789.1, the sum for January.

Month	Paasche Numerator	Paasche Denominator	Laspeyres Numerator	Paasche Index	Laspeyres Index
Jan	1079789.12	1079789.12	1079789.1	100.0	100.0
Feb	1115345.90	1112259.16	1083094.3	100.3	100.3
Mar	1263952.40	1251315.40	1092028.4	101.0	101.1
Apr	1240241.96	1223759.96	1097369.1	101.3	101.6
May	1262272.46	1255056.28	1086843.9	100.6	100.7
June	1128103.22	1125055.04	1084329.7	100.3	100.4
July	1066929.84	1067176.44	1081484.8	100.0	100.2
Aug	1008183.04	1009481.72	1079833.0	99.9	100.0
Sept	875434.26	880919.12	1073615.4	99.4	99.4
Oct	922140.12	930763.56	1069142.8	99.1	99.0
Nov	960478.70	961256.80	1078970.2	99.9	99.9
Dec	926191.26	933548.04	1070984.1	99.2	99.2

b.

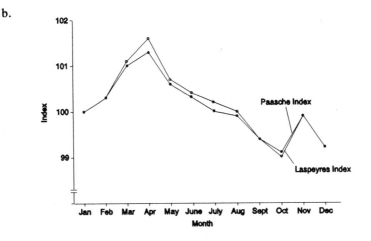

c. The Laspeyres index values for September and December are 99.4 and 99.2. The Paasche index values for the same months are 99.4 and 99.2.

The problem states that the price and production of metals is a measure of the strength of the industrial economy. Since the Paasche index takes into account the production values at each time period, it seems like it may be a more appropriate measure for describing the change in this 4-month period. The change is from 99.4 to 99.2—a very minor change.

14.19 a. The exponentially smoothed sales for the first period is equal to the sales for that period. For the rest of the time periods, the exponentially smoothed sales is found by multiplying the sales for the time period by .5 and adding to that $(1 - .5)$ times the exponentially smoothed value above it.

The exponential smoothed sales for time period 1980 is $.5(1.740) + (1 - .5)(2.237) = 1.989$. The rest of the values are shown below:

Year	Sales	Exponentially Smoothed Sales
1977	2.133	2.133
1978	2.349	2.241
1979	2.233	2.237
1980	1.740	1.989
1981	1.444	1.716
1982	0.986	1.351
1983	1.289	1.320
1984	1.455	1.388
1985	4.882	3.135

b.

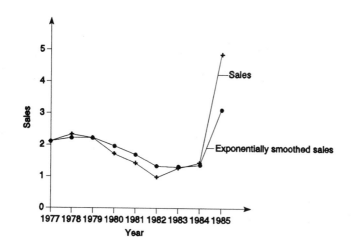

14.21 a. The exponentially smoothed gold price for the first period is equal to the gold price for that period. For the rest of the times periods, the exponentially smoothed gold price is found by multiplying the price for the time period by $w = .8$ and adding to that $(1 - .8)$ times the exponentially smoothed value from the previous time period. The exponentially smoothed value for the second period is $.8(125) + (1 - .8)(161) = 132.2$. The rest of the values are shown below.

Year	Price	$w = .8$ Exponentially Smoothed Price	Year	Price	$w = .8$ Exponentially Smoothed Price
1975	161	161.00	1983	424	418.46
1976	125	132.20	1984	361	372.49
1977	148	144.84	1985	318	328.90
1978	194	184.17	1986	368	360.18
1979	308	283.23	1987	448	430.44
1980	613	547.05	1988	438	436.49
1981	460	477.41	1989	383	393.70
1982	376	396.28	1990	387	388.34

b.

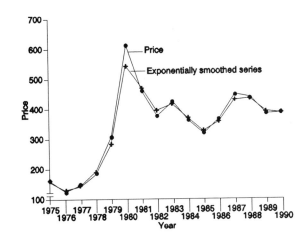

14.23 a.   The exponentially smoothed expenditure for the first time period is equal to the
            expenditure for that period. For the rest of the time periods, the exponentially smoothed
            expenditures are found by multiplying the expenditures for the time period by $w = .2$
            and adding to that $(1 - .2)$ times the exponentially smoothed value above it. The
            exponentially smoothed values for the year 1961 is $.2(44.8) + (1 - .2)(42.4) = 42.9$.
            The rest of the values appear in the table. The process is repeated with $w = .8$.

Year	Expendi-tures	$w = .2$ Exp. Smoothed Value	$w = .8$ Exp. Smoothed Value	Year	Expendi-tures	$w = .2$ Exp. Smoothed Value	$w = .8$ Exp. Snoothed Value
1960	42.4	42.4	42.4	1973	114.6	84.7	112.1
1961	44.8	42.9	44.3	1974	117.9	91.3	116.7
1962	47.4	43.8	46.8	1975	129.4	98.9	126.9
1963	49.5	44.9	49.0	1976	155.2	110.2	149.5
1964	54.3	46.8	53.2	1977	179.3	124.0	173.3
1965	58.4	49.1	57.4	1978	198.1	138.8	193.1
1966	60.4	51.4	59.8	1979	219.4	154.9	214.1
1967	63.3	53.8	62.6	1980	236.6	171.3	232.1
1968	69.3	56.9	68.0	1981	261.5	189.3	255.6
1969	75.7	60.6	74.2	1982	267.3	204.9	265.0
1970	80.6	64.6	79.3	1983	291.9	222.3	286.5
1971	92.3	70.2	89.7	1984	319.5	241.8	312.9
1972	105.4	77.2	102.3				

b. There is a much steeper increase in expenditures in the 1970s and early 1980s compared to the 1960s.

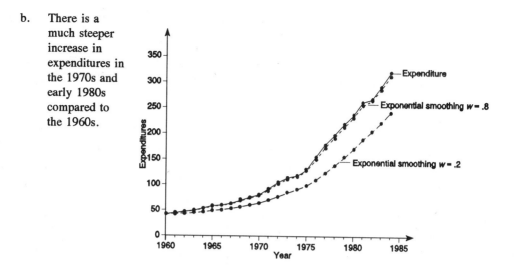

14.25 a. The simple composite index is found by summing the three steel prices, dividing by 64.25, the sum for the base period, 1980, and multiplying by 100. The values appear in the table.

Year	Cold Rolled Steel Price	Hot Rolled Steel Price	Galvanized Steel Price	Price Total	Index
1976	14.51	12.20	16.07	42.78	66.58
1977	16.44	13.79	18.10	48.33	75.22
1978	18.43	15.53	20.47	54.43	84.72
1979	20.25	17.05	22.32	59.62	92.79
1980	21.91	18.46	23.88	64.25	100.00
1981	23.90	20.15	26.88	70.93	110.40
1982	24.65	20.80	26.75	72.20	112.37
1983	26.36	22.23	28.43	77.02	119.88
1984	28.15	23.75	30.30	82.20	127.94
1985	28.15	23.75	30.30	82.20	127.94
1986	25.65	21.15	30.30	77.10	120.00
1987	27.38	21.64	30.49	79.51	123.75
1988	28.15	21.50	31.05	80.70	125.60
1989	28.15	21.50	31.05	80.70	125.60

b. This is a price index because it is based on the price of steel rather than quantity.

c. In order to compute the Laspeyres index, we need quantities of steel for the base year 1980. To compute the Paasche index, we need quantities of steel for each of the years.

14.27 To compute the simple index, divide each value by the value in the base year, 129.4, and multiply by 100. The index is:

Year	Expenditure	Simple Index	Year	Expenditure	Simple Index
1960	42.4	32.8	1973	114.6	88.6
1961	44.8	34.6	1974	117.9	91.1
1962	47.4	36.6	1975	129.4	100.0
1963	49.5	38.3	1976	155.2	119.9
1964	54.3	42.0	1977	179.3	138.6
1965	58.4	45.1	1978	198.1	153.1
1966	60.4	46.7	1979	219.4	169.6
1967	63.3	48.9	1980	236.6	182.8
1968	69.3	53.6	1981	261.5	202.1
1969	75.7	58.5	1982	267.3	206.6
1970	80.6	62.3	1983	291.9	225.6
1971	92.3	71.3	1984	319.5	246.9
1972	105.4	81.5			

14.29 a. The simple composite index is obtained from Exercise 14.28. The first value of the exponentially smoothed series is the same as the first value of the simple composite index. For all other time periods, the exponentially smoothed value is found by multiplying $w = .3$ times the corresponding simple composite index value and adding to it $(1 - .3)$ times the exponentially smoothed value from the previous time period. The exponentially smoothed series is shown in the table.

Year	Automobile	Mobile Home	Revolving	Total	Simple Composite Index(1980)	$w = .8$ Exponential Smoothed
1980	112.0	18.7	55.1	185.8	100.00	100.00
1981	119.0	20.1	61.1	200.2	107.75	102.33
1982	125.9	22.6	66.5	215.0	115.72	106.34
1983	143.6	23.6	79.1	246.3	132.56	114.21
1984	173.6	25.9	100.3	299.8	161.36	128.35
1985	210.2	26.8	121.8	358.8	193.11	147.78
1986	247.4	27.1	135.9	410.4	220.88	169.71
1987	265.9	25.9	153.1	444.9	239.45	190.63
1988	284.2	25.3	174.1	483.6	260.28	211.53
1989	290.7	22.5	199.1	512.3	275.73	230.79
1990	284.6	21.0	220.1	525.7	282.94	246.43
1991	267.9	19.1	234.5	521.5	280.68	256.71

b.

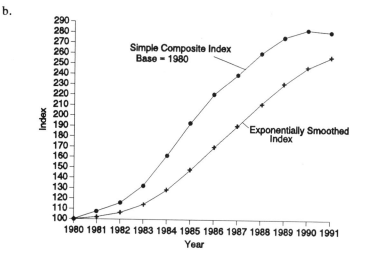

14.31 a. To compute the Paasche index, first compute the numerator by multiplying the quantity by the price for each category and summing the results for each year. To compute the denominator, multiply the price in the base year 1990 by the corresponding quantity for each category and each time period and summing the results for each year. The Paasche index is found by multiplying the numerator by 100 and dividing by the denominator. The index is:

Year	Auto.	Auto.	Mobile Home	Mobile Home	Revolv. Credit	Revolv. Credit	Numerator	Denominator	Paasche Imdex
1985	210.2	340,000	26.8	75,000	121.8	500,000	134,378,000	208,389,000	64.48
1986	247.4	350,000	27.1	80,000	135.9	600,000	170,298,000	233,350,000	72.98
1987	265.9	325,000	25.9	75,000	153.1	850,000	218,495,000	281,155,000	77.71
1988	284.2	350,000	25.3	90,000	174.1	900,000	258,437,000	299,590,000	86.26
1989	290.7	380,000	22.5	95,000	199.1	990,000	309,712,500	328,042,000	94.41
1990	284.6	400,000	21.0	100,000	220.1	1,000,000	336,040,000	336,040,000	100.00
1991	267.9	425,000	19.1	110,000	234.5	1,200,000	397,358,500	387,385,000	102.57

b.   From Exercises 14.28 and 14.30:

Year	Simple Composite Index(1980)	Simple Composite Index(1985)	Laspeyres Index	Paasche Index
1980	100.00	51.78	30.29	
1981	107.75	55.80	32.95	
1982	115.72	59.92	35.45	
1983	132.56	68.65	41.33	
1984	161.36	83.56	51.28	
1985	193.11	100.00	62.06	64.48
1986	220.88	114.38	70.70	72.98
1987	239.45	124.00	77.98	77.71
1988	260.28	134.78	86.39	86.26
1989	275.73	142.78	94.52	94.41
1990	282.94	146.52	100.00	100.00
1991	280.68	145.35	102.24	102.57

The simple composite index does not take quantity into account while each of the other two indexes do. The Paasche index uses the price only during the base period while the Laspeyres index uses both the price and quantity during the base period.

The 1985 simple composite index value is 193.11. This implies the price in 1985 is 93.11% higher than that in 1980. The 1985 Laspeyres index value is 62.06. This implies the value of the loans in 1985 are 37.94% lower than that in 1990. The 1985 Paasche index value is 64.48. This implies the price is 35.52% lower in 1985 than in 1990 assuming the quantities are at the 1985 levels for both periods.

14.33 a. The first value of the exponentially smoothed series is the same as the first value of the time series. All other values are found by multiplying w times the time series value and adding to it $(1 - w)$ times the previous exponentially smoothed value. The series for $w = .2$ and $w = .8$ are shown in the table that follows.

Year	Business Formation	Exponentially Smoothed Series	
		$w = .2$	$w = .8$
1970	106.4	106.4	106.4
1071	108.5	106.8	108.1
1972	115.9	108.6	114.3
1973	114.9	109.9	114.8
1974	109.2	109.8	110.3
1975	107.0	109.2	107.7
1976	115.6	110.5	114.0
1977	123.2	113.0	121.4
1978	128.2	116.1	126.8
1979	128.3	118.5	128.0
1980	122.4	119.3	123.5
1981	118.6	119.1	119.6
1982	113.2	118.0	114.5
1983	114.8	117.3	114.7
1984	117.1	117.3	116.6

 b.

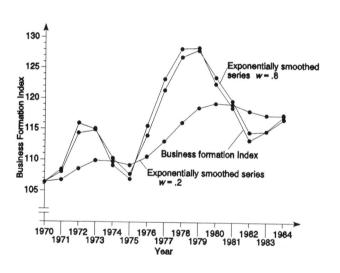

The exponentially smoothed series with $w = .2$ is the smoothest. The reason is that most of the weight is associated with the previous value and very little weight is associated with each yearly value.

CHAPTER FIFTEEN

Time Series: Models and Forecasting

15.1   Two measures of forecasting accuracy are:

    a.   Mean absolute deviation (MAD) – mean absolute difference between the forecast and actual values of the time series.

    b.   Root mean squared error (RMSE) – square root of the mean squared difference between the forecast and actual values of the time series.

15.3   a.   We first compute the exponentially smoothed values $E_1, E_2, \ldots, E_t$ for years 1970–1986.

$$E_1 = Y_1 = 133.1$$

For $w = .3$, $E_2 = wY_2 + (1 - w)E_1 = .3(137.4) + (1 - .3)(133.1) = 134.39$
$$E_3 = wY_3 + (1 - w)E_2 = .3(141.3) + (1 - .3)(134.39) = 136.46$$

The rest of the values appear in the table.

For $= .7$, $E_2 = wY_2 + (1 - w)E_1 = .7(137.4) + (1 - .7)(133.1) = 136.11$
$$E_3 = wY_3 + (1 - w)E_2 = .7(141.3) + (1 - .7)(136.11) = 139.74$$

The rest of the values appear in the table.

Year	Beer Production	Exponentially Smoothed Value $w = .3$	Exponentially Smoothed Value $w = .7$
1970	133.1	133.10	133.10
1971	137.4	134.39	136.11
1972	141.3	136.46	139.74
1973	148.6	140.10	145.94
1974	156.2	144.93	153.12
1975	160.6	149.63	158.36
1976	163.7	153.85	162.10
1977	170.5	158.85	167.98
1978	179.1	164.92	175.76
1979	184.2	170.71	181.67
1980	194.1	177.72	190.37
1981	193.7	182.52	192.70
1982	196.2	186.62	195.15
1983	195.4	189.26	195.33
1984	192.2	190.14	193.14
1985	194.3	191.39	193.95
1986	194.4	192.29	194.27
1987	195.9		
1988	197.4		
1989	197.8		

To forecast using exponentially smoothed values, we use the following:

For $w = .3$:

$$F_{1987} = F_{t+1} = E_t = 192.29 \qquad \text{Forecast error} = 195.9 - 192.29 = 3.61$$
$$F_{1988} = F_{t+2} = F_{t+1} = 192.29 \qquad \text{Forecast error} = 197.4 - 192.29 = 5.11$$
$$F_{1989} = F_{t+3} = F_{t+1} = 192.92 \qquad \text{Forecast error} = 197.8 - 192.29 = 5.51$$

For $w = .7$:

$$F_{1987} = F_{t+1} = E_t = 194.27 \qquad \text{Forecast error} = 195.9 - 194.27 = 1.63$$
$$F_{1988} = F_{t+2} = F_{t+1} = 194.27 \qquad \text{Forecast error} = 197.4 - 194.27 = 3.13$$
$$F_{1989} = F_{t+3} = F_{t+1} = 194.27 \qquad \text{Forecast error} = 197.8 - 194.27 = 3.53$$

b. We first compute the Holt-Winters values for years $1970-1986$.

With $w = .7$ and $v = .3$,

$$E_2 = Y_2 = 137.40$$
$$E_3 = wY_3 + (1 - w)(E_2 + T_2)$$
$$\quad = .7(141.3) + (1 - .7)(137.4 + 4.3)$$
$$\quad = 141.42$$

$$T_2 = Y_2 - Y_1 = 137.4 - 133.1 = 4.3$$
$$T_3 = v(E_3 - E_2) + (1 - v)T_2$$
$$\quad = .3(141.42 - 137.40) + (1 - .3)(4.3)$$
$$\quad = 4.22$$

The rest of the E_t's and T_t's appear in the table that follows.

With $w = .3$ and $v = .7$,

$$E_2 = Y_2 = 137.40$$
$$E_3 = .3(141.3) + (1 - .3)(137.4 + 4.3)$$
$$\quad = 141.58$$

$$T_2 = Y_2 - Y_1 = 137.4 - 133.1 = 4.3$$
$$T_3 = .7(141.58 - 137.40) + (1 - .7)(4.3)$$
$$\quad = 4.22$$

The rest of the E_t's and T_t's appear in the table below.

Year	Beer Production	E_t $w = .7$ $v = .3$	T_t $w = .7$ $v = .3$	E_t $w = .3$ $v = .7$	T_t $w = .3$ $v = .7$
1970	133.1				
1971	137.4	137.40	4.30	137.40	4.30
1972	141.3	141.42	4.22	141.58	4.22
1973	148.6	147.71	4.84	146.64	4.80
1974	156.2	155.10	5.61	152.87	5.80
1975	160.6	160.63	5.58	159.25	6.21
1976	163.7	164.45	5.05	164.93	5.84
1977	170.5	170.20	5.26	170.69	5.78
1978	179.1	178.01	6.03	177.26	6.33
1979	184.2	184.15	6.06	183.78	6.46
1980	194.1	192.93	6.88	191.40	7.27
1981	193.7	195.53	5.59	197.18	6.23
1982	196.2	197.68	4.56	201.24	4.72
1983	195.4	197.45	3.12	202.79	2.50
1984	192.2	194.71	1.36	201.36	−0.25
1985	194.3	194.83	0.99	199.07	−1.68
1986	194.4	194.83	0.69	196.49	−2.31
1987	195.9				
1988	197.4				
1989	197.8				

To forecast using the Holt-Winters Model:

For $w = .7$ and $v = .3$,

$$F_{1987} = F_{t+1} = E_t + T_t = 194.83 + .69 = 195.52$$
Forecast error $= 195.9 - 195.52 = .38$

$$F_{1988} = F_{t+2} = E_t + 2T_t = 194.83 + 2(.69) = 196.21$$
Forecast error $= 197.4 - 196.21 = 1.19$

$$F_{1989} = F_{t+3} = E_t + 3T_t = 194.83 + 3(.69) = 196.90$$
Forecast error $= 197.8 - 196.9 = .90$

For $w = .3$ and $v = .7$,

$$F_{1987} = F_{t+1} = E_t + T_t = 196.49 - 2.31 = 194.18$$
Forecast error $= 195.9 - 194.18 = 1.72$

$$F_{1988} = F_{t+2} = E_t + 2T_t = 196.49 - 2(2.31) = 191.87$$
Forecast error $= 197.4 - 191.87 = 5.53$

$$F_{1989} = F_{t+3} = E_t + 3T_t = 196.49 - 3(2.31) = 189.56$$
Forecast error $= 197.8 - 189.56 = 8.24$

c. For the exponential smoothed forecasts,
$w = .3$,

$$MAD = \frac{\sum_{t=1}^{n} |F_t - Y_t|}{N}$$

$$= \frac{|192.29 - 195.9| + |192.29 - 197.4| + |192.29 - 197.8|}{3} = \frac{14.23}{3}$$

$$= 4.7433$$

$$RMSE = \sqrt{\frac{\sum_{t=1}^{n} (F_t - Y_t)^2}{N}}$$

$$= \sqrt{\frac{(-3.61)^2 + (-5.11)^2 + (-5.51)^2}{3}} = 4.8133$$

$$MAD = \frac{\sum |F_t - Y_t|}{N}$$

$$= \frac{|194.27 - 195.9| + |194.27 - 197.4| + |194.27 - 197.8|}{3}$$

$$= \frac{8.29}{3} = 2.7633$$

$$RMSE = \sqrt{\frac{\sum (F_t - Y_t)^2}{N}} = \sqrt{\frac{(-1.63)^2 + (-3.13)^2 + (-3.53)^2}{3}} = 2.8818$$

For the Holt-Winters forecasts,
$w = .7$ and $v = .3$

$$MAD = \frac{\sum |F_t - Y_t|}{N}$$

$$= \frac{|195.52 - 195.9| + |196.21 - 197.4| + |196.9 - 197.8|}{3}$$

$$= \frac{2.47}{3} = .8233$$

$$RMSE = \sqrt{\frac{\sum (F_t - Y_t)^2}{N}} = \sqrt{\frac{(-.38)^2 + (-1.19)^2 + (-.90)^2}{3}} = .8889$$

$w = .3$ and $v = .7$,

$$MAD = \frac{\sum |F_t - Y_t|}{N}$$

$$= \frac{|194.18 - 195.9| + |191.87 - 197.4| + |189.56 - 197.8|}{3}$$

$$= \frac{15.49}{3} = 5.1633$$

$$RMSE = \sqrt{\frac{\sum (F_t - Y_t)^2}{N}} = \sqrt{\frac{(-1.72)^2 + (-5.53)^2 + (8.24)^2}{3}} = 5.1633$$

Both the MAD measures for the exponentially smoothed forecasts (4.7433 and 2.7633) are between the MAD measures for the Holt-Winters forecasts (.8233 and 5.8148). Similarly, the RMSE measures for the exponentially smoothed forecasts (4.8133 and 2.8818) are between the RMSE measures for the Holt-Winters forecasts (.8889 and 5.8148).

There does appear to be a trend in the beer production time series. The series increases for every year from 1970 to 1980, then goes up and down for a few years and then increases from 1985 on. It appears the Holt-Winters forecasting model with $w = .7$ and $v = .3$ is the best. It has the smallest error.

15.5   a.   We first compute the exponentially smoothed values $E_1, E_2, \ldots, E_t$ for 1978 through 1991.

$$E_1 = Y_1 = 88.82$$

For $w = .7$,

$$E_2 = wY_2 + (1 - w)E_1 = .7(97.66) + (1 - .7)(88.82) = 95.01$$
$$E_3 = wY_3 + (1 - w)E_2 = .7(103.9) + (1 - .7)(95.008) = 101.23$$

The rest of the values appear in the table:

Year	Quarter	S&P 500	Exponentially Smoothed Value ($w = .7$)	Year	Quarter	S&P 500	Exponentially Smoothed Value ($w = .7$)
1978	I	88.82	88.82	1986	I	232.3	222.73
	II	97.66	95.01		II	245.3	238.53
	III	103.9	101.23		III	238.3	238.37
	IV	96.11	97.65		IV	248.6	245.53
1979	I	100.1	99.36	1987	I	292.5	278.41
	II	101.7	101.00		II	301.4	294.50
	III	108.6	106.32		III	318.7	311.44
	IV	107.8	107.36		IV	241.0	262.13
1980	I	104.7	105.50	1988	I	265.7	264.63
	II	114.6	111.87		II	270.7	268.88
	III	126.5	122.11		III	268.0	268.26
	IV	133.5	130.08		IV	276.5	274.03
1981	I	133.2	132.26	1989	I	292.7	287.10
	II	132.3	132.29		II	323.7	312.72
	III	118.3	122.50		III	347.3	336.93
	IV	123.8	123.41		IV	348.6	345.10
1982	I	110.8	114.58	1990	I	338.5	340.48
	II	109.7	111.16		II	360.4	354.42
	III	122.4	119.03		III	315.4	327.11
	IV	139.4	133.29		IV	328.8	328.29
1983	I	151.9	146.32	1991	I	372.3	359.10
	II	166.4	160.37		II	378.3	372.54
	III	167.2	165.15		III	387.2	382.80
	IV	164.4	164.63		IV	388.5	386.79
1984	I	157.4	159.57	1992	I	407.36	
	II	153.1	155.04		II	408.27	
	III	166.1	162.78		III	418.48	
	IV	164.5	163.98		IV	435.64	
1985	I	179.4	174.78				
	II	188.9	184.66				
	III	184.1	184.27				
	IV	207.3	200.39				

The forecasts for the 4 quarters of 1992 based on 1978–1991 data are:

$$F_{1992,I} = F_{t+1} = E_t = 386.79$$
$$F_{1992,II} = F_{t+2} = F_{t+1} = 386.79$$
$$F_{1992,III} = F_{t+3} = F_{t+1} = 386.79$$
$$F_{1992,IV} = F_{t+4} = F_{t+1} = 386.79$$

b. The forecast errors are:

1992, I $407.36 - 386.79 = 20.57$
1992, II $408.27 - 386.79 = 21.48$
1992, III $418.48 - 386.79 = 31.69$
1992, IV $435.64 - 386.79 = 48.85$

c. We first compute the exponentially smoothed values, $E_1, E_2, \ldots, E_t$ for 1978–1991.

For $w = .3$,

$$E_1 = Y_1 = 88.82$$
$$E_2 = wY_2 + (1 - w)E_1 = .3(97.66) + (1 - .3)(88.82) = 91.472$$
$$E_3 = wY_3 + (1 - w)E_2 = .3(103.9) + (1 - .3)(91.472) = 95.20$$

The rest of the values appear in the table.

Year	Quarter	S&P 500	Exponentially Smoothed Value ($w = .7$)	Year	Quarter	S&P 500	Exponentially Smoothed Value ($w = .7$)
1978	I	88.82	88.82	1986	I	232.3	199.54
	II	97.66	91.47		II	245.3	213.27
	III	103.9	95.20		III	238.3	220.78
	IV	96.11	95.47		IV	248.6	229.12
1979	I	100.1	96.86	1987	I	292.5	248.14
	II	101.7	98.31		II	301.4	264.12
	III	108.6	101.40		III	318.7	280.49
	IV	107.8	103.32		IV	241.0	268.64
1980	I	104.7	103.73	1988	I	265.7	267.76
	II	114.6	106.99		II	270.7	268.64
	III	126.5	112.85		III	268.0	268.45
	IV	133.5	119.04		IV	276.5	270.86
1981	I	133.2	123.29	1989	I	292.7	277.42
	II	132.3	125.99		II	323.7	291.30
	III	118.3	123.68		III	347.3	308.10
	IV	123.8	123.72		IV	348.6	320.25
1982	I	110.8	119.84	1990	I	338.5	325.73
	II	109.7	116.80		II	360.4	336.13
	III	122.4	118.48		III	315.4	329.91
	IV	139.4	124.76		IV	328.8	329.58
1983	I	151.9	132.90	1991	I	372.3	342.39
	II	166.4	142.95		II	378.3	353.17
	III	167.2	150.22		III	387.2	363.38
	IV	164.4	154.48		IV	388.5	370.91
1984	I	157.4	155.35	1992	I	407.36	
	II	153.1	154.68		II	408.27	
	III	166.1	158.10		III	418.48	
	IV	164.5	160.02		IV	435.64	
1985	I	179.4	165.84				
	II	188.9	172.76				
	III	184.1	176.16				
	IV	207.3	185.50				

The forecasts are for the 4 quarters of 1992 based of 1978–1991 data are:

$$F_{1992,\text{I}} = F_{t+1} = E_t = 370.91$$
$$\text{Forecast error} = 407.36 - 370.91 = 36.45$$

$$F_{1992,\text{II}} = F_{t+2} = F_{t+1} = 370.91$$
$$\text{Forecast error} = 408.27 - 370.91 = 37.36$$

$$F_{1992,\text{III}} = F_{t+3} = F_{t+1} = 370.91$$
$$\text{Forecast error} = 418.48 - 370.91 = 47.57$$

$$F_{1992,\text{IV}} = F_{t+4} = F_{t+1} = 370.91$$
$$\text{Forecast error} = 435.64 - 370.91 = 64.73$$

d. We first compute the Holt-Winters values for years 1978–1991.

With $w = .7$ and $v = .5$,

$$E_2 = Y_2 = 97.66$$
$$\begin{aligned} E_3 &= wY_3 + (1 - w)(E_2 + T_2) \\ &= .7(103.9) + (1 - .7)(97.66 + 8.84) \\ &= 104.68 \end{aligned}$$

$$T_2 = Y_2 - Y_1 = 97.66 - 88.82 = 8.84$$
$$\begin{aligned} T_3 &= v(E_3 - E_2) + (1 - v)T_2 \\ &= .5(104.68 - 97.66) + (1 - .5)(8.84) \\ &= 7.93 \end{aligned}$$

The rest of the E_t's and T_t's appear in the table that follows.

With $w = .3$ and $v = .5$,

$$E_2 = Y_2 = 97.66$$
$$\begin{aligned} E_3 &= wY_3 + (1 - w)(E_2 + T_2) \\ &= .3(103.9) + (1 - .3)(97.66 + 8.84) \\ &= 105.72 \end{aligned}$$

$$T_2 = Y_2 - Y_1 = 97.66 - 88.82 = 8.84$$
$$\begin{aligned} T_3 &= v(E_3 - E_2) + (1 - v)T_2 \\ &= .5(105.72 - 97.66) + (1 - .5)(8.84) \\ &= 8.45 \end{aligned}$$

The rest of the E_t's and T_t's appear in the table.

| | | | Holt-Winters Model 1 | | Holt-Winters Model 2 | |
Year	Quarter	S&P 500	$E(w = .7)$	$T(v = .5)$	$E(w = .7)$	$T(v = .5)$
1978	I	88.82				
	II	97.66	97.66	8.84	97.66	8.84
	III	103.9	104.68	7.93	105.72	8.45
	IV	96.11	101.06	2.15	108.75	5.74
1979	I	100.1	101.03	1.06	110.18	3.58
	II	101.7	101.82	0.93	110.14	1.77
	III	108.6	106.84	2.97	110.92	1.28
	IV	107.8	108.41	2.27	110.88	0.62
1980	I	104.7	106.49	0.18	109.46	−0.40
	II	114.6	112.22	2.95	110.72	0.43
	III	126.5	123.10	6.92	115.75	2.73
	IV	133.5	132.46	8.14	122.99	4.98
1981	I	133.2	135.42	5.55	129.54	5.77
	II	132.3	134.90	2.52	134.41	5.32
	III	118.3	124.03	−4.17	133.30	2.10
	IV	123.8	122.62	−2.80	131.92	0.36
1982	I	110.8	113.51	−5.95	125.84	−2.86
	II	109.7	109.06	−5.20	119.00	−4.85
	III	122.4	116.84	1.29	116.62	−3.61
	IV	139.4	133.02	8.74	120.93	0.35
1983	I	151.9	148.86	12.29	130.46	4.94
	II	166.4	164.82	14.13	144.70	9.59
	III	167.2	170.72	10.01	158.16	11.53
	IV	164.4	169.30	4.30	168.10	10.73
1984	I	157.4	162.26	−1.37	172.41	7.52
	II	153.1	155.44	−4.10	171.88	3.49
	III	166.1	161.67	1.07	172.59	2.10
	IV	164.5	163.97	1.68	171.64	0.57
1985	I	179.4	175.28	6.49	174.37	1.65
	II	188.9	186.76	8.99	179.88	3.59
	III	184.1	187.60	4.91	183.66	3.68
	IV	207.3	202.86	10.09	193.33	6.67
1986	I	232.3	226.50	16.86	209.69	11.52
	II	245.3	244.72	17.54	228.44	15.13
	III	238.3	245.49	9.16	241.99	14.34
	IV	248.6	250.41	7.04	254.01	13.18
1987	I	292.5	281.99	19.31	274.79	16.98
	II	301.4	301.37	19.34	294.65	18.42
	III	318.7	319.30	18.64	314.77	19.27
	IV	241.0	270.08	−15.29	306.12	5.31

Year	Quarter	S&P 500	Holt-Winters Model 1		Holt-Winters Model 2	
			$E(w = .7)$	$T(v = .5)$	$E(w = .7)$	$T(v = .5)$
1988	I	265.7	262.43	−11.47	297.71	−1.55
	II	270.7	264.78	−4.56	288.53	−5.37
	III	268.0	265.66	−1.84	278.61	−7.64
	IV	276.5	272.70	2.60	272.63	−6.81
1989	I	292.7	287.48	8.69	273.88	−2.78
	II	323.7	315.44	18.33	286.88	5.11
	III	347.3	343.24	23.06	308.58	13.41
	IV	348.6	353.91	16.87	329.97	17.40
1990	I	338.5	348.18	5.57	344.71	16.07
	II	360.4	358.41	7.90	360.66	16.01
	III	315.4	330.67	-9.92	358.29	6.82
	IV	328.8	326.39	-7.10	354.22	1.37
1991	I	372.3	356.39	11.45	360.60	3.88
	II	378.3	375.16	15.11	368.63	5.95
	III	387.2	388.12	14.03	378.37	7.84
	IV	388.5	392.60	9.25	386.90	8.19
1992	I	407.36				
	II	408.27				
	III	418.48				
	IV	435.64				

The forecasts using the Holt-Winters model are:

Model 1

$F_{1992,\text{I}} = F_{t+1} = E_t + T_t = 392.60 + 9.25 = 401.86$

$F_{1992,\text{II}} = F_{t+2} = E_t + 2T_t = 392.6 + 2(9.25) = 411.10$

$F_{1992,\text{III}} = F_{t+3} = E_t + 3T_t = 392.6 + 3(9.25) = 420.35$

$F_{1992,\text{IV}} = F_{t+4} = E_t + 4T_t = 392.6 + 4(9.25) = 429.60$

Model 2

$F_{1992,\text{I}} = F_{t+1} = E_t + T_t = 386.9 + 8.19 = 395.09$

$F_{1992,\text{II}} = F_{t+2} = E_t + 2T_t = 386.9 + 2(8.19) = 403.28$

$F_{1992,\text{III}} = F_{t+3} = E_t + 3T_t = 386.9 + 3(8.19) = 411.47$

$F_{1992,\text{IV}} = F_{t+4} = E_t + 4T_t = 386.9 + 4(8.19) = 419.66$

The forecast errors are:

Model 1		Model 2	
1992, I	$407.36 - 401.86 = 5.50$	1992, I	$407.36 - 395.09 = 12.27$
1992, II	$408.27 - 411.10 = -2.83$	1992, II	$408.27 - 403.28 = 4.99$
1992, III	$418.48 - 420.35 = -1.87$	1992, III	$418.48 - 411.47 = 7.01$
1992, IV	$435.64 - 429.60 = 6.04$	1992, IV	$435.64 - 419.66 = 15.98$

15.7 **a.** We first compute the exponentially smoothed values $E_1, E_2, \ldots, E_t$ for 1987–1992.

For $w = .3$,

$$E_1 = Y_1 = 292.5$$
$$E_2 = wY_2 + (1 - w)E_1 = .3(301.4) + (1 - .3)(292.5) = 295.17$$
$$E_3 = wY_3 + (1 - w)E_2 = .3(318.7) + (1 - .3)(295.17) = 302.23$$

The rest of the values appear in the table.

For $w = .7$,

$$E_1 = Y_1 = 292.5$$
$$E_2 = wY_2 + (1 - w)E_1 = .3(301.4) + (1 - .7)(292.5) = 298.73$$
$$E_3 = wY_3 + (1 - w)E_2 = .3(318.7) + (1 - .7)(298.73) = 312.71$$

The rest of the values appear in the table.

Year	Quarter	S&P 500	Exponentially Smoothed Value ($w = .3$)	Exponentially Smoothed Value ($w = .7$)
1987	I	292.5	292.50	292.50
	II	301.4	295.17	298.73
	III	318.7	302.23	312.71
	IV	241.0	283.86	262.51
1988	I	265.7	278.41	264.74
	II	270.7	276.10	268.91
	III	268.0	273.67	268.27
	IV	276.5	274.52	274.03
1989	I	292.7	279.97	287.10
	II	323.7	293.09	312.72
	III	347.3	309.35	336.93
	IV	348.6	321.13	345.10
1990	I	338.5	326.34	340.48
	II	360.4	336.56	354.42
	III	315.4	330.21	327.11
	IV	328.8	329.79	328.29
1991	I	372.3	342.54	359.10
	II	378.3	353.27	372.54
	III	387.2	363.45	382.80
	IV	388.5	370.96	386.79
1992	I	407.36	381.88	401.19
	II	408.27	389.80	406.15
	III	418.48	398.40	414.78
	IV	435.64	409.57	429.38

The forecasts for the 4 quarters of 1993 based on 1987–1992 data are:

$w = .3$, $\qquad\qquad\qquad\qquad\qquad\qquad\qquad$ $w = .7$,

$$F_{1993,I} = F_{t+1} = E_t = 409.57 \qquad\qquad F_{t+1} = E_t = 429.38$$
$$F_{1993,II} = F_{t+2} = F_{t+1} = 409.57 \qquad\qquad F_{t+2} = F_{t+1} = 429.38$$
$$F_{1993,III} = F_{t+3} = F_{t+1} = 409.57 \qquad\qquad F_{t+3} = F_{t+1} = 429.38$$
$$F_{1993,IV} = F_{t+4} = F_{t+1} = 409.57 \qquad\qquad F_{t+4} = F_{t+1} = 429.38$$

b. We first compute the Holt-Winters values for the years 1987–1992.

With $w = .3$ and $v = .5$,

$$E_2 = Y_2 = 301.4$$
$$E_3 = wY_3 + (1 - w)(E_2 + T_2)$$
$$\quad = .3(318.7) + (1 - .3)(301.4 + 8.90)$$
$$\quad = 312.82$$

$$T_2 = Y_2 - Y_1 = 301.4 - 292.5 = 8.90$$
$$T_3 = v(E_3 - E_2) + (1 - v)T_2$$
$$\quad = .5(312.82 - 301.4) + .5(8.90)$$
$$\quad = 10.16$$

The rest of the E_t's and T_t's appear in the table.

With $w = .7$ and $v = .5$,

$$E_2 = Y_2 = 301.4$$
$$E_3 = .7(318.7) + (1 - .7)(301.4 + 8.90)$$
$$\quad = 316.18$$

$$T_2 = Y_2 - Y_1 = 301.4 - 292.5 = 8.9$$
$$T_3 = .5(316.18 - 301.4) + (1 - .5)(8.90)$$
$$\quad = 11.84$$

The rest of the E_t's and T_t's appear in the table.

Year	Quarter	S&P 500	Holt-Winters Model 1 $E(w = .3)$	$T(v = .5)$	Holt-Winters Model 2 $E(w = .7)$	$T(v = .5)$
1987	I	292.5				
	II	301.4	301.4	8.90	301.4	8.90
	III	318.7	312.82	10.16	316.18	11.84
	IV	241.0	298.39	−2.14	267.11	−18.62
1988	I	265.7	287.08	−6.72	260.54	−12.59
	II	270.7	277.47	−8.17	263.87	−4.63
	III	268.0	268.91	−8.36	265.37	−1.56
	IV	276.5	265.33	−5.97	272.69	2.88
1989	I	292.7	269.36	−0.97	287.56	8.87
	II	323.7	284.99	7.33	315.52	18.42
	III	347.3	308.81	15.57	343.29	23.09
	IV	348.6	331.65	19.21	353.94	16.87
1990	I	338.5	347.15	17.35	348.19	5.56
	II	360.4	363.27	16.74	358.41	7.89
	III	315.4	360.63	7.05	330.67	−9.92
	IV	328.8	356.01	1.22	326.38	−7.10
1991	I	372.3	361.75	3.48	356.39	11.45
	II	378.3	369.15	5.44	375.16	15.11
	III	387.2	378.37	7.33	388.12	14.04
	IV	388.5	386.54	7.75	392.60	9.25
1992	I	407.36	398.21	9.71	405.71	11.18
	II	408.27	408.03	9.76	410.86	8.17
	III	418.48	418.00	9.87	418.64	7.98
	IV	435.64	430.20	11.03	432.93	11.13

The forecasts using the Holt-Winters model are:

Model 1

$$F_{1993,I} = F_{t+1} = E_t + T_t = 430.2 + 11.03 = 441.23$$
$$F_{1993,II} = F_{t+2} = E_t + 2T_t = 430.2 + 2(11.03) = 452.26$$
$$F_{1993,III} = F_{t+3} = E_t + 3T_t = 430.2 + 3(11.03) = 463.29$$
$$F_{1993,IV} = F_{t+4} = E_t + 4T_t = 430.2 + 4(11.03) = 474.32$$

Model 2

$$F_{1993,I} = F_{t+1} = E_t + T_t = 432.93 + 11.13 = 444.06$$
$$F_{1993,II} = F_{t+2} = E_t + 2T_t = 432.93 + 2(11.13) = 455.19$$
$$F_{1993,III} = F_{t+3} = E_t + 3T_t = 432.93 + 3(11.13) = 466.32$$
$$F_{1993,IV} = F_{t+4} = E_t + 4T_t = 432.93 + 4(11.13) = 477.45$$

c. Since we do not know the S&P 500 values for 1993, there is no way to know which set of forecasts is best. In Exercise 15.6, the MAD and RMSE were computed for all four models. The values of MAD and RMSE for the Holt-Winters model with $w = .7$ and $v = .5$ are the smallest. Thus, we would choose this model to forecast 1993 values.

15.9   a. Using just the 1986–1991 data, the forecasts using the exponentially smoothed model with $w = .5$ and the differences between the forecasts and the actual values, $F_t - Y_t$ are listed in the table:

Year	Month	Gold Price, Y_t	Forecast F_t	Difference $F_t - Y_t$
1992	Jan	355.7	366.38	10.68
	Feb	355.2	366.38	11.18
	Mar	346.0	366.38	20.38
	Apr	339.8	366.38	26.58
	May	338.5	366.38	27.88
	Jun	342.0	366.38	24.38
	Jul	338.9	366.38	27.48
	Aug	344.2	366.38	22.18
	Sep	346.7	366.38	19.68
	Oct	345.6	366.38	20.78
	Nov	336.3	366.38	30.08
	Dec	335.5	366.38	30.88

$$\text{MAD} = \frac{\sum |F_t - Y_t|}{N} = \frac{|10.68| + |11.18| + \cdots + |30.88|}{12} = \frac{272.16}{12} = 22.68$$

$$\text{RMSE} = \sqrt{\frac{\sum (F_t - Y_t)^2}{N}} = \sqrt{\frac{(10.68)^2 + (11.18)^2 + \cdots + (30.88)^2}{12}}$$

$$= \sqrt{\frac{6657.1688}{12}} = 23.55$$

Using just the 1986–1991 data, the forecasts using the Holt-Winters model with $w = .5$ and $v = .5$ and the differences between the forecasts and the actual values, $F_t - Y_t$, are listed in the table:

Year	Month	Gold Price, Y_t		Forecast F_t	Difference $F_t - Y_t$
1992	Jan	355.7	1	369.12	13.42
	Feb	355.2	2	373.42	18.22
	Mar	346.0	3	377.72	31.72
	Apr	339.8	4	382.02	42.22
	May	338.5	5	386.32	47.82
	Jun	342.0	6	390.62	48.62
	Jul	338.9	7	394.92	56.02
	Aug	344.2	8	399.22	55.02
	Sep	346.7	9	403.52	56.82
	Oct	345.6	10	407.82	62.22
	Nov	336.3	11	412.12	75.82
	Dec	335.5	12	416.42	80.92

$$MAD = \frac{\sum |F_t - Y_t|}{N} = \frac{|13.42| + |18.22| + \cdots + |80.92|}{12} = \frac{588.84}{12} = 49.07$$

$$RMSE = \sqrt{\frac{\sum (F_t - Y_t)^2}{N}} = \sqrt{\frac{13.42^2 + 18.22^2 + \cdots + (80.92)^2}{12}}$$

$$= \sqrt{\frac{33513.4088}{12}} = 52.85$$

Using both the MAD and RMSE, forecasts based on the exponentially smoothed model appear to be better than those based on the Holt-Winters model. The MAD and RMSE for the exponentially smoothed forecasts are less than those for the Holt-Winters forecasts.

b. The one-step ahead forecasts based on the exponentially smoothed model with $w = .5$ and the differences between the forecasts and the actual values, $F_t - Y_t$, are listed in the table.

Year	Month	Gold Price, Y_t	Forecast F_t	Difference $F_t - Y_t$
1992	Jan	355.7	366.38	10.68
	Feb	355.2	361.04	5.84
	Mar	346.0	358.12	12.12
	Apr	339.8	352.06	12.26
	May	338.5	345.93	7.43
	Jun	342.0	342.21	0.21
	Jul	338.9	342.11	3.21
	Aug	344.2	340.50	3.70
	Sep	346.7	342.35	4.35
	Oct	345.6	344.53	1.07
	Nov	336.3	345.06	8.76
	Dec	335.5	340.68	5.18

$$\text{MAD} = \frac{\sum |F_t - Y_t|}{N} = \frac{|10.68| + |5.84| + \cdots + |5.18|}{12} = \frac{74.81}{12} = 6.23$$

$$\text{RMSE} = \sqrt{\frac{\sum (F_t - Y_t)^2}{N}} = \sqrt{\frac{(10.68)^2 + (5.84)^2 + \cdots + (5.18)^2}{12}}$$

$$= \sqrt{\frac{648.2505}{12}} = 7.35$$

The one-step ahead forecasts based on the Holt-Winters model with $w = .5$ and $v = .5$ and the differences between the forecasts and the actual values, $F_t - Y_t$ are listed in the table:

Year	Month	Gold Price, Y_t	Forecast F_t	Difference $F_t - Y_t$
1992	Jan	355.7	369.12	13.42
	Feb	355.2	363.35	8.15
	Mar	346.0	358.18	12.18
	Apr	339.8	347.95	8.15
	May	338.5	337.70	−0.80
	Jun	342.0	332.12	−9.88
	Jul	338.9	333.91	−4.99
	Aug	344.2	334.06	−10.14
	Sep	346.7	339.50	−7.20
	Oct	345.6	345.27	−0.33
	Nov	336.3	347.68	11.38
	Dec	335.5	341.40	5.90

$$\text{MAD} = \frac{\sum |F_t - Y_t|}{N} = \frac{|13.42| + |8.15| + \cdots + |5.90|}{12} = \frac{92.52}{12} = 7.71$$

$$\text{RMSE} = \sqrt{\frac{\sum (F_t - Y_t)^2}{N}} = \sqrt{\frac{(13.42)^2 + (8.15)^2 + \cdots + (5.90)^2}{12}}$$

$$= \sqrt{\frac{903.5312}{12}} = 8.68$$

Again, using both MAD and RMSE, the forecasts based on the exponentially smoothed model appear to be better than those based on the Holt-Winters model. The MAD and RMSE for the exponentially smoothed forecasts are less than those for the Holt-Winters forecasts.

15.11 a. The estimates of the parameters in the model, $E(Y_t) = \beta_0 + \beta_1 t$, are

$\hat{\beta}_0 = 6.609231.$ The price is estimated to be 6.609 for $t = 0$ or for 1970.

$\hat{\beta}_1 = 1.704198.$ The price is estimated to increase by 1.704 for each additional year.

Time Series: Models and Forecasting

b. The forecast for 1985 is:

$$\text{Using } t = 15, \hat{Y}_{1985} = 6.609231 + 1.704198(15) = 32.1722$$

The forecast for 1986 is:

$$\text{Using } t = 16, \hat{Y}_{1986} = 6.609231 + 1.704198(16) = 33.8764$$

c. From the printout, the 95% forecast intervals are:

1985 (30.0704, 34.2740)
1986 (31.7193, 36.0335)

15.13 a. The SAS printout for this problem is:

DEP VARIABLE: Y
ANALYSIS OF VARIANCE

SOURCE	DF	SUM OF SQUARES	MEAN SQUARE	F VALUE	PROB>F
MODEL	1	167208.30	167208.30	217.274	0.0001
ERROR	23	17700.17827	769.57297		
C TOTAL	24	184908.48			

ROOT MSE	27.74118	R-SQUARE	0.9043
DEP MEAN	133.38	ADJ R-SQ	0.9001
C.V.	20.7986		

PARAMETER ESTIMATES

VARIABLE	DF	PARAMETER ESTIMATE	STANDARD ERROR	T FOR H0: PARAMETER =0	PROB > |T|
INTERCEP	1	-2.71384615	10.77162590	-0.252	0.8033
T	1	11.34115385	0.76940185	14.740	0.0001

OBS	ACTUAL	PREDICT VALUE	STD ERR PREDICT	LOWER95% PREDICT	UPPER95% PREDICT	RESIDUAL
1	42.4000	-2.7138	10.7716	-64.2746	58.8469	45.1138
2	44.8000	8.6273	10.1199	-52.4584	69.7130	36.1727
3	47.4000	19.9685	9.4858	-40.6803	80.6172	27.4315
4	49.5000	31.3096	8.8732	-28.9410	91.5602	18.1904
5	54.3000	42.6508	8.2867	-17.2414	102.5	11.6492
6	58.4000	53.9919	7.7324	-5.5822	113.6	4.4081
7	60.4000	65.3331	7.2176	6.0360	124.6	-4.9331
8	63.3000	76.6742	6.7515	17.6126	135.7	-13.3742
9	69.3000	88.0154	6.3447	29.1471	146.9	-18.7154
10	75.7000	99.3565	6.0092	40.6391	158.1	-23.6565
11	80.6000	110.7	5.7577	52.0882	169.3	-30.0977
12	92.3000	122.0	5.6013	63.4942	180.6	-29.7388
13	105.4	133.4	5.5482	74.8570	191.9	-27.9800
14	114.6	144.7	5.6013	86.1765	203.3	-30.1212
15	117.9	156.1	5.7577	97.4528	214.7	-38.1623
16	129.4	167.4	6.0092	108.7	226.1	-38.0035
17	155.2	178.7	6.3447	119.9	237.6	-23.5446
18	179.3	190.1	6.7515	131.0	249.1	-10.7858
19	198.1	201.4	7.2176	142.1	260.7	-3.3269
20	219.4	212.8	7.7324	153.2	272.3	6.6319
21	236.6	224.1	8.2867	164.2	284.0	12.4908
22	261.5	235.5	8.8732	175.2	295.7	26.0496
23	267.3	246.8	9.4858	186.1	307.4	20.5085
24	291.9	258.1	10.1199	197.0	319.2	33.7673
25	319.5	269.5	10.7716	207.9	331.0	50.0262
26	.	280.8	11.4380	218.7	342.9	.
27	.	292.2	12.1166	229.5	354.8	.
28	.	303.5	12.8054	240.3	366.7	.

The fitted model is $\hat{Y}_t = -2.7138 + 11.3412t$

b.   The forecasts for 1985 to 1987 are:

1985 $\hat{Y}_{1985} = -2.7138 + 11.3412(25) = 280.8162$
1986 $\hat{Y}_{1986} = -2.7138 + 11.3412(26) = 292.1574$
1987 $\hat{Y}_{1987} = -2.7138 + 11.3412(27) = 303.4986$

The formula for the prediction interval is:

$$\hat{Y}_t \pm t_{\alpha/2} \sqrt{MSE \left[1 + \frac{1}{n} + \frac{(t - \bar{t})^2}{SS_{tt}} \right]}$$

$$\text{where } \bar{t} = \frac{0 + 1 + 2 + \cdots + 24}{25} = \frac{300}{25} = 12$$

$$SS_{tt} = \sum t^2 - \frac{(\sum t)^2}{n} = 4900 - \frac{300^2}{25} = 1300$$

For confidence coefficient .95, $\alpha = 1 - .95 = 05$ and $\alpha/2 = .05/2 = .025$. From Table VI, Appendix B, $t_{.025} = 2.069$ with df $= n - 2 = 25 - 2 = 23$.

The prediction intervals are:

1985 $280.8162 \pm 2.069 \sqrt{769.57297 \left[1 + \frac{1}{25} + \frac{(25 - 12)^2}{1300} \right]}$

$\Rightarrow 280.8162 \pm 62.0838 \Rightarrow (218.7324, 342.9000)$

1986 $292.1574 \pm 2.069 \sqrt{769.57297 \left[1 + \frac{1}{25} + \frac{(26 - 12)^2}{1300} \right]}$

$\Rightarrow 292.1574 \pm 62.6324 \Rightarrow (229.5250, 354.7898)$

1987 $303.4986 \pm 2.069 \sqrt{769.57297 \left[1 + \frac{1}{25} + \frac{(27 - 12)^2}{1300} \right]}$

$\Rightarrow 303.4986 \pm 63.2164 \Rightarrow (240.2822, 366.7150)$

15.15   a.   The SAS printout for this problem is shown as follows:

Dependent Variable: Y

Analysis of Variance

Source	DF	Sum of Squares	Mean Square	F Value	Prob>F
Model	1	3903056055.9	3903056055.9	2764.036	0.0001
Error	20	28241719.581	1412085.9791		
C Total	21	3931297775.5			

Root MSE	1188.31224	R-square	0.9928
Dep Mean	107685.54545	Adj R-sq	0.9925
C.V.	1.10350		

Parameter Estimates

Variable	DF	Parameter Estimate	Standard Error	T for H0: Parameter=0	Prob > $\mid$T$\mid$
INTERCEP	1	83542	524.48230116	159.284	0.0001
T	1	2099.462451	39.93339993	52.574	0.0001

Durbin-Watson D	0.302
(For Number of Obs.)	22
1st Order Autocorrelation	0.694

Obs	Dep Var Y	Predict Value	Std Err Predict	Lower95% Predict	Upper95% Predict	Residual
1	84889.0	85641.2	489.897	82960.0	88322.4	-752.2
2	86355.0	87740.7	456.186	85085.5	90395.8	-1385.7
3	88847.0	89840.1	423.558	87208.6	92471.6	-993.1
4	91203.0	91939.6	392.283	89329.2	94549.9	-736.6
5	93670.0	94039.0	362.713	91447.4	96630.7	-369.0
6	95453.0	96138.5	335.298	93562.9	98714.1	-685.5
7	97826.0	98238.0	310.609	95675.9	100800	-412.0
8	100665	100337	289.345	97786.2	102889	327.6
9	103882	102437	272.309	99893.9	104980	1445.1
10	106559	104536	260.334	101999	107074	2022.6
11	108544	106636	254.135	104101	109171	1908.2
12	110315	108735	254.135	106200	111270	1579.7
13	111872	110835	260.334	108297	113372	1037.3
14	113226	112934	272.309	110391	115477	291.8
15	115241	115034	289.345	112482	117585	207.3
16	117167	117133	310.609	114571	119695	33.8735
17	119540	119233	335.298	116657	121808	307.4
18	121602	121332	362.713	118740	123924	269.9
19	123378	123432	392.283	120821	126042	-53.5138
20	125557	125531	423.558	122899	128163	26.0237
21	126424	127630	456.186	124975	130286	-1206.4
22	126867	129730	489.897	127049	132411	-2862.9
23	.	131829	524.482	129120	134539	.
24	.	133929	559.780	131189	136669	.

Sum of Residuals	0
Sum of Squared Residuals	28241719.581
Predicted Resid SS (Press)	35491467.336

The fitted regression model is $\hat{Y}_t = 83{,}542 + 2099.46t$

b.   The forecasts for 1992 and 1993 are:

$$\hat{Y}_{1992} = 83{,}542 + 2099.46(23) + 131{,}829.58$$
$$\hat{Y}_{1993} = 83{,}542 + 2099.46(24) + 133{,}929.04$$

c.   The formula for the prediction interval is:

$$\hat{Y}_t \pm t_{\alpha/2}\sqrt{\text{MSE}\left[1 + \frac{1}{n} + \frac{(t - \bar{t})^2}{SS_{tt}}\right]}$$

where $\bar{t} = \dfrac{1 + 2 + \cdots + 22}{22} = 11.5$

$$SS_{tt} = \sum t^2 - \frac{\left(\sum t\right)^2}{n} = 3795 - \frac{253^2}{22} = 885.5$$

For confidence coefficient .95, $\alpha = 1 - .95 = .05$ and $\alpha/2 = .05/2 = .025$. From Table IV, Appendix B, $t_{.025} = 2.086$ with df $= n - 2 = 22 - 2 = 20$.

The prediction intervals are:

1985 $131{,}829.6 \pm 2.086\sqrt{1{,}412{,}085.98\left[1 + \dfrac{1}{22} + \dfrac{(23 - 11.5)^2}{885.5}\right]}$

$$\Rightarrow 131{,}829.6 \pm 2709.5 \Rightarrow (129{,}120.1, 134{,}538.1)$$

1986 $133{,}929.0 \pm 2.086\sqrt{1{,}412{,}085.98\left[1 + \dfrac{1}{22} + \dfrac{(24 - 11.5)^2}{885.5}\right]}$

$$\Rightarrow 133{,}929.0 \pm 2740.1 \Rightarrow (131{,}188.9, 136{,}669.1)$$

15.17 a. $d = 3.9$ indicates the residuals are very strongly negatively autocorrelated.

b. $d = .2$ indicates the residuals are very strongly positively autocorrelated.

c. $d = 1.99$ indicates the residuals are probably uncorrelated.

15.19 a.   The fitted model is $\hat{Y}_t = 3.3177 - .0844t$

The residuals are calculated using $\hat{R}_t = Y_t - \hat{Y}_t$, and appear in the table:

	Actual	Predicted Value	Residual
1	2.9940	3.2293	-0.2353
2	2.9940	3.1409	-0.1469
3	2.9850	3.0525	-0.0675
4	2.9940	2.9641	0.0299
5	2.9850	2.8757	0.1093
6	2.9330	2.7873	0.1457
7	2.8410	2.6989	0.1421
8	2.8090	2.6106	0.1984
9	2.7320	2.5222	0.2098
10	2.6320	2.4338	0.1982
11	2.5450	2.3454	0.1996
12	2.4690	2.2570	0.2120
13	2.3920	2.1686	0.2234
14	2.1930	2.0802	0.1128
15	1.9010	1.9918	-0.0908
16	1.7180	1.9034	-0.1854
17	1.6450	1.8150	-0.1700
18	1.5460	1.7266	-0.1806
19	1.4330	1.6382	-0.2052
20	1.2890	1.5498	-0.2608
21	1.1360	1.4615	-0.3255
22	1.0410	1.3731	-0.3321
23	1.0000	1.2847	-0.2847
24	0.9840	1.1963	-0.2123
25	0.9640	1.1079	-0.1439
26	0.9550	1.0195	-0.0645
27	0.9690	0.9311	0.0379
28	0.9490	0.8427	0.1063
29	0.9260	0.7543	0.1717
30	0.8800	0.6659	0.2141
31	0.8390	0.5775	0.2615
32	0.8220	0.4891	0.3329

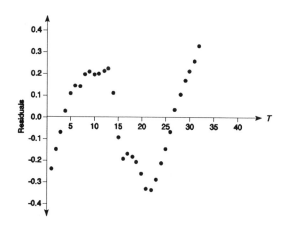

There is a tendency for the residuals to have long positive runs and negative runs. Residuals 4 through 14 are positive, while residuals 15 through 26 are negative. This indicates the error terms are correlated.

b. $$d = \frac{\sum_{t=2}^{n} (\hat{R}_t - \hat{R}_{t-1})^2}{\sum_{t=1}^{n} \hat{R}_t^2} \qquad \text{where } \hat{R}_t = \text{residual for time } t$$

$$= \frac{\left(-.1469 - (-.2353)\right)^2 + \left(-.0675 - (-.1469)\right)^2 + \left(.0299 - (-.0675)\right)^2 + \cdots + \left(.3329 - .2615\right)^2}{(-.2353)^2 + (-.1469)^2 + \cdots + (.3329)^2}$$

$$= \frac{.152931}{1.25455} = .122$$

To determine if the time series residuals are autocorrelated, we test:

H_0: No first-order autocorrelation of residuals
H_a: Positive or negative first-order autocorrelation of residuals

The test statistic is $d = .122$.

For $\alpha = .10$, the rejection region is $d < d_{L,\alpha/2} = d_{L,.05} = 1.37$ or $(4 - d) < d_{L,.05} = 1.37$. The value $d_{L,.05}$ is found in Table XIV, Appendix B, with $k = 1$, $n = 32$, and $\alpha = .10$.

Since the observed value of the test statistic falls in the rejection region ($d = .122 < 1.37$), H_0 is rejected. There is sufficient evidence to indicate the time series residuals are autocorrelated at $\alpha = .10$.

c. We must assume the residuals are normally distributed.

. . . .
Time Series: Models and Forecasting

15.21 a. We first compute the exponentially smoothed values for the years 1975–1991 for both series.

For the agricultural data with $w = .5$,

$$E_1 = Y_1 = 3408$$
$$E_2 = .5Y_2 + (1 - .5)E_1 = .5(3331) + .5(3408) = 3369.5$$
$$E_3 = .5Y_3 + (1 - .5)E_2 = .5(3283) + .5(3369.5) = 3326.25$$

The rest of the values appear in the table.

Similarly, we compute the values for the nonagricultural data:

$$E_1 = Y_1 = 82,438$$
$$E_2 = .5Y_2 + (1 - .5)E_1 = .5(85,421) + .5(82,438) = 83,929.5$$
$$E_3 = .5Y_3 + (1 - .5)E_2 = .5(88,734) + .5(83,929.5) = 86,331.75$$

Year	Agricultural	Exponentially Smoothed Values ($w = .5$)	Non-Agricultral	Exponentially Smoothed Values ($w = .5$)
1975	3408	3408.0	82,438.0	82,438.0
1976	3331	3369.5	85,421.0	83,929.5
1977	3283	3326.3	88,734.0	86,331.8
1978	3387	3356.6	92,661.0	89,496.4
1979	3347	3351.8	95,477.0	92,486.7
1980	3364	3357.9	95,938.0	94,212.3
1981	3368	3363.0	97,030.0	95,621.2
1982	3401	3382.0	96,125.0	95,873.1
1983	3383	3382.5	97,450.0	96,661.5
1984	3321	3351.7	101,685.0	99,173.3
1985	3179	3265.4	103,971.0	101,572.1
1986	3163	3214.2	106,434.0	104,003.1
1987	3208	3211.1	109,232.0	106,617.5
1988	3169	3190.0	111,800.0	109,208.8
1989	3199	3194.5	114,142.0	111,675.4
1990	3186	3190.3	114,728.0	113,201.7
1991	3233	3211.6	113,644.0	113,422.8

The forecasts for 1992 are:

Agricultural $\quad F_{1992} = F_{t+1} = E_t = 3211.6$

Nonagricultural $\quad F_{1992} = F_{t+1} = E_t = 113,422.8$

b.   We first compute the Holt-Winters values for years 1975–1991.

For the agricultural data with $w = .5$ and $v = .5$,

$$E_2 = Y_2 = 3331$$
$$E_3 = .5Y_3 + (1 - .5)(E_2 + T_2)$$
$$= .5(3283) + .5(3331 - 77)$$
$$= 3268.5$$

$$T_2 = Y_2 - Y_1 = 3331 - 3408 = -77$$
$$T_3 = .5(E_3 - E_2) + (1 - .5)(T_2)$$
$$= .5(3268.5 - 3331) + .5(-77)$$
$$= -69.75$$

The rest of the values appear in the table.

For the nonagricultural data with $w = .5$ and $v = .5$,

$$E_2 = Y_2 = 85,421$$
$$E_3 = .5Y_3 + (1 - .5)(E_2 + T_2)$$
$$= .5(88,734) + .5(85,421 + 2983)$$
$$= 88,569$$

$$T_2 = Y_2 - Y_1 = 85,421 - 82,438 = 2983$$
$$T_3 = .5(E_3 - E_2) + (1 - .5)T_2$$
$$= .5(88,569 - 85,421) + .5(2983) = 3065.5$$
$$= 88,569$$

The rest of the values appear in the table.

Year	Agricultural	E_t $w = .5$ $v = .5$	T_t $w = .5$ $v = .5$	Non-Agricultural	E_t $w = .5$ $v = .5$	T_t $w = .5$ $v = .5$
1975	3408			82,438.0		
1976	3331	3331.0	−77.0	85,421.0	85,421.0	2,983.0
1977	3283	3268.5	−69.8	88,734.0	88,569.0	3,065.5
1978	3387	3292.9	−22.7	92,661.0	92,147.8	3,322.1
1979	3347	3308.6	−3.5	95,477.0	95,473.4	3,323.9
1980	3364	3334.6	11.2	95,938.0	97,367.7	2,609.1
1981	3368	3356.9	16.8	97,030.0	98,503.4	1,872.4
1982	3401	3387.3	23.6	96,125.0	98,250.4	809.7
1983	3383	3397.0	16.6	97,450.0	98,255.0	407.2
1984	3321	3367.3	−6.5	101,685.0	100,173.6	1,162.9
1985	3179	3269.9	−52.0	103,971.0	102,653.7	1,821.5
1986	3163	3190.5	−65.7	106,434.0	105,454.6	2,311.2
1987	3208	3166.4	−44.9	109,232.0	108,498.9	2,677.7
1988	3169	3145.2	−33.0	111,800.0	111,488.3	2,833.6
1989	3199	3155.6	−11.3	114,142.0	114,232.0	2,788.6
1990	3186	3165.1	−0.9	114,728.0	115,874.3	2,215.5
1991	3233	3198.6	16.3	113,644.0	115,866.9	1,104.0

The forecasts for 1992 are:

Agricultural $F_{1992} = F_{t+1} = E_t + T_t = 3198.6 + 16.3 = 3214.9$
Nonagricultural $F_{1992} = F_{t+1} = E_t + T_t = 115,866.9 + 1,104.0 = 116,970.9$

15.23   a.   We first calculate the exponentially smoothed values for 1973–1986.

$$E_1 = Y_1 = 49.875$$
$$E_2 = .8Y_1 + (1 - .8)E_1 = .8(50.750) + .2(49.875) = 50.575$$
$$E_3 = .8Y_2 + (1 - .8)E_2 = .8(41.250) + .2(50.575) = 43.115$$

The rest of the values appear in the table.

Year	Closing Price	Exponentially Smoothed Value $(w = .8)$
1973	49.875	49.875
1974	50.750	50.575
1975	41.250	43.115
1976	49.125	47.923
1977	56.500	54.785
1978	33.750	37.957
1979	41.125	40.491
1980	56.500	53.298
1981	27.000	32.260
1982	38.750	37.452
1983	45.250	43.690
1984	41.750	42.138
1985	68.375	63.128
1986	45.625	49.126

The forecasts for 1987 and 1988 are:

$$F_{1987} = F_{t+1} = E_t = 49.126$$
$$F_{1988} = F_{t+2} = F_{t+1} = 49.126$$

The expected gain is $F_{1988} - Y_{1986} = 49.126 - 45.625 = 3.501$

b.   We first calculate the Holt-Winters values for 1973–1986.

For $w = .8$ and $v = .5$,

$$E_2 = Y_2 = 50.750$$
$$E_3 = .8Y_3 + (1 - .8)(E_2 + T_2)$$
$$= .8(41.25) + .2(50.750 + .875)$$
$$= 43.325$$

$$T_2 = Y_2 - Y_1 = 50.750 - 49.875 = .875$$
$$T_3 = .5(E_3 - E_2) + (1 - .5)(T_2)$$
$$= .5(43.325 - 50.750) + .5(.875)$$
$$= -3.275$$

The rest of the values appear in the table.

Year	Closing Price	E_t $w = .8$ $v = .5$	T_t $w = .8$ $v = .5$
1973	49.875		
1974	50.750	50.750	0.875
1975	41.250	43.325	-3.275
1976	49.125	47.310	0.355
1977	56.500	54.733	3.889
1978	33.750	38.724	-6.060
1979	41.125	39.433	-2.676
1980	56.500	52.551	5.221
1981	27.000	33.155	-7.088
1982	38.750	36.213	-2.014
1983	45.250	43.040	2.406
1984	41.750	42.489	0.928
1985	68.375	63.383	10.911
1986	45.625	51.359	-0.557

The forecasts for 1987 and 1988 are:

$$F_{1987} = F_{t+1} = E_t + T_t = 51.359 - .557 = 50.802$$
$$F_{1988} = F_{t+2} = E_t + 2T_t = 51.359 + 2(-.557) = 50.245$$

The expected gain is $F_{1988} - Y_{1986} = 50.245 - 45.625 = 4.62$

15.31 The fitted regression model is

$$\hat{Y} = 45.7473 + .0492t$$

The residuals are:

	Actual	Predict Value	Residual
1	49.8750	45.7964	4.0786
2	50.7500	45.8456	4.9044
3	41.2500	45.8948	-4.6448
4	49.1250	45.9440	3.1810
5	56.5000	45.9931	10.5069
6	33.7500	46.0423	-12.2923
7	41.1250	46.0915	-4.9665
8	56.5000	46.1407	10.3593
9	27.0000	46.1898	-19.1898
10	38.7500	46.2390	-7.4890
11	45.2500	46.2882	-1.0382
12	41.7500	46.3374	-4.5874
13	68.3750	46.3865	21.9885
14	45.6250	46.4357	-0.8107

The Durbin-Watson statistic is

$$d = \frac{\sum\limits_{t=2}^{n} (\hat{R}_t - \hat{R}_{t-1})^2}{\sum \hat{R}_t^2}$$

$$= \frac{(4.9044 - 4.0786)^2 + (-4.6448 - 4.9044)^2 + \cdots + (-.8107 - 21.9855)^2}{4.0786^2 + 4.9044^2 + \cdots + (-.8107)^2}$$

$$= \frac{3305.482}{1396.469} = 2.367$$

To determine if the time series residuals are autocorrelated, we test:

H_0: No first-order autocorrelation of residuals
H_a: Positive or negative first-order autocorrelation of residuals

The test statistic is $d = 2.367$.

For $\alpha = .10$, the rejection region is $d < d_{L,\alpha/2} = d_{L,.05} \approx 1.08$ or $(4 - d) < d_{L,.05} \approx 1.08$. The value $d_{L,.05}$ is found in Table XIV, Appendix B, with $k = 1$, $n = 14$ and $\alpha = .10$. Since $n = 14$ is not in the table, we will use $n = 15$.

Since the observed value of the test statistic does not fall in the rejection region ($d = 2.367$ $\not< 1.08$ and $4 - d = 4 - 2.367 = 1.633 \not< 1.08$), H_0 is not rejected. There is insufficient evidence to indicate positive or negative first-order autocorrelation of the residuals at $\alpha = .10$.

15.27 a. The following model is fit: $E(Y_t) = \beta_0 + \beta_1 t$. The SAS output is:

```
Dependent Variable: Y
                  Analysis of Variance
                        Sum of        Mean
    Source      DF      Squares      Square      F Value     Prob>F

    Model        1  80032458.232  80032458.232   12760.880   0.0001
    Error       58   363758.80211   6271.70348
    C Total     59  80396217.034

        Root MSE      79.19409    R-square    0.9955
        Dep Mean    3535.99000    Adj R-sq    0.9954
        C.V.            .23966


                        Parameter Estimates
                      Parameter    Standard   T for H0:
    Variable  DF      Estimate       Error   Parameter=0    Prob > |T|

    INTERCEP   1    1501.961695   20.70612401     72.537      0.0001
    T          1      66.689453    0.59035988    112.964      0.0001
```

Obs	Dep Var Y	Predict Value	Std Err Predict	Lower95% Predict	Upper95% Predict	Residual
1	1717.8	1568.7	20.195	1405.1	1732.2	149.1
2	1746.4	1635.3	19.688	1472.0	1798.7	111.1
3	1779.9	1702.0	19.186	1538.9	1865.1	77.8699
4	1829.6	1768.7	18.689	1605.8	1931.6	60.8805
5	1881.7	1835.4	18.198	1672.8	1998.1	46.2910
6	1952.9	1902.1	17.712	1739.7	2064.5	50.8016
.	.	.	.	.	.	.
.	.	.	.	.	.	.
.	.	.	.	.	.	.
49	4752.4	4769.7	14.960	4608.4	4931.1	-17.3449
50	4857.2	4836.4	15.397	4674.9	4997.9	20.7657
51	4947.3	4903.1	15.843	4741.5	5064.8	44.1762
52	5044.6	4969.8	16.298	4808.0	5131.7	74.7868
53	5139.9	5036.5	16.762	4874.5	5198.5	103.4
54	5218.5	5103.2	17.234	4941.0	5265.4	115.3
55	5277.3	5169.9	17.712	5007.4	5332.3	107.4
56	5340.4	5236.6	18.198	5073.9	5399.2	103.8
57	5422.4	5303.3	18.689	5140.4	5466.1	119.1
58	5504.7	5369.9	19.186	5206.8	5533.1	134.8
59	5570.5	5436.6	19.688	5273.3	5600.0	133.9
60	5557.5	5503.3	20.195	5339.7	5666.9	54.1711
61	.	5570.0	20.706	5406.2	5733.9	.
62	.	5636.7	21.222	5472.6	5800.8	.
63	.	5703.4	21.741	5539.0	5867.8	.
64	.	5770.1	22.263	5605.4	5934.8	.

The estimated regression line is $\hat{Y}_t = 1501.9617 + 66.6895t$

From the printout, the 1991 quarterly GDP forecasts are:

		Forecast	95% Lower Limit	95% Upper Limit
1991	Q1	5570.0	5406.2	5733.9
	Q2	5636.7	5472.6	5800.8
	Q3	5703.4	5539.0	5867.8
	Q4	5770.1	5605.4	5934.8

b. The following model is fit: $E(Y_t) = \beta_0 + \beta_1 t + \beta_2 Q_1 + \beta_3 Q_2 + \beta_4 Q_3$

where $Q_1 = \begin{cases} 1 & \text{if quarter 1} \\ 0 & \text{otherwise} \end{cases}$ $Q_2 = \begin{cases} 1 & \text{if quarter 2} \\ 0 & \text{otherwise} \end{cases}$

$Q_3 = \begin{cases} 1 & \text{if quarter 3} \\ 0 & \text{otherwise} \end{cases}$

The SAS printout is:

Dependent Variable: Y

Analysis of Variance

Source	DF	Sum of Squares	Mean Square	F Value	Prob>F
Model	4	80033235.398	20008308.849	3031.715	0.0001
Error	55	362981.63643	6599.66612		
C Total	59	80396217.034			

Root MSE	81.23833	R-square	0.9955
Dep Mean	3535.99000	Adj R-sq	0.9952
C.V.	2.29747		

Time Series: Models and Forecasting

Parameter Estimates

Variable	DF	Parameter Estimate	Standard Error	T for H0: Parameter=0	Prob > |T|
INTERCEP	1	1495.579048	28.58498765	52.320	0.0001
T	1	66.701696	0.60686484	109.912	0.0001
Q1	1	9.131756	29.71985947	0.307	0.7598
Q2	1	8.443393	29.68886360	0.284	0.7772
Q3	1	6.461696	29.67025054	0.218	0.8284

Durbin-Watson D	0.155
(For Number of Obs.)	60
1st Order Autocorrelation	0.888

Obs	Dep Var Y	Predict Value	Std Err Predict	Lower95% Predict	Upper95% Predict	Residual
1	1717.8	1571.4	26.995	1399.9	1743.0	146.4
2	1746.4	1637.4	26.995	1465.9	1809.0	109.0
3	1779.9	1702.1	26.995	1530.6	1873.7	77.7542
4	1829.6	1762.4	26.995	1590.8	1933.9	67.2142
5	1881.7	1838.2	25.536	1667.6	2008.9	43.4807
6	1952.9	1904.2	25.536	1733.6	2074.9	48.6674
.	.	.	.	.	.	.
.	.	.	.	.	.	.
.	.	.	.	.	.	.
49	4752.4	4773.1	24.234	4603.2	4943.0	-20.6939
50	4857.2	4839.1	24.234	4669.2	5009.0	18.0927
51	4947.3	4903.8	24.234	4733.9	5073.7	43.4727
52	5044.6	4964.1	24.234	4794.2	5134.0	80.5327
53	5139.9	5039.9	25.536	4869.2	5210.6	99.9993
54	5218.5	5105.9	25.536	4935.3	5276.6	112.6
55	5277.3	5170.6	25.536	5000.0	5341.3	106.7
56	5340.4	5230.9	25.536	5060.2	5401.5	109.5
57	5422.4	5306.7	26.995	5135.1	5478.3	115.7
58	5504.7	5372.7	26.995	5201.2	5544.3	132.0
59	5570.5	5437.4	26.995	5265.9	5609.0	133.1
60	5557.5	5497.7	26.995	5326.1	5669.2	59.8192
61	.	5573.5	28.585	5400.9	5746.1	.
62	.	5639.5	28.585	5466.9	5812.1	.
63	.	5704.2	28.585	5531.7	5876.8	.
64	.	5764.5	28.585	5591.9	5937.1	.

The fitted model is

$$\hat{Y} = 1495.5790 + 66.7017t + 9.1318Q_1 + 8.4434Q_2 + 6.4617Q_3$$

To determine whether the data indicate a significant seasonal component, we test:

H_0: $\beta_2 = \beta_3 = \beta_4 = 0$
H_a: At least one $\beta_i \neq 0$, $i = 2, 3, 4$

The test statistic is $F = \dfrac{(SSE_r - SSE_c)/(k - g)}{SSE_c/[n - (k + 1)]}$

$$= \frac{(363758.80211 - 362981.63643)/(4 - 1)}{362981.63643/[60 - (4 + 1)]} = \frac{259.0552333}{6599.666116} = .039$$

The rejection region requires $\alpha = .05$ in the upper tail of the F distribution with numerator df $= k - g = 4 - 1 = 3$ and denominator df $= n - (k + 1) = 60 - (4 + 1) = 55$. From Table VIII, Appendix B, $F_{.05} \approx 3.23$. The rejection region is $F > 3.23$.

Since the observed value of the test statistic does not fall in the rejection ($F = .039 \not> 3.23$), H_0 is not rejected. There is insufficient evidence to indicate a seasonal component at $\alpha = .05$. This supports the assertion that the data have been seasonally adjusted.

c. Using the printout, the 1991 quarterly GDP forecasts are:

		Forecast
1991	Q_1	5573.5
	Q_2	5639.5
	Q_3	5704.2
	Q_4	5764.5

d. From the printout provided in part **b**, part of the residuals are reported. The Durbin-Watson d is .155 (from the printout).

To determine if the time series residuals are autocorrelated, we test:

H_0: No first-order autocorrelation of residuals
H_a: Positive or negative first-order autocorrelation of residuals

The test statistic is $d = .155$.

For $\alpha = .02$, the rejection region is $d < d_{L,\alpha/2} = d_{L,.01} = 1.28$ or $(4 - d) < d_{L,.01} = 1.28$. The value $d_{L,.01}$ is found in Table XIV, Appendix B, with $k = 4$, $n = 60$, and $\alpha = .02$.

Since the observed value of the test statistic falls in the rejection region ($d = .155 < 1.28$), H_0 is rejected. There is sufficient evidence to indicate the time series residuals are autocorrelated at $\alpha = .02$. (If we reject H_0 for $\alpha = .02$, we will reject H_0 for $\alpha = .05$).

15.29 a. The following model is fit: $E(Y_t) = \beta_0 + \beta_1 t$,

where $t = 0$ corresponds to 1945,
        $t = 5$ corresponds to 1950, etc.

The SAS printout is:

DEP VARIABLE: YP
ANALYSIS OF VARIANCE

SOURCE	DF	SUM OF SQUARES	MEAN SQUARE	F VALUE	PROB>F
MODEL	1	326.56177	326.56177	136.393	0.0001
ERROR	8	19.15422711	2.39427839		
C TOTAL	9	345.71600			

ROOT MSE	1.547346	R-SQUARE	0.9446
DEP MEAN	48.08	ADJ R-SQ	0.9377
C.V.	3.218273		

PARAMETER ESTIMATES

VARIABLE	DF	PARAMETER ESTIMATE	STANDARD ERROR	T FOR HO: PARAMETER =0	PROB > ¦T¦
INTERCEP	1	36.66659341	1.09293555	33.549	0.0001
T	1	0.69172161	0.05922922	11.679	0.0001

OBS	ACTUAL	PREDICT VALUE	STD ERR PREDICT	LOWER95% PREDICT	UPPER95% PREDICT	RESIDUAL
1	34.3000	36.6666	1.0929	32.2980	41.0352	-2.3666
2	43.4000	40.1252	0.8387	36.0666	44.1838	3.2748
3	42.5000	43.5838	0.6226	39.7376	47.4301	-1.0838
4	48.1000	47.0424	0.4973	43.2944	50.7904	1.0576
5	50.9000	50.5010	0.5314	46.7282	54.2738	0.3990
6	51.0000	51.1927	0.5572	47.4002	54.9853	-0.1927
7	51.5000	51.8845	0.5878	48.0674	55.7015	-0.3845
8	52.4000	52.5762	0.6226	48.7299	56.4224	-0.1762
9	52.9000	53.2679	0.6609	49.3879	57.1480	-0.3679
10	53.8000	53.9596	0.7021	50.0413	57.8780	-0.1596
11	.	54.6514	0.7457	50.6904	58.6123	.
12	.	55.3431	0.7913	51.3353	59.3508	.
13	.	56.0348	0.8387	51.9762	60.0934	.
14	.	56.7265	0.8875	52.6131	60.8399	.
15	.	57.4182	0.9374	53.2463	61.5902	.

The fitted model is $\hat{Y}_t = 36.67 + .692t$.

b.   The plot of the data and fitted regression line is shown using SAS.

Plot of YP*T   Symbol used is *
Plot of YHAT*T   Symbol used is +

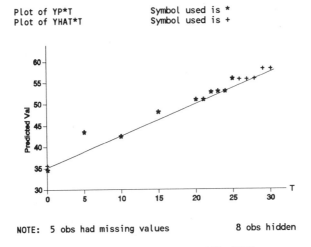

NOTE: 5 obs had missing values   8 obs hidden

c.   Using the printout, the forecasts for 1971–1975 are:

	Forecasts
1971	54.6514
1972	55.3431
1973	56.0348
1974	56.7265
1975	57.4182

d.   The following model is fit: $E(Y_t) = \beta_0 + \beta_1 t$ where Y_t is the market share of secondary lenders. The SAS output is:

```
DEP VARIABLE: YS
ANALYSIS OF VARIANCE
```

SOURCE	DF	SUM OF SQUARES	MEAN SQUARE	F VALUE	PROB>F
MODEL	1	351.65130	351.65130	145.245	0.0001
ERROR	8	19.36870330	2.42108791		
C TOTAL	9	371.02000			

ROOT MSE	1.555985	R-SQUARE	0.9478	
DEP MEAN	50.6	ADJ R-SQ	0.9413	
C.V.	3.075068			

```
PARAMETER ESTIMATES
```

VARIABLE	DF	PARAMETER ESTIMATE	STANDARD ERROR	T FOR HO: PARAMETER =0	PROB > |T|
INTERCEP	1	62.44373626	1.09903750	56.817	0.0001
T	1	-0.71780220	0.05955990	-12.052	0.0001

The fitted model is $\hat{Y}_t = 62.444 - .178t$.

The forecast for market share of secondary lenders for 1975 is

$$\hat{Y}_{1975} = 62.444 - .718(30) = 40.904$$

The forecasts are:

Year	Primary	Secondary
1955 (actual)	42.5	56.4
1975 (forecast)	57.42	40.90

The projections yield the same conclusions.

15.31 a.

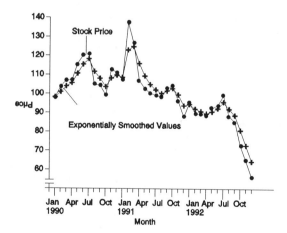

b.   The forecasts are:

$$Y_{1992,J} = F_{t+1} = E_t = 64.491$$
$$Y_{1992,F} = F_{t+2} = F_{t+1} = 64.491$$
$$Y_{1992,M} = F_{t+3} = F_{t+1} = 64.491$$

Time Series: Models and Forecasting

CHAPTER SIXTEEN

· ·

Design of Experiments and Analysis of Variance

16.1   The treatments are the factor levels A, B, C, and D.

16.3   a.   College GPA's are measured on college students. The experimental units are college students.

   b.   Unemployment rate is measured by counting the number of unemployed workers and dividing by the total number of workers. The experimental units are the workers of the state.

   c.   Gasoline mileage is measured on automobiles. The experimental units are the automobiles of a particular model.

   d.   The experimental units are the sectors on a computer diskette.

16.5   a.   The response is the debt-to-equity ratio measured on each company.

   b.   The factor is the type of company, which is qualitative.

   c.   The treatments are the 4 types of companies: insurance, publishing, electric utility, and banking.

   d.   The experimental units are the companies.

16.7   a.   The response is the quality of a steel ingot on a scale from 0 to 10.

   b.   There are two factors—temperature and pressure. Both factors are quantitative.

   c.   There are $3 \times 5 = 15$ treatments. Each treatment is a combination of a temperature and a pressure. One example is temperature 1100 and pressure 500.

   d.   The experimental units are the steel ingots.

16.9   a.   From Table VIII with $\nu_1 = 4$ and $\nu_2 = 4$, $F_{.05} = 6.39$.

   b.   From Table X with $\nu_1 = 4$ and $\nu_2 = 4$, $F_{.01} = 15.98$.

   c.   From Table VII with $\nu_1 = 30$ and $\nu_2 = 40$, $F_{.10} = 1.54$.

   d.   From Table IX with $\nu_1 = 15$, and $\nu_2 = 12$, $F_{.025} = 3.18$.

· ·

16.11 In dot diagram **b**, the difference between the sample means is small relative to the variability within the sample observations. The observations for Sample 1 overlap those of Sample 2. In diagram **a**, no observations from Sample 1 overlap any observations from Sample 2, implying the variation within the sample observations is small relative to the difference between the sample means.

16.13 Some preliminary calculations are:

For diagram **a**:

$$\bar{y}_1 = \frac{\sum y_1}{n_1} = \frac{54}{6} = 9 \qquad s_1^2 = \frac{\sum y_1^2 - \frac{\left(\sum y_1\right)^2}{n_1}}{n_1 - 1} = \frac{496 - \frac{54^2}{6}}{6 - 1} = \frac{10}{5} = 2$$

$$\bar{y}_2 = \frac{\sum y_2}{n_2} = \frac{84}{6} = 14 \qquad s_2^2 = \frac{\sum y_2^2 - \frac{\left(\sum y_2\right)^2}{n_2}}{n_2 - 1} = \frac{1186 - \frac{84^2}{6}}{6 - 1} = \frac{10}{5} = 2$$

$$s_p^2 = \frac{(n_1 - 1)s_1^2 + (n_2 - 1)s_2^2}{n_1 + n_2 - 2} = \frac{(6 - 1)2 + (6 - 1)2}{6 + 6 - 2} = \frac{20}{10} = 2$$

From Exercise 16.12, MSE $= \dfrac{20}{10} = 2$

To determine if the two treatment means are different, we test:

H_0: $\mu_1 - \mu_2 = 0$
H_a: $\mu_1 - \mu_2 \neq 0$

The test statistic is:

$$t = \frac{(\bar{y}_1 - \bar{y}_2) - 0}{\sqrt{s_p^2\left[\frac{1}{n_1} + \frac{1}{n_2}\right]}} = \frac{9 - 14 - 0}{\sqrt{2\left[\frac{1}{6} + \frac{1}{6}\right]}} = \frac{-5}{.8165} = -6.12$$

From Exercise 16.12, $F = 37.5$, $t^2 = (-6.12)^2 = 37.5$

The rejection region requires $\alpha/2 = .05/2 = .025$ in each tail of the t distribution with df $= n_1 + n_2 - 2 = 6 + 6 - 2 = 10$. From Table VI, Appendix B, $t_{.025} = 2.228$. The rejection region is $t > 2.228$ or $t < -2.228$. The critical value from Exercise 16.12 is $F_{.05} = 4.96$, $t^2 = 2.228^2 = 4.96$.

Since the observed value of the test statistic falls in the rejection region ($t = -6.12 < -2.228$), H_0 is rejected. There is sufficient evidence to indicate the population means are different at $\alpha = .05$. The conclusion in Exercise 16.12 is reject H_0 at $\alpha = .05$.

The assumptions necessary for the t test are:

1. Both populations are normal.
2. Samples are independent.
3. The population variances are equal.

These are the same assumptions necessary for the F test.

For diagram **b**:

$$\bar{y}_1 = \frac{\sum y_1}{n_1} = \frac{54}{6} = 9 \quad s_1^2 = \frac{\sum y_1^2 - \frac{(\sum y_1)^2}{n_1}}{n_1 - 1} = \frac{558 - \frac{54^2}{6}}{6 - 1} = \frac{72}{5} = 14.4$$

$$\bar{y}_2 = \frac{\sum y_2}{n_2} = \frac{84}{6} = 14 \quad s_2^2 = \frac{\sum y_2^2 - \frac{(\sum y_2)^2}{n_2}}{n_2 - 1} = \frac{1248 - \frac{84^2}{6}}{6 - 1} = \frac{72}{5} = 14.4$$

$$s_p^2 = \frac{(n_1 - 1)s_1^2 + (n_2 - 1)s_2^2}{n_1 + n_2 - 2} = \frac{(6 - 1)14.4 + (6 - 1)14.4}{6 + 6 - 2} = \frac{144}{10} = 14.4$$

From Exercise 16.12, MSE $= \dfrac{144}{10} = 14.4$

To determine if the two treatment means are different, we test:

$H_0: \mu_1 - \mu_2 = 0$
$H_a: \mu_1 - \mu_2 \neq 0$

The test statistic is:

$$t = \frac{(\bar{y}_1 - \bar{y}_2) - 0}{\sqrt{s_p^2 \left[\frac{1}{n_1} + \frac{1}{n_2} \right]}} = \frac{9 - 14 - 0}{\sqrt{14.4 \left[\frac{1}{6} + \frac{1}{6} \right]}} = \frac{-5}{2.1909} = -2.282$$

From Exercise 16.12, $F = 5.21$, $t^2 = (-2.282)^2 = 5.21$.

The rejection region requires $\alpha/2 = .05/2 = .025$ in each tail of the t distribution with df $= n_1 + n_2 - 2 = 6 + 6 - 2 = 10$. From Table VI, Appendix B, $t_{.025} = 2.228$. The rejection region is $t > 2.228$ or $t < -2.228$. The critical value from Exercise 16.12 is $F_{.05} = 4.96$, $t^2 = 2.228^2 = 4.96$.

Since the observed value of the test statistic falls in the rejection region ($t = -2.282 < -2.228$), H_0 is rejected. There is sufficient evidence to indicate the population means are different at $\alpha = .05$. The conclusion in Exercise 16.12 is reject H_0 at $\alpha = .05$.

The assumptions for both tests are the same as those stated in diagram **a**.

16.15 For all parts, the hypotheses are:

$H_0: \mu_1 = \mu_2 = \mu_3 = \mu_4 = \mu_5 = \mu_6$
$H_a:$ At least two treatment means differ

The rejection region for all parts is the same.

The rejection region requires $\alpha = .10$ in the upper tail of the F distribution with $\nu_1 = p - 1 = 6 - 1 = 5$ and $\nu_2 = n - p = 36 - 6 = 30$. From Table VII, Appendix B, $F_{.10} = 2.05$. The rejection region is $F > 2.05$.

a. $SST = .2(500) = 100$ $SSE = SS(Total) - SST = 500 - 100 = 400$

$$MST = \frac{SST}{p - 1} = \frac{100}{6 - 1} = 20 \qquad MSE = \frac{SSE}{n - p} = \frac{400}{36 - 6} = 13.333$$

$$F = \frac{MST}{MSE} = \frac{20}{13.333} = 1.5$$

Since the observed value of the test statistic does not fall in the rejection region ($F = 1.5 \not> 2.05$), H_0 is not rejected. There is insufficient evidence to indicate differences among the treatment means at $\alpha = .10$.

b. $SST = .5(500) = 250$ $SSE = SS(Total) - SST = 500 - 250 = 250$

$$MST = \frac{SST}{p - 1} = \frac{250}{6 - 1} = 50 \qquad MSE = \frac{SSE}{n - p} = \frac{250}{36 - 6} = 8.333$$

$$F = \frac{MST}{MSE} = \frac{50}{8.333} = 6$$

Since the observed value of the test statistic falls in the rejection region ($F = 6 > 2.05$), H_0 is rejected. There is sufficient evidence to indicate differences among the treatment means at $\alpha = .10$.

c. $SST = .8(500) = 400$ $SSE = SS(Total) - SST = 500 - 400 = 100$

$$MST = \frac{SST}{p - 1} = \frac{400}{6 - 1} = 80 \qquad MSE = \frac{SSE}{n - p} = \frac{100}{36 - 6} = 3.333$$

$$F = \frac{MST}{MSE} = \frac{80}{3.333} = 24$$

Since the observed value of the test statistic falls in the rejection region ($F = 24 > 2.05$), H_0 is rejected. There is sufficient evidence to indicate differences among the treatment means at $\alpha = .10$.

d.   The F ratio increases as the treatment sum of squares increases.

16.17   a.   The number of treatments is $p = 3 + 1 = 4$. The total sample size is $37 + 1 = 38$.

b.   To determine if the treatment means differ, we test:

H_0: $\mu_1 = \mu_2 = \mu_3 = \mu_4$
H_a: At least two treatment means differ

The test statistic is $F = 14.80$.

The rejection region requires $\alpha = .01$ in the upper tail of the F distribution with $\nu_1 = p - 1 = 4 - 1 = 3$ and $\nu_2 = n - p = 38 - 4 = 34$. From Table X, Appendix B, $F_{.01} \approx 4.51$. The rejection region is $F > 4.51$.

Since the observed value of the test statistic falls in the rejection region ($F = 14.80 >$ 4.51), H_0 is rejected. There is sufficient evidence to indicate differences among the treatment means at $\alpha = .01$.

c. We need the sample means to compare specific pairs of treatment means.

16.19 a. Some preliminary calculations are:

$$CM = \frac{\left(\sum y_i\right)^2}{n} = \frac{37.1^2}{12} = 114.701$$

$$SS(Total) = \sum y_i^2 - CM = 145.89 - 114.701 = 31.189$$

$$SST = \sum \frac{T_i^2}{n_i} - CM = \frac{16.9^2}{5} + \frac{16.0^2}{4} + \frac{4.2^2}{3} - 114.701$$

$$= 127.002 - 114.701 = 12.301$$

$$SSE = SS(Total) - SST = 31.189 - 12.301 = 18.888$$

$$MST = \frac{SST}{p-1} = \frac{12.301}{3-1} = 6.1505 \qquad MSE = \frac{SSE}{n-p} = \frac{18.888}{12-3} = 2.0987$$

$$F = \frac{MST}{MSE} = \frac{6.1505}{2.0987} = 2.931$$

Source	df	SS	MS	F
Treatments	2	12.301	6.1505	2.931
Error	9	18.888	2.0987	
Total	11	31.189		

b. H_0: $\mu_1 = \mu_2 = \mu_3$
H_a: At least two treatment means differ

The test statistic is $F = 2.931$.

The rejection region requires $\alpha = .01$ in the upper tail of the F distribution with $\nu_1 = p - 1 = 3 - 1 = 2$ and $\nu_2 = n - p = 12 - 3 = 9$. From Table X, Appendix B, $F_{.01} = 8.02$. The rejection region is $F > 8.02$.

Since the observed value of the test statistic does not fall in the rejection region ($F = 2.931 \ngtr 8.02$), H_0 is not rejected. There is insufficient evidence to indicate a difference in the treatment means at $\alpha = .01$.

c. There are 3 pairs of treatments to compare, implying $c = 3$. For $\alpha/2c = .05/2(3) = .0083 \approx .005$ and df $= n - p = 12 - 3 = 9$, $t_{.005} = 3.250$ from Table VI, Appendix B.

We now form confidence intervals for the difference between each pair of treatments using the formula.

$$(\bar{y}_i - \bar{y}_j) \pm t_{\alpha/2c} s \sqrt{\frac{1}{n_i} + \frac{1}{n_j}} \quad \text{where } s = \sqrt{MSE} = \sqrt{2.0987} = 1.4487$$

$$\bar{y}_1 = \frac{\sum y_1}{n_1} = \frac{16.9}{5} = 3.38, \; \bar{y}_2 = \frac{\sum y_2}{n_2} = \frac{16.0}{4} = 4, \; \bar{y}_3 = \frac{\sum y_3}{n_3} = \frac{4.2}{3} = 1.4$$

Pair

| 1-2 | $(3.38 - 4) \pm 3.25(1.4487)\sqrt{\frac{1}{5} + \frac{1}{4}} \Rightarrow -.62 \pm 3.158 \Rightarrow (-3.778, 2.538)$ |

| 1-3 | $(3.38 - 1.4) \pm 3.25(1.4487)\sqrt{\frac{1}{5} + \frac{1}{3}} \Rightarrow 1.98 \pm 3.438 \Rightarrow (-1.458, 5.418)$ |

| 2-3 | $(4 - 1.4) \pm 3.25(1.4487)\sqrt{\frac{1}{4} + \frac{1}{3}} \Rightarrow 2.6 \pm 3.596 \Rightarrow (-.996, 6.196)$ |

All the intervals contain 0. This indicates there is insufficient evidence to say any two treatment means are different.

16.21 a. $SS = SST + SSE = 509.87 + 12259.96 = 12769.83$

$$MST = \frac{SST}{p - 1} = \frac{509.87}{3 - 1} = 254.935 \qquad MSE = \frac{SSE}{n - p} = \frac{12259.96}{93 - 3} = 136.222$$

$$F = \frac{MST}{MSE} = \frac{254.935}{136.222} = 1.871$$

Source	df	SS	MS	F
Treatments	2	509.87	254.935	1.871
Error	90	12,259.96	136.222	
Total	92	12,769.83		

b. To determine if differences in mean risk-taking propensities exist among the three groups, we test:

H_0: $\mu_1 = \mu_2 = \mu_3$
H_a: At least two treatment means differ

The test statistic is $F = 1.871$.

The rejection region requires $\alpha = .05$ in the upper tail of the F distribution with $\nu_1 = p - 1 = 3 - 1 = 2$ and $\nu_2 = n - p = 93 - 3 = 90$. From Table VIII, Appendix B, $F_{.05} \approx 3.15$. The rejection region is $F > 3.15$.

Since the observed value of the test statistic does not fall in the rejection region ($F = 1.871 \ngtr 3.15$), H_0 is not rejected. There is insufficient evidence to indicate differences in the mean risk-raking propensities among the three groups at $\alpha = .05$.

c. We must assume:

    1. All 3 population probability distributions are normal.
    2. The 3 population variances are equal.
    3. Samples are selected randomly and independently from the 3 populations.

d. No. No differences were found using ANOVA.

e. The experiment is observational. The experimenter "observed" from which group the individuals came.

16.23   a.   Some preliminary calculations are:

$$CM = \frac{\left(\sum y_i\right)^2}{n} = \frac{5369^2}{12} = 2,402,180.083$$

$$SS(Total) = \sum y_i^2 - CM = 2,419,553 - 2,402,180.083 = 17,372.917$$

$$SST = \sum \frac{T_i^2}{n_i} - CM = \frac{1693^2}{5} + \frac{1695^2}{4} + \frac{1981^2}{4} - 2,402,180.083$$

$$= 2,415,908.75 - 2,402,180.083 = 13,728.667$$

$$SSE = SS(Total) - SST = 17,372.917 - 13,728.667 = 3644.25$$

$$MST = \frac{SST}{p-1} = \frac{13728.667}{3-1} = 6864.3335$$

$$MSE = \frac{SSE}{n-p} = \frac{3644.25}{12-3} = 404.9167$$

$$F = \frac{MST}{MSE} = \frac{6864.3335}{404.9167} = 16.952$$

Source	df	SS	MS	F
Treatments	2	13,728.667	6864.3335	16.952
Error	9	3,664.25	404.9167	
Total	11	17,372.917		

To determine whether there is a difference among the mean sales at the three locations, we test:

H_0: $\mu_1 = \mu_2 = \mu_3$
H_a: At least two treatment means differ

The test statistic is $F = 16.952$.

The rejection region requires $\alpha = .05$ in the upper tail of the F distribution with $\nu_1 = p - 1 = 3 - 1 = 2$ and $\nu_2 = n - p = 12 - 3 = 9$. From Table VIII, Appendix B, $F_{.05} = 4.26$. The rejection region is $F > 4.26$.

Since the observed value of the test statistic falls in the rejection region ($F = 16.952 > 4.26$), H_0 is rejected. There is sufficient evidence to indicate a difference among the mean sales at the three locations at $\alpha = .05$.

b. The confidence interval for $\mu_1 - \mu_3$ is:

$$(\bar{y}_1 - \bar{y}_3) \pm t_{\alpha/2} s \sqrt{\frac{1}{n_1} + \frac{1}{n_3}} \quad \text{where } s = \sqrt{MSE} = \sqrt{404.9167} = 20.1225$$

For confidence coefficient .90, $\alpha = 1 - .90 = .10$ and $\alpha/2 = .10/2 = .05$. From Table VI, Appendix B, $t_{.05} = 1.833$ with df $= n - p = 12 - 3 = 9$.

$$\bar{y}_1 = \frac{\sum y_1}{n_1} = \frac{1693}{4} = 423.25 \qquad \bar{y}_3 = \frac{\sum y_3}{n_3} = \frac{1981}{4} = 495.25$$

The 90% confidence interval is

$$(423.25 - 495.25) \pm 1.833(20.1225)\sqrt{\frac{1}{4} + \frac{1}{4}} \Rightarrow (-72) \pm 26.081$$
$$\Rightarrow (-98.081, -45.919)$$

16.25 a. The response is the job satisfaction score; the factor is the type of work scheduling which is qualitative; the treatments are the three types of work scheduling—flextime, staggered starting hours, and fixed hours; and the experimental units are the workers.

b. $SST = \sum n_i(\bar{y}_i - \bar{y})^2$ $\qquad\qquad$ $\bar{y} = \dfrac{\sum n_i \bar{y}_i}{n}$

$\quad = 27(35.22 - 31.563)^2$ $\qquad\qquad\quad = \dfrac{27(35.22) + 59(31.05) + 24(28.71)}{110}$

$\qquad + 59(31.05 - 31.563)^2$ $\qquad\qquad = 31.563$

$\qquad + 24(28.71 - 31.563)^2$

$\quad = 571.96611$

c. $SSE = \sum (n_i - 1)s_i^2 = (27 - 1)10.22^2 + (59 - 1)7.22^2 + (24 - 1)9.28^2$
$\qquad\qquad\qquad = 7719.8288$

d. $SS(Total) = SSE + SST = 7719.8288 + 571.96611 = 8291.79491$

Treatment $\qquad$ df $= p - 1 = 3 - 1 = 2$
Error $\qquad\qquad$ df $= n - p = 110 - 3 = 107$
Total $\qquad\qquad$ df $= n - 1 = 110 - 1 = 109$

$$MST = \frac{SST}{p - 1} = \frac{571.96611}{2} = 285.983055$$

$$MSE = \frac{SSE}{n - p} = \frac{7719.8288}{107} = 72.1479$$

$$F = \frac{MST}{MSE} = \frac{285.983055}{72.1479} = 3.96$$

The ANOVA table is:

Source	df	SS	MS	F
Treatments	2	571.96611	285.983055	3.96
Error	107	7719.8288	72.1479	
Total	109	8291.79491		

e. To determine whether the three groups differ with respect to their mean job satisfaction, we test:

H_0: $\mu_1 = \mu_2 = \mu_3$
H_a: At least two treatment means differ

The test statistic is $F = 3.96$.

The rejection region requires $\alpha = .05$ in the upper tail of the F distribution with $\nu_1 = p - 1 = 3 - 1 = 2$ and $\nu_2 = n - p = 110 - 3 = 107$. From Table VIII, Appendix B, $F_{.05} \approx 3.15$. The rejection region is $F > 3.15$.

Since the observed value of the test statistic falls in the rejection region ($F = 3.96 > 3.15$), H_0 is rejected. There is sufficient evidence to indicate the three groups differ with respect to their mean job satisfaction at $\alpha = .05$.

f. There are 3 pairs of treatments to compare, implying $c = 3$. For $\alpha/2c = .05/2(3) = .0083$ and df $= n - p = 110 - 3 = 107$, $t_{.0083} = 2.40$ from Table IV, Appendix B. (Since the df is so large, we use the z distribution instead of the t.)

We now form confidence intervals for the difference between each pair of means using the formula:

$$(\bar{y}_i - \bar{y}_j) \pm t_{.0083} \, s \sqrt{\frac{1}{n_i} + \frac{1}{n_j}} \quad \text{where } s = \sqrt{MSE} = \sqrt{72.1479} = 8.494$$

Pair

1-2 $(35.22 - 31.05) \pm 2.4(8.494)\sqrt{\dfrac{1}{27} + \dfrac{1}{59}} \Rightarrow 4.17 \pm 4.737 \Rightarrow (-.567, 8.907)$

1-3 $(35.22 - 28.71) \pm 2.4(8.494)\sqrt{\dfrac{1}{27} + \dfrac{1}{24}} \Rightarrow 6.51 \pm 5.719 \Rightarrow (.791, 12.229)$

2-3 $(31.05 - 28.71) \pm 2.4(8.494)\sqrt{\dfrac{1}{59} + \dfrac{1}{24}} \Rightarrow 2.34 \pm 4.935$
$\Rightarrow (-2.595, 7.275)$

The mean job satisfaction scores are significantly different between flextime and fixed. No other pair of means are significantly different.

16.27 a. Some preliminary calculations are:

$$MST = \frac{SST}{p-1} = \frac{273}{3} = 91 \qquad MSE = \frac{SSE}{n-p} = \frac{494}{94} = 5.2553$$

$$F = \frac{MST}{MSE} = \frac{91}{5.2553} = 17.32$$

To determine whether the mean level of confidence differs among the four forms of auditor association, we test:

$H_0: \mu_1 = \mu_2 = \mu_3 = \mu_4$
$H_a:$ At least two treatment means differ

The test statistic is $F = 17.32$.

The rejection region requires $\alpha = .05$ in the upper tail of the F distribution with $\nu_1 = p - 1 = 4 - 1 = 3$ and $\nu_2 = n - p = 98 - 4 = 94$. From Table VIII, Appendix B, $F_{.05} \approx 2.76$. The rejection region is $F > 2.76$.

Since the observed value of the test statistic falls in the rejection region ($F = 17.32 > 2.76$), H_0 is rejected. There is sufficient evidence to indicate the mean level of confidence differs among the four forms of auditor association at $\alpha = .05$.

b. The p-value $= P(F \geq 17.32)$. Using Table X, Appendix B, with $\nu_1 = 3$ and $\nu_2 = 94$, $P(F \geq 13.32) < .01$.

c. To determine if the mean level of confidence associated with the audit is significantly higher than the mean for review, we test:

$H_0: \mu_4 - \mu_3 = 0$
$H_a: \mu_4 - \mu_3 > 0$

The test statistic is $t = \dfrac{\bar{y}_4 - \bar{y}_3 - 0}{\sqrt{MSE\left[\dfrac{1}{n_4} + \dfrac{1}{n_3}\right]}} = \dfrac{8.8 - 6.1 - 0}{\sqrt{5.2553\left[\dfrac{1}{27} + \dfrac{1}{25}\right]}} = \dfrac{2.7}{.6363}$

$= 4.24$

The rejection region requires $\alpha = .05$ in the upper tail of the t distribution with df $= n - p = 94$. From Table VI, Appendix B, $t_{.05} \approx 1.671$. The rejection region is $t > 1.671$.

Since the observed value of the test statistic falls in the rejection region ($t = 4.24 > 1.671$), H_0 is rejected. There is sufficient evidence to indicate the mean level of confidence associated with the audit is significantly higher than the mean for review at $\alpha = .05$.

d. There are 6 pairs of treatments to compare, implying $c = 6$. For $\alpha/2c = .05/2(6) = .0042$ and df $= n - p = 98 - 4 = 94$, $t_{.0042} = 2.635$ from Table IV, Appendix B. (Since the df is so large, we use the z distribution instead of the t).

We now form confidence intervals for the difference between each pair of means using the formula:

$$(\bar{y}_i - \bar{y}_j) \pm t_{.0042}\, s \sqrt{\frac{1}{n_i} + \frac{1}{n_j}} \quad \text{where } s = \sqrt{\text{MSE}} = \sqrt{5.2553} = 2.2924$$

Pair

1−2 $(3.9 - 5.5) \pm 2.635(2.2924)\sqrt{\frac{1}{15} + \frac{1}{31}} \Rightarrow -1.6 \pm 1.900 \Rightarrow (-3.500, .300)$

1−3 $(3.9 - 6.1) \pm 2.635(2.2924)\sqrt{\frac{1}{15} + \frac{1}{25}} \Rightarrow -2.2 \pm 1.973$

                                           $\Rightarrow (-4.173, -.227)$

1−4 $(3.9 - 8.8) \pm 2.635(2.2924)\sqrt{\frac{1}{15} + \frac{1}{27}} \Rightarrow -4.9 \pm 1.945$

                                           $\Rightarrow (-6.845, -2.955)$

2−3 $(5.5 - 6.1) \pm 2.635(2.2924)\sqrt{\frac{1}{31} + \frac{1}{25}} \Rightarrow -.6 \pm 1.624 \Rightarrow (-2.224, 1.024)$

2−4 $(5.5 - 8.8) \pm 2.635(2.2924)\sqrt{\frac{1}{31} + \frac{1}{27}} \Rightarrow -3.3 \pm 1.59$

                                           $\Rightarrow (-4.890, -1.710)$

3−4 $(6.1 - 8.8) \pm 2.635(2.2924)\sqrt{\frac{1}{25} + \frac{1}{27}} \Rightarrow -2.7 \pm 1.677$

                                           $\Rightarrow (-4.377, -1.023)$

The pairs of means that are significantly different are: 1 and 3, 1 and 4, 2 and 4, and 3 and 4.

e.   Since there are differences among the treatment means, there is evidence that the auditors were not accepting the same degree of responsibility for each form of financial report.

16.29   a.   The ANOVA table is:

Source	df	SS	MS	F
A	2	.8	.4000	3.69
B	3	5.3	1.7667	16.31
AB	6	9.6	1.6000	14.77
Error	12	1.3	.1083	
Total	23	17.0		

df for A is $a - 1 = 3 - 1 = 2$
df for $B = b - 1 = 4 - 1 = 3$
df for AB is $(a - 1)(b - 1) = 2(3) = 6$
df for Error is $n - ab = 24 - 3(4) = 12$
df for Total is $n - 1 = 24 - 1 = 23$

$$SSE = SS(Total) - SSA - SSB - SSAB = 17.0 - .8 - 5.3 - 9.6 = 1.3$$

$$MSA = \frac{SSA}{a - 1} = \frac{.8}{3 - 1} = .40 \qquad MSB = \frac{SSB}{b - 1} = \frac{5.3}{4 - 1} = 1.7667$$

$$MSAB = \frac{SSAB}{(a - 1)(b - 1)} = \frac{9.6}{(3 - 1)(4 - 1)} = 1.60$$

$$MSE = \frac{SSE}{n - ab} = \frac{1.3}{24 - 3(4)} = .1083$$

$$F_A = \frac{MSA}{MSE} = \frac{.4000}{.1083} = 3.69 \qquad F_B = \frac{MSB}{MSE} = \frac{1.7667}{.1083} = 16.31$$

$$F_{AB} = \frac{MSAB}{MSE} = \frac{1.6000}{.1083} = 14.77$$

b. Sum of Squares for Treatment $= SSA + SSB + SSAB = .8 = 5.3 + 2.6 = 15.7$

$$MST = \frac{SST}{ab - 1} = \frac{15.7}{3(4) - 1} = 1.4273 \qquad F_T = \frac{MST}{MSE} = \frac{1.4273}{.1083} = 13.18$$

To determine if the treatment means differ, we test:

H_0: $\mu_1 = \mu_2 = \cdots = \mu_{12}$
H_a: At least two treatments means differ

The test statistic is $F = 13.18$.

The rejection region requires $\alpha = .05$ in the upper tail of the F distribution with $\nu_1 = ab - 1 = 3(4) - 1 = 11$ and $\nu_2 = n - ab = 24 - 3(4) = 12$. From Table VIII, Appendix B, $F_{.05} \approx 2.75$. The rejection region is $F > 2.75$.

Since the observed value of the test statistic falls in the rejection region ($F = 13.18 > 2.75$), H_0 is rejected. There is sufficient evidence to indicate the treatment means differ at $\alpha = .05$.

c. Yes. We need to partition the Treatment Sum of Squares into the Main Effects and Interaction Sum of Squares. Then we test whether factors A and B interact. Depending on the conclusion of the test for interaction, we either test for main effects or compare the treatment means.

d. Two factors are said to interact if the effect of one factor on the dependent variable is not the same at different levels of the second factor. If the factors interact, then tests for main effects are not necessary. We need to compare the treatment means for one factor at each level of the second.

e. To determine if the factors interact, we test:

H_0: Factors A and B do not interact to affect the response mean
H_a: Factors A and B do interact to affect the response mean

The test statistic is $F = \dfrac{\text{MS}AB}{\text{MSE}} = 14.77$

The rejection region requires $\alpha = .05$ in the upper tail of the F distribution with $\nu_1 = (a - 1)(b - 1) = (3 - 1)(4 - 1) = 6$ and $\nu_2 = n - ab = 24 - 3(4) = 12$. From Table VIII, Appendix B, $F_{.05} = 3.00$. The rejection region is $F > 3.00$.

Since the observed value of the test statistic falls in the rejection region ($F = 14.77 > 3.00$), H_0 is rejected. There is sufficient evidence to indicate the two factors interact to affect the response mean at $\alpha = .05$.

f. No. Testing for main effects is not warranted because interaction is present. Instead, we compare the treatment means of one factor at each level of the second factor.

16.31 a. The treatments for this experiment consist of a level for factor A and a level for factor B. There are 6 treatments—(1, 1), (1, 2), (1, 3), (2, 1), (2, 2), and (2, 3) where the first number represents the level of factor A and the second number represents the level of factor B.

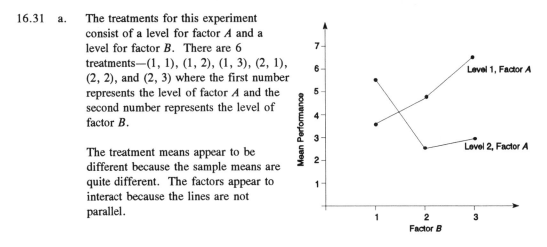

The treatment means appear to be different because the sample means are quite different. The factors appear to interact because the lines are not parallel.

b. $\text{SST} = \text{SS}A + \text{SS}B + \text{SS}AB = 4.441 + 4.127 + 18.007 = 26.575$

$\text{MST} = \dfrac{\text{SST}}{ab - 1} = \dfrac{26.575}{2(3) - 1} = 5.315$ $F_\text{T} = \dfrac{\text{MST}}{\text{MSE}} = \dfrac{5.315}{.246} = 21.61$

To determine whether the treatment means differ, we test:

H_0: $\mu_1 = \mu_2 = \mu_3 = \mu_4 = \mu_5 = \mu_6$
H_a: At least two treatment means differ

The test statistic is $F = \dfrac{\text{MST}}{\text{MSE}} = 21.61$

The rejection region requires $\alpha = .05$ in the upper tail of the F distribution with $\nu_1 = ab - 1 = 2(3) - 1 = 5$ and $\nu_2 = n - ab = 12 - 2(3) = 6$. From Table VIII, Appendix B, $F_{.05} = 4.39$. The rejection region is $F > 4.39$.

Since the observed value of the test statistic falls in the rejection region ($F = 21.61 > 4.39$), H_0 is rejected. There is sufficient evidence to indicate that the treatment means differ at $\alpha = .05$. This supports the plot in a.

c. Yes. Since there are differences among the treatment means, we test for interaction. To determine whether the factors A and B interact, we test:

H_0: Factors A and B do not interact to affect the mean response
H_a: Factors A and B do interact to affect the mean response

The test statistic is $F = \dfrac{MSAB}{MSE} = \dfrac{9.003}{.246} = 36.60$

The rejection region requires $\alpha = .05$ in the upper tail of the F distribution with $v_1 = (a - 1)(b - 1) = (2 - 1)(3 - 1) = 2$ and $v_2 = n - ab = 12 - 2(3) = 6$. From Table VIII, Appendix B, $F_{.05} = 5.14$. The rejection region is $F > 5.14$.

Since the observed value of the test statistic falls in the rejection region ($F = 36.60 > 5.14$), H_0 is rejected. There is sufficient evidence to indicate that factors A and B interact to affect the response mean at $\alpha = .05$.

d. No. Because interaction is present, the tests for main effects are not warranted.

e. The results of the tests in parts b and c support the visual interpretation in part a.

f. There are 15 pairs of treatment means to compare, implying $c = 15$. For $\alpha/2c = .10/2(15) = .0033 \approx .005$ and df $= n - ab = 12 - 2(3) = 6$, $t_{.005} = 3.707$ from Table VI, Appendix B. We now form confidence intervals for the difference between each pair of means using the formula:

$$(\bar{y}_i - \bar{y}_j) \pm t_{\alpha/2c} s \sqrt{\frac{1}{n_i} + \frac{1}{n_j}} \quad \text{where } s = \sqrt{MSE} = \sqrt{.246} = .4959$$

$$\bar{y}_{11} = \frac{7.1}{2} = 3.55 \quad \bar{y}_{12} = \frac{8.8}{2} = 4.4 \quad \bar{y}_{13} = \frac{13.5}{2} = 6.75$$

$$\bar{y}_{21} = \frac{11.2}{2} = 5.6 \quad \bar{y}_{22} = \frac{5.1}{2} = 2.55 \quad \bar{y}_{23} = \frac{5.8}{2} = 2.9$$

Pair

$A_1B_1 - A_1B_2 \quad (3.55 - 4.4) \pm 3.707(.4959)\sqrt{\dfrac{1}{2} + \dfrac{1}{2}} \Rightarrow -.85 \pm 1.838$

$\Rightarrow (-2.688, .988)$

$A_1B_1 - A_1B_3 \quad (3.55 - 6.75) \pm 1.838 \Rightarrow -3.2 \pm 1.838 \Rightarrow (-5.038, -1.362)$

$A_1B_1 - A_2B_1 \quad (3.55 - 5.6) \pm 1.838 \Rightarrow -2.05 \pm 1.838 \Rightarrow (-3.888, -.212)$

$A_1B_1 - A_2B_2 \quad (3.55 - 2.55) \pm 1.838 \Rightarrow 1.00 \pm 1.838 \Rightarrow (-.838, 2.838)$

$A_1B_1 - A_2B_3 \quad (3.55 - 2.9) \pm 1.838 \Rightarrow .65 \pm 1.838 \Rightarrow (-1.188, 2.488)$

$A_1B_2 - A_1B_3 \quad (4.4 - 6.75) \pm 1.838 \Rightarrow -2.35 \pm 1.838 \Rightarrow (-4.188, -.512)$

$A_1B_2 - A_2B_1 \quad (4.4 - 5.6) \pm 1.838 \Rightarrow -1.2 \pm 1.838 \Rightarrow (-3.038, .638)$

$A_1B_2 - A_2B_2 \quad (4.4 - 2.55) \pm 1.838 \Rightarrow 1.85 \pm 1.838 \Rightarrow (.012, 3.638)$

$A_1B_2 - A_2B_3 \quad (4.4 - 2.9) \pm 1.838 \Rightarrow 1.5 \pm 1.838 \Rightarrow (-.338, 3.338)$

$A_1B_3 - A_2B_1$ $(6.75 - 5.6) \pm 1.838 \Rightarrow 1.15 \pm 1.838 \Rightarrow (-.688, 2.988)$

$A_1B_3 - A_2B_2$ $(6.75 - 2.55) \pm 1.838 \Rightarrow 4.2 \pm 1.838 \Rightarrow (2.362, 6.038)$

$A_1B_3 - A_2B_3$ $(6.75 - 2.9) \pm 1.838 \Rightarrow 3.85 \pm 1.838 \Rightarrow (2.012, 5.688)$

$A_2B_1 - A_2B_2$ $(5.6 - 2.55) \pm 1.838 \Rightarrow 3.05 \pm 1.838 \Rightarrow (1.212, 4.888)$

$A_2B_1 - A_2B_3$ $(5.6 - 2.9) \pm 1.838 \Rightarrow 2.7 \pm 1.838 \Rightarrow (.862, 4.538)$

$A_2B_2 - A_2B_3$ $(2.55 - 2.9) \pm 1.838 \Rightarrow -.35 \pm 1.838 \Rightarrow (-2.188, 1.488)$

The pairs that are significantly different are those that do not have 0 within their confidence interval. They are: A_1B_1 and A_1B_3, A_1B_1 and A_2B_1, A_1B_2 and A_1B_3, A_1B_2 and A_2B_2, A_1B_3 and A_2B_2, A_1B_3 and A_2B_3, A_2B_1 and A_2B_2, and A_2B_1 and A_2B_3.

16.33 a. $SSA = .2(1000) = 200$, $SSB = .1(1000) = 100$, $SSAB = .1(1000) = 100$

$SSE = SS(\text{Total}) - SSA - SSB - SSAB = 1000 - 200 - 100 - 100 = 600$

$SST = SSA + SSB + SSAB = 200 + 100 + 100 = 400$

$MSA = \dfrac{SSA}{a - 1} = \dfrac{200}{3 - 1} = 100$ $MSB = \dfrac{SSB}{b - 1} = \dfrac{100}{3 - 1} = 50$

$MSAB = \dfrac{SSAB}{(a - 1)(b - 1)} = \dfrac{100}{(3 - 1)(3 - 1)} = 25$

$MSE = \dfrac{SSE}{n - ab} = \dfrac{600}{27 - 3(3)} = 33.333$ $MST = \dfrac{SST}{ab - 1} = \dfrac{400}{3(3) - 1} = 50$

$F_A = \dfrac{MSA}{MSE} = \dfrac{100}{33.333} = 3.00$ $F_B = \dfrac{MSB}{MSE} = \dfrac{50}{33.333} = 1.50$

$F_{AB} = \dfrac{MSAB}{MSE} = \dfrac{25}{33.333} = .75$ $F_T = \dfrac{MST}{MSE} = \dfrac{50}{33.333} = 1.50$

Source	df	SS	MS	F
A	2	200	100	3.00
B	2	100	50	1.50
AB	4	100	25	.75
Error	18	600	33.333	
Total	26	1000		

To determine whether the treatment means differ, we test:

H_0: $\mu_1 = \mu_2 = \cdots = \mu_9$

H_a: At least two treatments means differ

The test statistic is $F = \dfrac{MST}{MSE} = 1.50$

The rejection region requires $\alpha = .05$ in the upper tail of the F distribution with $\nu_1 = ab - 1 = 3(3) - 1 = 8$ and $\nu_2 = n - ab = 27 - 3(3) = 18$. From Table VIII, Appendix B, $F_{.05} = 2.51$. The rejection region is $F > 2.51$.

Since the observed value of the test statistic does not fall in the rejection region ($F = 1.50 \not> 2.51$), H_0 is not rejected. There is insufficient evidence to indicate the treatment means differ at $\alpha = .05$. Since there are no treatment mean differences, we have nothing more to do.

b. $SSA = .1(1000) = 100$, $SSB = .1(1000) = 100$, $SSAB = .5(1000) = 500$
$SSE = SS(\text{Total}) - SSA - SSB - SSAB = 1000 - 100 - 100 - 500 = 300$
$SST = SSA + SSB + SSAB = 100 + 100 + 500 = 700$

$$MSA = \frac{SSA}{a - 1} = \frac{100}{3 - 1} = 50 \qquad MSB = \frac{SSB}{b - 1} = \frac{100}{3 - 1} = 50$$

$$MSAB = \frac{SSAB}{(a - 1)(b - 1)} = \frac{500}{(3 - 1)(3 - 1)} = 125$$

$$MSE = \frac{SSE}{n - ab} = \frac{300}{27 - 3(3)} = 16.667$$

$$MST = \frac{SST}{ab - 1} = \frac{700}{9 - 1} = 87.5$$

$$F_A = \frac{MSA}{MSE} = \frac{50}{16.667} = 3.00 \qquad F_B = \frac{MSB}{MSE} = \frac{50}{16.667} = 3.00$$

$$F_{AB} = \frac{MSAB}{MSE} = \frac{125}{16.667} = 7.50 \qquad F_T = \frac{MST}{MSE} = \frac{87.5}{16.667} = 5.25$$

Source	df	SS	MS	F
A	2	100	50	3.00
B	2	100	50	3.00
AB	4	500	125	7.50
Error	18	300	16.667	
Total	26	1000		

To determine whether the treatment means differ, we test:

H_0: $\mu_1 = \mu_2 = \cdots = \mu_9$
H_a: At least two treatments means differ

The test statistic is $F = \dfrac{MST}{MSE} = 5.25$

The rejection region requires $\alpha = .05$ in the upper tail of the F distribution with $\nu_1 = ab - 1 = 3(3) - 1 = 8$ and $\nu_2 = n - ab = 27 - 3(3) = 18$. From Table VIII, Appendix B, $F_{.05} = 2.51$. The rejection region is $F > 2.51$.

Since the observed value of the test statistic falls in the rejection region ($F = 5.25 > 2.51$), H_0 is rejected. There is sufficient evidence to indicate the treatment means differ at $\alpha = .05$.

Since the treatment means differ, we next test for interaction between factors A and B.

To determine if factors A and B interact, we test:

H_0: Factors A and B do not interact to affect the mean response
H_a: Factors A and B do interact to affect the mean response

The test statistic is $F = \dfrac{MSAB}{MSE} = 7.50$

The rejection region requires $\alpha = .05$ in the upper tail of the F distribution with $\nu_1 = (a - 1)(b - 1) = (3 - 1)(3 - 1) = 4$ and $\nu_2 = n - ab = 27 - 3(3) = 18$. From Table VIII, Appendix B, $F_{.05} = 2.93$. The rejection region is $F > 2.93$.

Since the observed value of the test statistic falls in the rejection region ($F = 7.50 > 2.93$), H_0 is rejected. There is sufficient evidence to indicate the factors A and B interact at $\alpha = .05$. Since interaction is present, no tests for main effects are necessary.

c. $SSA = .4(1000) = 400$, $SSB = .1(1000) = 100$, $SSAB = .2(1000) = 200$

$SSE = SS(\text{Total}) - SSA - SSB - SSAB = 1000 - 400 - 100 - 200 = 300$

$SST = SSA + SSB + SSAB = 400 + 100 + 200 = 700$

$MSA = \dfrac{SSA}{a - 1} = \dfrac{400}{3 - 1} = 200 \qquad MSB = \dfrac{SSB}{b - 1} = \dfrac{100}{3 - 1} = 50$

$MSAB = \dfrac{SSAB}{(a - 1)(b - 1)} = \dfrac{200}{(3 - 1)(3 - 1)} = 50$

$MSE = \dfrac{SSE}{n - ab} = \dfrac{300}{27 - 3(3)} = 16.667$

$MST = \dfrac{SST}{ab - 1} = \dfrac{700}{3(3) - 1} = 87.5$

$F_A = \dfrac{MSA}{MSE} = \dfrac{200}{16.667} = 12.00 \qquad F_B = \dfrac{MSB}{MSE} = \dfrac{50}{16.667} = 3.00$

$F_{AB} = \dfrac{MSAB}{MSE} = \dfrac{50}{16.667} = 3.00 \qquad F_T = \dfrac{MST}{MSE} = \dfrac{87.5}{16.667} = 5.25$

Source	df	SS	MS	F
A	2	400	200	12.00
B	2	100	50	3.00
AB	4	200	50	3.00
Error	18	300	16.667	
Total	26	1000		

To determine whether the treatment means differ, we test:

H_0: $\mu_1 = \mu_2 = \cdots = \mu_9$

H_a: At least two treatments means differ

The test statistic is $F = \dfrac{MST}{MSE} = 5.25$

The rejection region requires $\alpha = .05$ in the upper tail of the F distribution with $\nu_1 = ab - 1 = 3(3) - 1 = 8$ and $\nu_2 = n - ab = 27 - 3(3) = 18$. From Table VIII, Appendix B, $F_{.05} = 2.51$. The rejection region is $F > 2.51$.

Since the observed value of the test statistic falls in the rejection region ($F = 5.25 > 2.51$), H_0 is rejected. There is sufficient evidence to indicate the treatment means differ at $\alpha = .05$.

Since the treatment means differ, we next test for interaction between factors A and B. To determine if factors A and B interact, we test:

H_0: Factors A and B do not interact to affect the mean response
H_a: Factors A and B do interact to affect the mean response

The test statistic is $F = \dfrac{\text{MS}AB}{\text{MSE}} = 3.00$

The rejection region requires $\alpha = .05$ in the upper tail of the F distribution with $\nu_1 = (a - 1)(b - 1) = (3 - 1)(3 - 1) = 4$ and $\nu_2 = n - ab = 27 - 3(3) = 18$. From Table VIII, Appendix B, $F_{.05} = 2.93$. The rejection region is $F > 2.93$.

Since the observed value of the test statistic falls in the rejection region ($F = 3.00 > 2.93$), H_0 is rejected. There is sufficient evidence to indicate the factors A and B interact at $\alpha = .05$. Since interaction is present, no tests for main effects are necessary.

d. $\text{SS}A = .4(1000) = 400$, $\text{SS}B = .4(1000) = 400$, $\text{SS}AB = .1(1000) = 100$
$\text{SSE} = \text{SS(Total)} - \text{SS}A - \text{SS}B - \text{SS}AB = 1000 - 400 - 400 - 100 = 100$
$\text{SST} = \text{SS}A + \text{SS}B + \text{SS}AB = 400 + 400 + 100 = 900$

$\text{MS}A = \dfrac{\text{SS}A}{a - 1} = \dfrac{400}{3 - 1} = 200 \qquad \text{MS}B = \dfrac{\text{SS}B}{b - 1} = \dfrac{400}{3 - 1} = 200$

$\text{MS}AB = \dfrac{\text{SS}AB}{(a - 1)(b - 1)} = \dfrac{100}{(3 - 1)(3 - 1)} = 25$

$\text{MSE} = \dfrac{\text{SSE}}{n - ab} = \dfrac{100}{27 - 3(3)} = 5.556$

$\text{MST} = \dfrac{\text{SST}}{ab - 1} = \dfrac{900}{3(3) - 1} = 112.5$

$F_A = \dfrac{\text{MS}A}{\text{MSE}} = \dfrac{200}{5.556} = 36.00 \qquad F_B = \dfrac{\text{MS}B}{\text{MSE}} = \dfrac{200}{5.556} = 36.00$

$F_{AB} = \dfrac{\text{MS}AB}{\text{MSE}} = \dfrac{25}{5.556} = 4.50 \qquad F_T = \dfrac{\text{MST}}{\text{MSE}} = \dfrac{112.5}{5.556} = 20.25$

Source	df	SS	MS	F
A	2	400	200	36.00
B	2	400	200	36.00
AB	4	100	25	4.50
Error	18	100	5.556	
Total	26	1000		

To determine whether the treatment means differ, we test:

H_0: $\mu_1 = \mu_2 = \cdots = \mu_9$
H_a: At least two treatments means differ

The test statistic is $F = \dfrac{\text{MST}}{\text{MSE}} = 20.25$

The rejection region requires $\alpha = .05$ in the upper tail of the F distribution with $\nu_1 = ab - 1 = 3(3) - 1 = 8$ and $\nu_2 = n - ab = 27 - 3(3) = 18$. From Table VIII, Appendix B, $F_{.05} = 2.51$. The rejection region is $F > 2.51$.

Since the observed value of the test statistic falls in the rejection region ($F = 20.25 > 2.51$), H_0 is rejected. There is sufficient evidence to indicate the treatment means differ at $\alpha = .05$.

Since the treatment means differ, we next test for interaction between factors A and B. To determine if factors A and B interact, we test:

H_0: Factors A and B do not interact to affect the mean response
H_a: Factors A and B do interact to affect the mean response

The test statistic is $F = \dfrac{\text{MS}AB}{\text{MSE}} = 4.50$

The rejection region requires $\alpha = .05$ in the upper tail of the F distribution with $\nu_1 = (a - 1)(b - 1) = (3 - 1)(3 - 1) = 4$ and $\nu_2 = n - ab = 27 - 3(3) = 18$. From Table VIII, Appendix B, $F_{.05} = 2.93$. The rejection region is $F > 2.93$.

Since the observed value of the test statistic falls in the rejection region ($F = 4.50 > 2.93$), H_0 is rejected. There is sufficient evidence to indicate the factors A and B interact at $\alpha = .05$. Since interaction is present, no tests for main effects are necessary.

16.35 a. The response variable is the dollar increases in sales per advertising dollar. There are two factors: advertising medium at 3 levels, and agency at 2 levels. Both factors are qualitative—the levels of each are not measured on a numerical scale. The treatments are the combinations of levels of the 2 factors. There are $2 \times 3 = 6$ treatments consisting of an agency type and an advertising medium. The experimental units are the twelve small towns. The experiment is a complete factorial experiment and is completely randomized.

b. $\text{SST} = \text{SS}A + \text{SS}B + \text{SS}AB = 39.967 + 198.332 + 77.345 = 315.644$

$\text{MST} = \dfrac{\text{SST}}{ab - 1} = \dfrac{315.644}{3(2) - 1} = 63.1288 \qquad F_T = \dfrac{\text{MST}}{\text{MSE}} = \dfrac{63.1288}{5.701} = 11.074$

To determine if there is a difference among the treatment means, we test:

H_0: $\mu_1 = \mu_2 = \mu_3 = \mu_4 = \mu_5 = \mu_6$
H_a: At least two treatment means differ

The test statistic is $F = \dfrac{\text{MST}}{\text{MSE}} = 11.074$

The rejection region requires $\alpha = .10$ in the upper tail of the F distribution with $\nu_1 = ab - 1 = 3(2) - 1 = 5$ and $\nu_2 = n - ab = 12 - 3(2) = 6$. From Table VII, Appendix B, $F_{.10} = 3.11$. The rejection region is $F > 3.11$.

Since the observed value of the test statistic falls in the rejection region ($F = 11.074 > 3.11$), H_0 is rejected. There is sufficient evidence to indicate a difference among the treatment means at $\alpha = .10$.

Since differences exist among the treatment means, we continue to test. To determine if an interaction between agency and advertising medium exist, we test:

H_0: Factors A and B do not interact to affect the response mean
H_a: Factors A and B do interact to affect the response mean

The test statistic is $F = \dfrac{\text{MSAB}}{\text{MSE}} = 6.784$

The rejection region requires $\alpha = .10$ in the upper tail of the F distribution with $\nu_1 = (a - 1)(b - 1) = (2 - 1)(3 - 1) = 2$ and $\nu_2 = n - ab = 12 - 3(2) = 6$. From Table VII, Appendix B, $F_{.10} = 3.46$. The rejection region is $F > 3.46$.

Since the observed value of the test statistic falls in the rejection region ($F = 6.784 > 3.46$), H_0 is rejected. There is sufficient evidence to indicate interaction between the two factors is present at $\alpha = .10$.

Since interaction is present, we use the Bonferroni multiple comparisons procedure to compare all pairs of treatment means. There are 15 pairs of treatments, so $c = 15$. For $\alpha/2c = .10/2(15) = .0033 \approx .005$ and df $= n - ab = 12 - 3(2) = 6$, $t_{.005} = 3.707$ from Table VI, Appendix B. We now form confidence intervals for the difference between each pair of means using the formula:

$$(\bar{y}_i - \bar{y}_j) \pm t_{\alpha/2c}\, s \sqrt{\frac{1}{n_i} + \frac{1}{n_j}} \quad \text{where } s = \sqrt{\text{MSE}} = \sqrt{5.701} = 2.3877$$

$$\bar{y}_{1N} = \frac{28.0}{2} = 14.0 \qquad \bar{y}_{1R} = \frac{37.5}{2} = 18.75 \qquad \bar{y}_{1T} = \frac{28.9}{2} = 14.45$$

$$\bar{y}_{2N} = \frac{41.3}{2} = 20.65 \qquad \bar{y}_{2R} = \frac{53.1}{2} = 26.55 \qquad \bar{y}_{2T} = \frac{21.9}{2} = 10.95$$

Pair

$A_1N - A_1R \quad (14 - 18.75) \pm 3.707(2.3877)\sqrt{\dfrac{1}{2} + \dfrac{1}{2}} \Rightarrow -4.75 \pm 8.85$

$$\Rightarrow (-13.6, 4.1)$$

$A_1N - A_1T \quad (14 - 14.45) \pm 8.85 \Rightarrow -.45 \pm 8.85 \Rightarrow (-9.30, 8.40)$

$A_1N - A_2N \quad (14 - 20.65) \pm 8.85 \Rightarrow -6.65 \pm 8.85 \Rightarrow (-15.5, 2.2)$

$A_1N - A_2R \quad (14 - 26.55) \pm 8.85 \Rightarrow -12.55 \pm 8.85 \Rightarrow (-21.4, -3.7)$

$A_1N - A_2T \quad (14 - 10.95) \pm 8.85 \Rightarrow 3.05 \pm 8.85 \Rightarrow (-5.8, 11.9)$

$A_1R - A_1T \quad (18.75 - 14.45) \pm 8.85 \Rightarrow 4.3 \pm 8.85 \Rightarrow (-4.55, 13.15)$

$A_1R - A_2N \quad (18.75 - 20.65) \pm 8.85 \Rightarrow -1.9 \pm 8.85 \Rightarrow (-10.75, 6.95)$

$A_1R - A_2R \quad (18.75 - 26.55) \pm 8.85 \Rightarrow -7.8 \pm 8.85 \Rightarrow (-16.65, 1.05)$

$A_1R - A_2T \quad (18.75 - 10.95) \pm 8.85 \Rightarrow 7.8 \pm 8.85 \Rightarrow (-1.05, 16.65)$

$$A_1T - A_2N \quad (14.45 - 20.65) \pm 8.85 \Rightarrow -6.2 \pm 8.85 \Rightarrow (-15.05, 2.65)$$
$$A_1T - A_2R \quad (14.45 - 26.55) \pm 8.85 \Rightarrow -12.1 \pm 8.85 \Rightarrow (-20.95, -3.25)$$
$$A_1T - A_2T \quad (14.45 - 10.95) \pm 8.85 \Rightarrow 3.5 \pm 8.85 \Rightarrow (-5.35, 12.35)$$
$$A_2N - A_2R \quad (20.65 - 26.55) \pm 8.85 \Rightarrow -5.9 \pm 8.85 \Rightarrow (-14.75, 2.95)$$
$$A_2N - A_2T \quad (20.65 - 10.95) \pm 8.85 \Rightarrow 9.7 \pm 8.85 \Rightarrow (.85, 18.55)$$
$$A_2R - A_2T \quad (26.55 - 10.95) \pm 8.85 \Rightarrow 15.6 \pm 8.85 \Rightarrow (6.75, 24.45)$$

The following treatment pairs are significantly different because 0 is not in the confidence interval:

$$A_1N \text{ and } A_2R, \ A_1T \text{ and } A_2R, \ A_2N \text{ and } A_2T, \text{ and } A_2R \text{ and } A_2T$$

There are no significant differences among the other pairs.

c. The lines are not parallel which implies interaction is present. Also, because the sample mean responses appear quite different, it appears that there are differences among the treatment means.

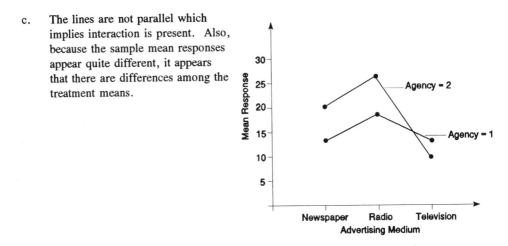

16.37 a. We first calculate the total of $n = 10$ pulling force measurements for each of the four categories by multiplying the means in each category by 10.

	Light	Heavy	Totals
Females	462.6	627.2	1089.8
Males	880.7	862.9	1743.6
Totals	1343.3	1490.1	

$$\sum x_i = 462.6 + 627.2 + 880.7 + 862.9 = 2833.4$$

$$CM = \frac{\left(\sum x_i\right)^2}{n} = \frac{2833.4^2}{40} = 200{,}703.889$$

$$SS(Sex) = \frac{\sum A_i^2}{br} - CM = \frac{1089.8^2}{2(10)} + \frac{1743.6^2}{2(10)} - 200{,}703.889$$

$$= 211{,}390.25 - 200{,}703.889 = 10{,}686.361$$

$$SS(\text{Weight}) = \frac{\sum B_i^2}{ar} - CM = \frac{1343.3^2}{2(10)} + \frac{1490.1^2}{2(10)} - 200{,}703.889$$

$$= 201{,}242.645 - 200{,}703.889 = 538.756$$

$$SS(\text{Sex} \times \text{Weight}) = \frac{\sum\sum AB_{ij}^2}{r} - SS(\text{Sex}) - SS(\text{Weight}) - CM$$

$$= \frac{462.6^2}{10} + \frac{627.2^2}{10} + \frac{880.7^2}{10} + \frac{862.9^2}{10}$$

$$- 10686.361 - 538.756 - 200{,}703.889$$

$$= 212{,}760.75 - 211{,}929.006 = 831.744$$

The sum of squares of deviations within each sample are found by multiplying the variance by $n - 1$.

	Standard Deviation	Variance	SS
Female, Light	14.23	202.4929	1822.4361
Female, Heavy	13.97	195.1609	1756.4481
Male, Light	8.32	69.2224	623.0016
Male, Heavy	12.45	155.0025	1395.0225
			5596.9083

$SSE = 5596.9083$

$SS(\text{Total}) = SS(\text{Sex}) + SS(\text{Weight}) + SS(\text{Sex} \times \text{Weight}) + SSE$

$= 10686.361 + 538.756 + 831.744 + 5596.9083$

$= 17653.7693$

Source	df	SS	MS	F
Sex	1	10,686.361	10,686.361	68.74
Weight	1	538.756	538.756	3.47
Sex × Weight	1	831.744	831.744	5.35
Error	36	5,596.9083	155.4697	
Total	39	17,653.7693		

b. $SST = SS(\text{Sex}) + SS(\text{Weight}) + SS(\text{Sex} \times \text{Weight})$

$= 10{,}686.361 + 538.756 + 831.744$

$= 12{,}056.861$

$$MST = \frac{SST}{ab - 1} = \frac{12{,}056.861}{2(2) - 1} = 4018.954 \qquad F_T = \frac{MST}{MSE} = \frac{4018.954}{155.4697} = 25.85$$

To determine if differences exist among the treatment means, we test:

H_0: $\mu_1 = \mu_2 = \mu_3 = \mu_4$
H_a: At least one treatment mean is different

The test statistic is $F = \dfrac{\text{MST}}{\text{MSE}} = 25.85$

The rejection region requires $\alpha = .05$ in the upper tail of the F distribution with $\nu_1 = ab - 1 = 2(2) - 1 = 3$ and $\nu_2 = n - ab = 40 - 2(2) = 36$. From Table VIII, Appendix B, $F_{.05} \approx 2.92$. The rejection region is $F > 2.92$.

Since the observed value of the test statistic falls in the rejection region ($F = 25.85 > 2.92$), H_0 is rejected. There is sufficient evidence to indicate the treatment means differ at $\alpha = .05$.

To determine if sex and weight interact, we test:

H_0: Sex and weight do not interact to affect the response mean
H_a: Sex and weight do interact to affect the response mean

The test statistic is $F = \dfrac{\text{MS(Sex} \times \text{Weight)}}{\text{MSE}} = 5.35$

The rejection region requires $\alpha = .05$ in the upper tail of the F distribution with $\nu_1 = (a - 1)(b - 1) = (2 - 1)(2 - 1) = 1$ and $\nu_2 = n - ab = 40 - 2(2) = 36$. From Table VIII, Appendix B, $F_{.05} \approx 4.17$. The rejection region is $F > 4.17$.

Since the observed value of the test statistic falls in the rejection region ($F = 5.35 > 4.17$), H_0 is rejected. There is sufficient evidence to indicate sex and weight interact at $\alpha = .05$.

Since interaction is present, we use the Bonferroni multiple comparisons procedure to compare all pairs of treatment means for one factor at each level of the second factor. Suppose we compare the sexes at each level of weight. There is one pair of means to compare at each level of weight, so $c = 2 \times 1 = 2$. For $\alpha/2c = .05/2(2) = .0125 \approx .01$ and df $= n - ab = 40 - 2(2) = 36$, $t_{.01} \approx 2.457$ from Table VI, Appendix B. We now form confidence intervals for the difference between each pair of means using the formula:

$$(\bar{y}_i - \bar{y}_j) \pm t_{\alpha/2c} s \sqrt{\frac{1}{n_i} + \frac{1}{n_j}} \quad \text{where } s = \sqrt{\text{MSE}} = \sqrt{155.4697} = 12.4687$$

F, L $-$ M, L $\quad (46.26 - 88.07) \pm 2.457(12.4687)\sqrt{\dfrac{1}{10} + \dfrac{1}{10}} \Rightarrow -41.81 \pm 13.701$
$$\Rightarrow (-55.511, -28.109)$$

F, H $-$ M, H $\quad (62.72 - 86.29) \pm 13.701 \Rightarrow -23.57 \pm 13.701 \Rightarrow (-37.271, -9.869)$

Both pairs are significantly different.

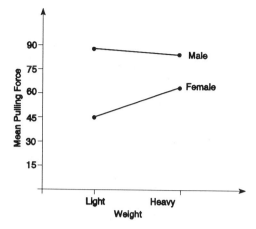

c. These standard deviations are sample standard deviations. Even though the sample standard deviations are different, it is not necessarily true that the population standard deviations are different. We would need to run a test to determine if the sample standard deviations are different enough to infer the population standard deviations are different.

16.39 a. This model is a 3×2 factorial model.

b. To determine if the treatment means are equal, we test:

H_0: $\beta_1 = \beta_2 = \beta_3 = \beta_4 = \beta_5 = 0$
H_a: At least one $\beta_i \neq 0$, $i = 1, 2, 3, 4, 5$

The test statistic is $F = \dfrac{MST}{MSE}$

The rejection region requires $\alpha = .05$ in the upper tail of the F distribution with $\nu_1 = k = 5$ and $\nu_2 = n - (k + 1) = 18 - (5 + 1) = 12$. From Table VIII, Appendix B, $F_{.05} = 3.11$. The rejection region is $F > 3.11$.

c. To test for interaction, we would test:

H_0: $\beta_4 = \beta_5 = 0$
H_a: At least one $\beta_i \neq 0$, $i = 4, 5$

To find the test statistic, we would have to fit the reduced model, $E(y) = \beta_0 + \beta_1 x_1 + \beta_2 x_2 + \beta_3 x_3$. Then

$$F = \frac{(SSE_r - SSE_c)/(k - g)}{SSE_c / [n - (k + 1)]}$$

where SSE_r is the error sum of squares for the reduced model and SSE_c is the error sum of squares for the complete model.

The rejection region requires $\alpha = .05$ in the upper tail of the F distribution with $\nu_1 = k - g = 5 - 3 = 2$ and $\nu_2 = n - (k + 1) = 18 - (5 + 1) = 12$. From Table VIII, Appendix B, $F_{.05} = 3.89$. The rejection region is $F > 3.89$.

16.41   a.

Treatment Price, Display	Estimate of Mean Response
Regular, Normal	$\hat{\beta}_0 + \hat{\beta}_1 + \hat{\beta}_3 + \hat{\beta}_5 = 1828.67 - 626 - 250.67 + 62.67$ $= 1014.67$
Regular, Normal Plus	$\hat{\beta}_0 + \hat{\beta}_1 + \hat{\beta}_4 + \hat{\beta}_6 = 1828.67 - 626 + 681.33 - 669$ $= 1215$
Regular, Twice Normal	$\hat{\beta}_0 + \hat{\beta}_1 = 1828.67 - 626 = 1202.67$
Reduced, Normal	$\hat{\beta}_0 + \hat{\beta}_2 + \hat{\beta}_3 + \hat{\beta}_7 = 1828.67 - 323.67 - 250.67 + 51.67 = 1202.66$
Reduced, Normal Plus	$\hat{\beta}_0 + \hat{\beta}_2 + \hat{\beta}_4 + \hat{\beta}_8 = 1828.67 - 323.67 + 681.33 - 287.67 = 1898.66$
Reduced, Twice Normal	$\hat{\beta}_0 + \hat{\beta}_2 = 1828.67 - 323.67 = 1505$
Cost, Normal	$\hat{\beta}_0 + \hat{\beta}_3 = 1828.67 - 250.67 = 1578$
Cost, Normal Plus	$\hat{\beta}_0 + \hat{\beta}_4 = 1828.67 + 681.33 = 2510$
Cost, Twice Normal	$\hat{\beta}_0 = 1828.67$

b.   The estimate of the standard deviation is $\sqrt{\text{MSE}} = \sqrt{495} = 22.2486$.

We expect most of the observed values to fall within $\pm 2s$ or $\pm 2(22.2486)$ or ± 44.4972 of their predicted values.

$R^2 = .998$. 99.8% of the sample variation in unit sales is explained by the 9 different display and price combinations.

c.   To determine if the mean unit sales differ for the 9 treatments, we test:

H_0: $\beta_1 = \beta_2 = \beta_3 = \cdots = \beta_8 = 0$
H_a: At least one $\beta_i \neq 0$, $i = 1, 2, \ldots, 8$

The test statistic is $F = \dfrac{\text{MSR}}{\text{MSE}} = \dfrac{661,394}{495} = 1336.15$

The rejection region requires $\alpha = .10$ in the upper tail of the F distribution with $\nu_1 = 8$ and $\nu_2 = 18$. From Table VII, Appendix B, $F_{.10} = 2.04$. The rejection region is $F > 2.04$.

Since the observed value of the test statistic falls in the rejection region ($F = 1336.15 > 2.04$), H_0 is rejected. There is sufficient evidence to indicate there are differences in mean unit sales among the 9 treatments at $\alpha = .10$. This is the same result as in Exercise 16.34.

d. The null hypothesis used to test for interaction is:

$$H_0: \beta_5 = \beta_6 = \beta_7 = \beta_8 = 0$$

The test statistic is $F = \dfrac{(SSE_r - SSE_c)/(k - g)}{SSE_c / [n - (k + 1)]}$

$$= \frac{(519{,}610 - 8905)/(8 - 4)}{8905/[27 - (8 + 1)]} = \frac{127{,}676.25}{494.7222} = 258.08$$

The rejection region requires $\alpha = .10$ in the upper tail of the F distribution with $\nu_1 = 4$ and $\nu_2 = 18$. From Table VII, Appendix B, $F_{.10} = 2.29$. The rejection region is $F > 2.29$.

e. Since the observed value of the test statistic falls in the rejection region ($F = 258.08 > 2.29$), H_0 is rejected. There is sufficient evidence to indicate interaction is present at $\alpha = .10$. This is the same result as in Exercise 16.34.

f. No further testing is necessary. Our next step would be to compare the treatment means using Bonferroni's multiple comparisons procedure.

16.43 a. The type of experimental design is completely randomized and a 2 factor complete factorial.

b. The degrees of freedom associated with SSE $= n - ab = 120 - 2(2) = 116$.

c. To determine whether the salesperson's territory and level of effort interact, we test:

H_0: Territory and level of effort do not interact to affect the perception of performance

H_a: Territory and level of effort do interact to affect the perception of performance

The test statistic is $F = 1.95$.

The rejection region requires $\alpha = .05$ in the upper tail of the F distribution with $\nu_1 = (a - 1)(b - 1) = (2 - 1)(2 - 1) = 1$ and $\nu_2 = n - ab = 120 - 2(2) = 116$. From Table VIII, Appendix B, $F_{.05} \approx 3.92$. The rejection region is $F > 3.92$.

Since the observed value of the test statistic does not fall in the rejection region ($F = 1.95 \not> 3.92$), H_0 is not rejected. There is insufficient evidence to indicate territory and level of effort interact at $\alpha = .05$.

d. To determine whether the salesperson's territory influences performance ratings, we test:

$$H_0: \mu_1 = \mu_2$$
$$H_a: \mu_1 \neq \mu_2$$

The test statistic is $F = .39$.

The rejection region requires $\alpha = .05$ in the upper tail of the F distribution with $\nu_1 = (a - 1) = 2 - 1 = 1$ and $\nu_2 = n - ab = 120 - 2(2) = 116$. From Table VIII, Appendix B, $F_{.05} \approx 3.92$. The rejection region is $F > 3.92$.

Since the observed value of the test statistic does not fall in the rejection region ($F = .39 \not> 3.92$), H_0 is not rejected. There is insufficient evidence to indicate a salesperson's territory influences performance ratings at $\alpha = .05$.

e. To determine whether the salesperson's level of effort influences performance ratings, we test:

H_0: $\mu_1 = \mu_2$
H_a: $\mu_1 \neq \mu_2$

The test statistic is $F = 53.27$.

The rejection region requires $\alpha = .05$ in the upper tail of the F distribution with $\nu_1 = (b - 1) = 2 - 1 = 1$ and $\nu_2 = n - ab = 120 - 2(2) = 116$. From Table VIII, Appendix B, $F_{.05} \approx 3.92$. The rejection region is $F > 3.92$.

Since the observed value of the test statistic falls in the rejection region ($F = 53.27 > 3.92$), H_0 is rejected. There is sufficient evidence to indicate a salesperson's level of effort influences performance ratings at $\alpha = .05$.

f. The results confirm the hypothesis of Mowen, et al. They suggested that managers underutilize data on the salesperson's territory and rely on data associated with how much effort the salesperson expended. The tests showed that territory did not affect the perceived performance, implying the managers do not take territory into account.

g. We must assume:

1. The response distribution for each factor level combination is normal.
2. The response variance is constant for all treatments.
3. Random and independent samples of experimental units are associated with each treatment.

16.45 a. The type of experiment is completely randomized. The response is the increase in "haze." There is one factor, type of treatment, which is qualitative with 4 levels. There are 4 treatments corresponding to the 4 factor levels. The experimental units are the 28 castings.

b. Some preliminary calculations are:

$$CM = \frac{\left(\sum x_i\right)^2}{n} = \frac{345.92^2}{28} = 4273.5945$$

$$SS(Total) = \sum x_i^2 - CM = 4370.6264 - 4273.5945 = 97.0319$$

$$SST = \sum \frac{T_i^2}{n_i} - CM = \frac{83.90^2}{7} + \frac{98.65^2}{7} + \frac{87.00^2}{7} + \frac{76.37^2}{7}$$

$$- 4273.5945 = 4310.3442 - 4273.5945 = 36.7497$$

$$SSE = SS(Total) - SST = 97.0319 - 36.7497 = 60.2822$$

$$MST = \frac{SST}{p-1} = \frac{36.7497}{4-1} = 12.2499 \qquad MSE = \frac{SSE}{n-p} = \frac{60.2822}{28-4} = 2.5118$$

$$F = \frac{MST}{MSE} = \frac{12.2499}{2.5118} = 4.88$$

Source	df	SS	MS	F
Treatments	3	36.7497	12.2499	4.88
Error	24	60.2822	2.5118	
Total	27	97.0319		

To determine if a difference in mean wear exists among the treatments, we test:

H_0: $\mu_1 = \mu_2 = \mu_3 = \mu_4$
H_a: At least one treatment mean is different

The test statistic is $F = 4.88$.

The rejection region requires $\alpha = .05$ in the upper tail of the F distribution with $\nu_1 = p - 1 = 4 - 1 = 3$ and $\nu_2 = n - p = 28 - 4 = 24$. From Table VIII, Appendix B, $F_{.05} = 3.01$. The rejection region is $F > 3.01$.

Since the observed value of the test statistic falls in the rejection region ($F = 4.88 > 3.01$), H_0 is rejected. There is sufficient evidence to indicate a difference in mean wear exists among the four treatments at $\alpha = .05$.

c.   The observed significance level is $P(F \geq 4.88) < .01$ from Table X, Appendix B, with $\nu_1 = 3$ and $\nu_2 = 24$.

d.   There are 6 pairs of treatments, so $c = 6$. For $\alpha/2c = .10/2(6) = .0083 \approx .005$, and $df = n - p = 28 - 4 = 24$, $t_{.005} = 2.797$ from Table VI, Appendix B. We form confidence intervals for the difference between each pair of means using the formula:

$$(\bar{y}_i - \bar{y}_j) \pm t_{\alpha/2c}s\sqrt{\frac{1}{n_i} + \frac{1}{n_j}} \qquad \text{where } s = \sqrt{MSE} = \sqrt{2.5118} = 1.5849$$

$$\bar{y}_A = \frac{83.90}{7} = 11.99 \qquad\qquad \bar{y}_B = \frac{98.65}{7} = 14.09$$

$$\bar{y}_C = \frac{87.00}{7} = 12.43 \qquad\qquad \bar{y}_D = \frac{76.37}{7} = 10.91$$

Pair

A – B $\quad (11.99 - 14.09) \pm 2.797(1.5849)\sqrt{\dfrac{1}{7} + \dfrac{1}{7}} \Rightarrow -2.1 \pm 2.369$

$\qquad\qquad\qquad\qquad\qquad\qquad\qquad\qquad\qquad\qquad \Rightarrow (-4.469, .269)$

A – C $\quad (11.99 - 12.43) \pm 2.369 \Rightarrow -.44 \pm 2.369 \Rightarrow (-2.809, 1.929)$

A – D $\quad (11.99 - 10.91) \pm 2.369 \Rightarrow 1.08 \pm 2.369 \Rightarrow (-1.289, 3.449)$

B – C $\quad (14.09 - 12.43) \pm 2.369 \Rightarrow 1.66 \pm 2.369 \Rightarrow (-.709, 4.029)$

B – D $\quad (14.09 - 10.91) \pm 2.369 \Rightarrow 3.18 \pm 2.369 \Rightarrow (.811, 5.549)$

C – D $\quad (12.43 - 10.91) \pm 2.369 \Rightarrow 1.52 \pm 2.369 \Rightarrow (-.849, 3.889)$

Only treatments B and D have different mean wears.

e. The confidence interval for μ_1 is

$$\bar{y}_A \pm t_{\alpha/2}s\sqrt{\dfrac{1}{n_A}} \quad \text{where } s = \sqrt{\text{MSE}} = 1.5849$$

For confidence coefficient .90, $\alpha = 1 - .90 = .10$ and $\alpha/2 = .10/2 = .05$. From Table VI, Appendix B, $t_{.05} = 1.711$ with df $= n - p = 28 - 4 = 24$. The 90% confidence interval is:

$$11.99 \pm 1.711(1.5849)\sqrt{\dfrac{1}{7}} \Rightarrow 11.99 \pm 1.025 \Rightarrow (10.965, 13.015)$$

16.47 a. df(Companies) $= p - 1 = 2 - 1 = 1$
       df(Error) $= n - p = 100 - 2 = 98$

$$\text{MST} = \dfrac{\text{SST}}{p - 1} = \dfrac{3237.2}{2 - 1} = 3237.2 \qquad \text{MSE} = \dfrac{\text{SSE}}{n - p} = \dfrac{16167.7}{100 - 2} = 164.9765$$

$$F = \dfrac{\text{MST}}{\text{MSE}} = \dfrac{3237.2}{164.9765} = 19.62$$

Source	df	SS	MS	F
Companies	1	3237.2	3237.2	19.62
Error	98	16167.7	164.9765	
Total	99	19404.9		

b. To determine if the mean number of hours missed differs for employees of the two companies, we test:

$$H_0: \ \mu_1 = \mu_2$$
$$H_a: \ \mu_1 \neq \mu_2$$

The test statistic is $F = \dfrac{\text{MST}}{\text{MSE}} = 19.62$

The rejection region requires $\alpha = .05$ in the upper tail of the F distribution with $\nu_1 = p - 1 = 2 - 1 = 1$ and $\nu_2 = n - p = 100 - 2 = 98$. From Table VIII, Appendix B, $F_{.05} \approx 4.00$. The rejection region is $F > 4.00$.

Since the observed value of the test statistic falls in the rejection region ($F = 19.62 > 4.00$), H_0 is rejected. There is sufficient evidence to indicate the mean number of hours missed differs for employees of the two companies at $\alpha = .05$.

c. No. We need the sample means for the two companies.

16.49 a. The experiment is completely randomized. The response is the attitude test score after 1 month. The two factors are scheduling (2 levels) and payment (2 levels). Both factors are qualitative. There are 4 different treatments, with the experimental units the workers.

b. To determine if the treatment means differ, we test:

H_0: $\mu_1 = \mu_2 = \mu_3 = \mu_4$
H_a: At least one treatment mean differs

The test statistic is $F = \dfrac{MST}{MSE} = 12.29$

The rejection region requires $\alpha = .05$ in the upper tail of the F distribution with $\nu_1 = ab - 1 = 2(2) - 1 = 3$ and $\nu_2 = n - ab = 16 - 2(2) = 24$. From Table VIII, Appendix B, $F_{.05} = 3.49$. The rejection region is $F > 3.49$.

Since the observed value of the test statistic falls in the rejection region ($F = 12.29 > 3.49$), H_0 is rejected. There is sufficient evidence to indicate the treatment means differ at $\alpha = .05$.

c. To determine if the factors interact, we test:

H_0: Factor A and factor B do not interact to affect the response mean
H_a: Factors A and B do interact to affect the response mean

The test statistic is $F = \dfrac{MSAB}{MSE} = .02$

The rejection region requires $\alpha = .05$ in the upper tail of the F distribution with $\nu_1 = (a - 1)(b - 1) = (2 - 1)(2 - 1) = 1$ and $\nu_2 = n - ab = 16 - 2(2) = 12$. From Table VIII, Appendix B, $F_{.05} = 4.75$. The rejection region is $F > 4.75$.

Since the observed value of the test statistic does not fall in the rejection region ($F = .02 \not> 4.75$), H_0 is not rejected. There is insufficient evidence to indicate the factors interact at $\alpha = .05$.

To determine if the mean attitude test scores differ for the two types of scheduling, we test:

H_0: $\mu_1 = \mu_2$
H_a: $\mu_1 \neq \mu_2$

The test statistic is $F = \dfrac{MS(\text{Schedule})}{MSE} = 7.37$

The rejection region requires $\alpha = .05$ in the upper tail of the F distribution with $\nu_1 = (a - 1) = 2 - 1 = 1$ and $\nu_2 = n - ab = 16 - 2(2) = 12$. From Table VIII, Appendix B, $F_{.05} = 4.75$. The rejection region is $F > 4.75$.

Since the observed value of the test statistic falls in the rejection region ($F = 7.37 > 4.75$), H_0 is rejected. There is sufficient evidence to indicate the mean attitude test scores differ for the two types of scheduling at $\alpha = .05$.

To determine if the mean attitude test scores differ for the two types of payments, we test:

H_0: $\mu_1 = \mu_2$
H_a: $\mu_1 \neq \mu_2$

The test statistic is $F = \dfrac{\text{MS(Payment)}}{\text{MSE}} = 29.47$

The rejection region requires $\alpha = .05$ in the upper tail of the F distribution with $\nu_1 = b - 1 = 2 - 1 = 1$ and $\nu_2 = n - ab = 16 - 2(2) = 12$. From Table VIII, Appendix B, $F_{.05} = 4.75$. The rejection region is $F > 4.75$.

Since the observed value of the test statistic falls in the rejection region ($F = 29.47 > 4.75$), H_0 is rejected. There is sufficient evidence to indicate the mean attitude test scores differ for the two types of payment at $\alpha = .05$.

Since the mean attitude test scores for $8 - 5$ is $558/8 = 69.75$ and the mean for worker-modified schedules is $634/8 = 79.25$, the mean attitude test scores for those on worker-modified schedules is significantly higher than for those on $8 - 5$ schedules.

Since the mean attitude test scores for those on hourly rate is $520/8 = 65$ and the mean for those on hourly and piece rate is $672/8 = 84$, the mean attitude test scores for those on hourly and piece rate is significantly higher than for those on hourly rate.

d. The necessary assumptions are:

1. The probability distributions for each schedule-payment combination is normal.
2. The variances for each distribution are equal.
3. The samples are random and independent.

16.51 $\text{MST} = \dfrac{\text{SST}}{p - 1} = \dfrac{421.74}{3 - 1} = 210.87$ $\text{MSE} = \dfrac{\text{SSE}}{n - p} = \dfrac{3574.06}{53 - 3} = 71.4812$

$F = \dfrac{\text{MST}}{\text{MSE}} = \dfrac{210.87}{71.4812} = 2.95$

To determine if there is a difference in mean length of service among the three factories, we test:

H_0: $\mu_1 = \mu_2 = \mu_3$
H_a: At least one treatment mean is different

The test statistic is $F = \dfrac{\text{MST}}{\text{MSE}} = 2.95$

The rejection region requires $\alpha = .05$ in the upper tail of the F distribution with $\nu_1 = p - 1 = 3 - 1 = 2$ and $\nu_2 = n - p = 53 - 3 = 50$. From Table VIII, Appendix B, $F_{.05} \approx 3.23$. The rejection region is $F > 3.23$.

Since the observed value of the test statistic does not fall in the rejection region ($F = 2.95 \not> 3.23$), H_0 is not rejected. There is insufficient evidence to indicate the mean lengths of service differ among the three factories at $\alpha = .05$.

16.53 a. Some preliminary calculations are:

$$\bar{x}_1 = \frac{\sum x_1}{n_1} = \frac{426}{8} = 53.25 \qquad \bar{x}_2 = \frac{\sum x_2}{n_2} = \frac{451.5}{8} = 56.4375$$

$$s_1^2 = \frac{\sum x_1^2 - \dfrac{\left(\sum x_1\right)^2}{n_1}}{n_1 - 1} = \frac{24207.25 - \dfrac{426^2}{8}}{8 - 1} = 217.5714$$

$$s_2^2 = \frac{\sum x_2^2 - \dfrac{\left(\sum x_2\right)^2}{n_2}}{n_2 - 1} = \frac{26607.25 - \dfrac{451.5^2}{8}}{8 - 1} = 160.8170$$

$$s_p^2 = \frac{(n_1 - 1)s_1^2 + (n_2 - 1)s_2^2}{n_1 + n_2 - 2} = \frac{(8 - 1)(217.5714) + (8 - 1)(160.8170)}{8 + 8 - 2} = 189.1942$$

H_0: $\mu_A = \mu_B$
H_a: $\mu_A \neq \mu_B$

The test statistic is $t = \dfrac{(\bar{x}_1 - \bar{x}_2) - D_0}{\sqrt{s_p^2\left[\dfrac{1}{n_1} + \dfrac{1}{n_2}\right]}} = \dfrac{53.25 - 56.4375}{\sqrt{189.1942\left[\dfrac{1}{8} + \dfrac{1}{8}\right]}} = \dfrac{-3.1875}{6.8774} = -.46$

The rejection region requires $\alpha/2 = .05/2 = .025$ in each tail of the t distribution with df $= n_1 + n_2 - 2 = 8 + 8 - 2 = 14$. From Table VI, Appendix B, $t_{.025} = 2.145$. The rejection region is $t < -2.145$ or $t > 2.145$.

Since the observed value of the test statistic does not fall in the rejection region ($t = -.46 \not< -2.145$), H_0 is not rejected. There is insufficient evidence to indicate the average home value is different in the two communities at $\alpha = .05$.

b. H_0: $\beta_1 = 0$
H_a: $\beta_1 \neq 0$

Design of Experiments and Analysis of Variance

The test statistic is $t = .46$.

The rejection region is $t < -2.145$ or $t > 2.145$ (as in part a). The conclusion is the same as in part a.

c. H_0: $\mu_A = \mu_B$
H_a: $\mu_A \neq \mu_B$

The test statistic is $F = \dfrac{MST}{MSE} = .21$

The rejection region requires $\alpha = .05$ in the upper tail of the F distribution with $\nu_1 = p - 1 = 2 - 1 = 1$ and $\nu_2 = n - p = 16 - 2 = 14$. From Table VIII, Appendix B, $F_{.05} = 4.60$. The rejection region is $F > 4.60$.

Since the observed value of the test statistic does not fall in the rejection region ($F = .21$ $\not> 4.60$), H_0 is not rejected. There is insufficient evidence to indicate the average home value is different in the two communities at $\alpha = .05$.

d. The test statistics in parts a and b are the same except for the sign, and the rejection regions are identical. We can also see that $t^2 = .46^2 = .21 = F$ in part c. Also,

$t^2_{.025} = 2.145^2 = 4.60 = F$ in part c.

16.55   a. From Exercise 16.54, $\bar{y}_1 = 26.16$, $\bar{y}_2 = 14.42$, and $\bar{y}_3 = 15.56$

Let $x_1 = \begin{cases} 1 & \text{if NYSE} \\ 0 & \text{otherwise} \end{cases}$ $x_2 = \begin{cases} 1 & \text{if ASE} \\ 0 & \text{otherwise} \end{cases}$

The model is $E(y) = \beta_0 + \beta_1 x_1 + \beta_2 x_2$

b. The fitted model is $\hat{y} = 15.56 + 10.60 x_1 - 1.14 x_2$

$\hat{\beta}_0 = 15.56$ This is the estimated mean closing price for stocks traded OTC

$\hat{\beta}_1 = 10.60$ This is the estimated difference in mean closing prices between stocks traded on the NYSE and OTC.

$\hat{\beta}_2 = -1.14$ This is the estimated difference in mean closing prices between stocks traded on the ASE and OTC.

Note: $\bar{y}_3 = 15.56 = \hat{\beta}_0$, $\bar{y}_1 = 26.16 = \hat{\beta}_0 + \hat{\beta}_1$, $\bar{y}_2 = 14.42 = \hat{\beta}_0 + \hat{\beta}_2$

c. To determine whether the three markets had the same mean closing prices, we test:

H_0: $\beta_1 = \beta_2 = 0$
H_a: At least one $\beta_i \neq 0$, $i = 1, 2$

The test statistic is $F = 4.114$.

The rejection region requires $\alpha = .10$ in the upper tail of the F distribution with $\nu_1 = k = 2$ and $\nu_2 = n - (k + 1) = 87 - (2 + 1) = 84$. From Table VII, Appendix B, $F_{.10} \approx 2.39$. The rejection region is $F > 2.39$.

Since the observed value of the test statistic falls in the rejection region ($F = 4.114 > 2.39$), H_0 is rejected. There is sufficient evidence to indicate the three markets had different mean closing prices at $\alpha = .10$.

The observed significance level is $P(F \geq 4.114) = .0197$.

16.57   a.   Some preliminary calculations are:

$$CM = \frac{(\sum y)^2}{n} = \frac{28,381^2}{18} = 44,748,953.39$$

$$SS(Total) = \sum y^2 - CM = 44,829,635 - 44,748,953.39 = 80,681.61$$

$$SSA = \frac{\sum A_i^2}{br} - CM = \frac{13,971^2}{3(3)} + \frac{14,410^2}{3(3)} - 44,748,953.39$$

$$= 44,759,660.11 - 44,748,953.39 = 10,706.72$$

$$SSB = \frac{\sum B_i^2}{ar} - CM = \frac{9013^2}{2(3)} + \frac{9607^2}{2(3)} + \frac{9761^2}{2(3)} - 44,748,953.39$$

$$= 44,800,956.5 - 44,748,953.39 = 52,003.11$$

$$SSAB = \frac{\sum \sum AB_{ij}^2}{r} - SSA - SSB - CM$$

$$= \frac{4438^2}{3} + \frac{4722^2}{3} + \cdots + \frac{4950^2}{3} - 10,706.72 - 52,003.11$$

$$- 44,748,953.39 = 44,811,733 - 44,811,663.22 = 69.78$$

$SSE = SS(Total) - SSA - SSB - SSAB$

$= 80,681.62 - 10,706.72 - 52,003.11 - 69.78$

$= 17,902.00$

$$MSA = \frac{SSA}{a - 1} = \frac{10,706.72}{2 - 1} = 10,706.72$$

$$MSB = \frac{SSB}{b - 1} = \frac{52,003.11}{3 - 1} = 26,001.56$$

$$MSAB = \frac{SSAB}{(a - 1)(b - 1)} = \frac{69.78}{(2 - 1)(3 - 1)} = 34.89$$

$$MSE = \frac{SSE}{n - ab} = \frac{17,902}{18 - 2(3)} = 1491.8333$$

$$F_A = \frac{MSA}{MSE} = \frac{10,706.72}{1491.8333} = 7.18 \qquad F_B = \frac{MSB}{MSE} = \frac{26,001.56}{1491.8333} = 17.43$$

$$F_{AB} = \frac{MSAB}{MSE} = \frac{34.89}{1491.8333} = .023$$

The ANOVA table is:

Source	df	SS	MS	F
A (Type Plant)	1	10,706.72	10,706.72	7.18
B (Incentive)	2	52,003.11	26,001.56	17.43
AB	2	69.78	34.89	.023
Error	12	17,902.00	1491.8333	
Total	17	80,681.61		

b. To determine whether interaction between incentive level and type of plant exists, we test:

H_0: Incentive level and type of plant do not interact to affect productivity
H_a: Incentive level and type of plant do interact to affect productivity

The test statistic is $F = .023$.

The rejection region requires $\alpha = .05$ in the upper tail of the F distribution with $\nu_1 = (a - 1)(b - 1) = (2 - 1)(3 - 1) = 2$ and $\nu_2 = n - ab = 18 - 2(3) = 12$. From Table VIII, Appendix B, $F_{.05} = 3.89$. The rejection region is $F > 3.89$.

Since the observed value of the test statistic does not fall in the rejection region ($F = .023 \not> 3.89$), H_0 is not rejected. There is insufficient evidence to indicate interaction exists between incentive level and type of plant at $\alpha = .05$.

c. No interaction implies that the difference in the value of the incentive affects union and nonunion workers the same.

d. From Figure 12.21, SSE = 23,349.2222, while SSE = 17,902.00 for this problem. The difference between the two is because different models were used to fit the data. In Figure 12.21, $x_1 = $ incentive, has only one degree of freedom. This implies only a linear relationship is expressed between incentive and productivity. In this problem, incentive has two degrees of freedom. This implies the relationship between incentive and productivity could be linear or quadratic.

e. If the independent variables are quantitative, multiple regression provides more practical information than analysis of variance. Analysis of variance will just provide information about whether a relationship exists between the independent and dependent variables. Regression analysis specifies the type of relationship between the independent and dependent variables.

CHAPTER SEVENTEEN

. . . .

Nonparametric Statistics

17.1   The sign test is preferred to the t test when the population from which the sample is selected is not normal.

17.3   a. $P(x \geq 7) = 1 - P(x \leq 6) = 1 - .965 = .035$

     b. $P(x \geq 5) = 1 - P(x \leq 4) = 1 - .637 = .363$

     c. $P(x \geq 8) = 1 - P(x \leq 7) = 1 - .996 = .004$

     d. $P(x \geq 10) = 1 - P(x \leq 9) = 1 - .849 = .151$

$$\mu = np = 15(.5) = 7.5 \text{ and } \sigma = \sqrt{npq} = \sqrt{15(.5)(.5)} = 1.9365$$

$$P(x \geq 10) \approx P\left[z \geq \frac{(10 - .5) - 7.5}{1.9365}\right] = P(z \geq 1.03) = .5 - .3485 = .1515$$

     e. $P(x \geq 15) = 1 - P(x \leq 14) = 1 - .788 = .212$

$$\mu = np = 25(.5) = 12.5 \text{ and } \sigma = \sqrt{npq} = \sqrt{25(.5)(.5)} = 2.5$$

$$P(x \geq 15) \approx P\left[z \geq \frac{(15 - .5) - 12.5}{2.5}\right] = P(z \geq .80) = .5 - .2881 = .2119$$

17.5   To determine if the median is greater than 75, we test:

H_0: $M = 75$
H_a: $M > 75$

The test statistic is S = number of measurements greater than $75 = 17$.

The p-value $= P(x \geq 17)$ where x is a binomial random variable with $n = 25$ and $p = .5$. From Table II,

$$p\text{-value} = P(x \geq 17) = 1 - P(x \leq 16) = 1 - .946 = .054$$

Since the p-value $= .054 < \alpha = .10$, H_0 is rejected. There is sufficient evidence to indicate the median is greater than 75 at $\alpha = .10$.

We must assume the sample was randomly selected from a continuous probability distribution.

Note: Since $n \geq 10$, we could use the large-sample approximation.

17.7   a.   I would recommend the sign test because 5 of the sample measurements are of similar magnitude, but the 6th is about three times as large as the others. It would be very unlikely to observe this sample if the population were normal.

b. To determine if the airline is meeting the requirement, we test:

H_0: $M = 30$
H_a: $M < 30$

c. The test statistic is S = number of measurements less than $30 = 5$.

H_0 will be rejected if the p-value $< \alpha = .01$.

d. The test statistic is $S = 5$.

The p-value $= P(x \geq 5)$ where x is a binomial random variable with $n = 6$ and $p = .5$. From Table II,

$$p\text{-value} = P(x \geq 5) = 1 - P(x \leq 4) = 1 - .891 = .109$$

Since the p-value $= .109$ is not less than $\alpha = .01$, H_0 is not rejected. There is insufficient evidence to indicate the airline is meeting the maintenance requirement at $\alpha = .01$.

17.9 a. The test statistic is T_B, the rank sum of population B (because $n_B < n_A$).

The rejection region is $T_B \leq 35$ or $T_B \geq 67$, from Table XI, Appendix B, with $n_A = 10$, $n_B = 6$, and $\alpha = .10$.

b. The test statistic is T_A, the rank sum of population A (because $n_A < n_B$).

The rejection region is $T_A \geq 43$, from Table XI, Appendix B, with $n_A = 5$, $n_B = 7$, and $\alpha = .05$.

c. The test statistic is T_B, the rank sum of population B (because $n_B < n_A$).

The rejection region is $T_B \geq 93$, from Table XI, Appendix B, with $n_A = 9$, $n_B = 8$, and $\alpha = .025$.

d. Since $n_A = n_B = 15$, the test statistic is

$$z = \frac{T_A - \dfrac{n_1(n_1 + n_2 + 1)}{2}}{\sqrt{\dfrac{n_1 n_2 (n_1 + n_2 + 1)}{12}}}$$

The rejection region is $z < -z_{\alpha/2}$ or $z > z_{\alpha/2}$. For $\alpha = .05$ and $\alpha/2 = .05/2 = .025$, $z_{.025} = 1.96$ from Table IV, Appendix B. The rejection region is $z < -1.96$ or $z > 1.96$.

17.11 The alternative hypotheses differ for one- and two-tailed versions of the Wilcoxon rank sum test. For a two-tailed test, the alternative hypothesis is H_a: The probability distribution for one population is shifted to the right or left of the other distribution. For a one-tailed test, the alternative hypothesis is H_a: The probability distribution for one population is shifted to the right (left) of the other distribution.

The rejection regions are also different. For a two-tailed test, the rejection region is $T \leq T_L$ or $T \geq T_U$ where T is the rank sum of the sample with the smallest sample size. For a one-tailed test, the rejection region is $T \geq T_U$ (or $T \leq T_L$).

17.13 a. We first rank all observations:

Neighborhood A		Neighborhood B	
Observation	**Rank**	**Observation**	**Rank**
.850	11	.911	16
1.060	18	.770	3
.910	15	.815	8
.813	7	.748	2
.787	1	.835	9
.880	13	.800	6
.895	14	.793	4
.844	10	.769	5
.965	17		
.875	12		
	$T_A = 118$		$T_B = 53$

To determine the fairness of the assessments between the two neighborhoods, we test:

H_0: Two sampled populations have identical probability distributions
H_a: The probability distribution for Neighborhood A is shifted to the right or left of that for Neighborhood B

The test statistic is $T_B = 53$ because $n_B < n_A$.

The rejection region is $T_B \leq 54$ or $T_B \geq 98$ from Table XI, Appendix B, with $n_A = 10$, $n_B = 8$, and $\alpha = .05$.

Since the observed value of the test statistic falls in the rejection region ($T = 53 \leq 54$), H_0 is rejected. There is sufficient evidence to indicate the fairness of the assessments are not the same for the two neighborhoods at $\alpha = .05$.

b. In order to use the two-sample t test, we have to have normal distributions for both neighborhoods A and B, the samples must be independent, and the population variances must be equal.

c. We must assume the two samples are random and independent, and the two probability distributions are continuous.

17.15 We first rank all data:

Before Right-Turn Law	Rank	After Right-Turn Law	Rank
150	3	145	2
500	11	390	8
250	5	680	13
301	7	560	12
242	4	899	14
435	10	1250	16
100	1	290	6
402	9	963	15
$T_{Before} = 50$		$T_{After} = 86$	

To determine whether the damages tended to increase after the enactment of the law, we test:

H_0: The distributions before and after the right-turn law are identical
H_a: The distribution after the right-turn law is shifted to the right of that before the right-turn law

The test statistic is $T_{After} = 86$.

The rejection region is $T \geq 84$ from Table XI, Appendix B, with $n_A = n_B = 8$ and $\alpha = .05$.

Since the observed value of the test statistic falls in the rejection region ($T = 86 \geq 84$), H_0 is rejected. There is sufficient evidence to indicate the damages tended to increase after the enactment of the law at $\alpha = .05$.

17.17 a. Some preliminary calculations are:

$$\bar{x}_A = \frac{\sum x_A}{n_A} = \frac{279}{6} = 46.5 \qquad \bar{x}_B = \frac{\sum x_B}{n_B} = \frac{442}{8} = 55.25$$

$$s_A^2 = \frac{\sum x_A^2 - \frac{\left(\sum x_A\right)^2}{n_A}}{n_A - 1} = \frac{13247 - \frac{279^2}{6}}{6 - 1} = \frac{273.5}{5} = 54.7$$

$$s_B^2 = \frac{\sum x_B^2 - \frac{\left(\sum x_B\right)^2}{n_B}}{n_B - 1} = \frac{26136 - \frac{442^2}{8}}{8 - 1} = \frac{1715.5}{7} = 245.0714$$

$$s_p^2 = \frac{(n_A - 1)s_A^2 + (n_B - 1)s_B^2}{n_A + n_B - 2} = \frac{(6 - 1)(54.7) + (8 - 1)(245.0714)}{6 + 8 - 2} = \frac{1989}{12}$$
$$= 165.75$$

To determine whether the mean prices per house differ in the two subdivisions, we test:

H_0: $\mu_1 = \mu_2$
H_a: $\mu_1 \neq \mu_2$

The test statistic is

$$t = \frac{(\bar{x}_A - \bar{x}_B) - 0}{\sqrt{s_p^2 \left[\frac{1}{n_A} + \frac{1}{n_B} \right]}} = \frac{46.5 - 55.25}{\sqrt{165.75 \left[\frac{1}{6} + \frac{1}{8} \right]}} = \frac{-8.75}{6.9530} = -1.26$$

The rejection region requires $\alpha/2 = .05/2 = .025$ in each tail of the t distribution with df $= n_A + n_B - 2 = 6 + 8 - 2 = 12$. From Table VI, Appendix B, $t_{.025} = 2.179$. The rejection region is $t < -2.179$ or $t > 2.179$.

Since the observed value of the test statistic does not fall in the rejection region ($t = -1.26 \nless -2.179$), H_0 is not rejected. There is insufficient evidence to indicate the mean prices differ for the two subdivisions at $\alpha = .05$.

We must assume:

1. Both populations are normal
2. The samples are independent
3. $\sigma_1^2 = \sigma_2^2$

Because $s_A^2 = 54.7$ and $s_B^2 = 245.0714$, it appears the assumption $\sigma_1^2 = \sigma_2^2$ may not be reasonable.

b. The data and their ranks are:

A	Rank	B	Rank
43	5	57	11
48	8	39	1.5
42	4	55	10
60	12	52	9
39	1.5	88	14
47	7	46	6
		41	3
		64	13
	$T_A = 37.5$		$T_B = 67.5$

To determine whether there is a shift in the locations of the probability distributions of house prices in the two subdivisions, we test:

H_0: The 2 sampled populations have identical probability distributions
H_a: The probability distribution for subdivision A is shifted to the right or left of that for subdivision B

The test statistic is $T_A = 37.5$.

The rejection region is $T_A \leq 29$ or $T_A \geq 61$ from Table XI, Appendix B, with $n_A = 6$, $n_B = 8$, and $\alpha = .05$.

Since the observed value of the test statistic does not fall in the rejection region ($T_A = 37.5 \not\leq 29$ and $\not\geq 61$), H_0 is not rejected. There is insufficient evidence to indicate a shift in the locations of the distributions of house prices in the two subdivisions.

17.19 a.   The test statistic is T_- or T_+, the smaller of the two.

The rejection region is $T \leq 152$, from Table XII, Appendix B, with $n = 30$, $\alpha = .10$, and two-tailed.

b.   The test statistic is T_-.

The rejection region is $T_- \leq 60$, from Table XII, Appendix B, with $n = 20$, $\alpha = .05$, and one-tailed.

c.   The test statistic is T_+.

The rejection region is $T_+ \leq 0$, from Table XII, Appendix B, with $n = 8$, $\alpha = .005$, and one-tailed.

17.21 The difference between a one- and two-tailed Wilcoxon signed rank test is the following:

A one-tailed test is used to test if one population distribution is shifted in a specified direction. A two-tailed test is used to test if one population distribution is shifted in either direction.

17.23 a. H_0: The two sampled populations have identical probability distributions
H_a: The probability distribution for population A is located to the right of that for population B

b.   The test statistic is

$$z = \frac{T_+ - \dfrac{n(n+1)}{4}}{\sqrt{\dfrac{n(n+1)(2n+1)}{24}}} = \frac{354 - \dfrac{30(30+1)}{4}}{\sqrt{\dfrac{30(30+1)(60+1)}{24}}} = \frac{121.5}{48.6184} = 2.499$$

The rejection region requires $\alpha = .05$ in the upper tail of the z distribution. From Table IV, Appendix B, $z_{.05} = 1.645$. The rejection region is $z > 1.645$.

Since the observed value of the test statistic falls in the rejection region ($z = 2.499 >$ 1.645), H_0 is rejected. There is sufficient evidence to indicate population A is located to the right of that for population B at $\alpha = .05$.

c.   The p-value $= P(z \geq 2.499) = .5 - .4938 = .0062$.

d.   The necessary assumptions are:

    1.   The sample of differences is randomly selected from the population of differences.

    2.   The probability distribution from which the sample of paired differences is drawn is continuous.

17.25   Some preliminary calculations are:

Employee	Before Flextime	After Flextime	Difference (B − A)	Difference
1	54	68	−14	7
2	25	42	−17	9
3	80	80	0	(Eliminated)
4	76	91	−15	8
5	63	70	−7	5
6	82	88	−6	3.5
7	94	90	4	2
8	72	81	−9	6
9	33	39	−6	3.5
10	90	93	−3	$\underline{1}$
				$T_+ = 2$

To determine if the pilot flextime program is a success, we test:

H_0:   The two probability distributions are identical
H_a: The probability distribution before is shifted to the left of that after

The test statistic is $T_+ = 2$.

The rejection region is $T_+ \leq 8$, from Table XII, Appendix B, with $n = 9$ and $\alpha = .05$.

Since the observed value of the test statistic falls in the rejection region ($T_+ = 2 \leq 8$), H_0 is rejected. There is sufficient evidence to indicate the pilot flextime program has been a success at $\alpha = .05$.

17.27 Some preliminary calculations are:

Location	A	B	Difference A − B	Rank of Absolute Differences
1	879	1085	−206	6
2	445	325	120	2
3	692	848	−156	5
4	1565	1421	144	4
5	2326	2778	−452	8
6	857	992	−135	3
7	1250	1303	−53	1
8	773	1215	−442	7
				$T_+ = 6$

To determine whether one of the chains tends to have more customers than the other, we test:

H_0: The two sampled populations have identical probability distributions

H_a: The probability distribution for chain A is shifted to the right or left of that for chain B

The test statistic is $T_+ = 6$.

The rejection region is $T_+ \le 4$ from Table XII, Appendix B, with $n = 8$ and $\alpha = .05$.

Since the observed value of the test statistic does not fall in the rejection region ($T_+ = 6 \not\le 4$), H_0 is not rejected. There is insufficient evidence to indicate one of the chains tends to have more customers than the other at $\alpha = .05$.

17.29 a. Some preliminary calculations are:

Product Category	Jan 1985	Jan 1986	1985−1986	Rank of Absolute Differences
Processed poultry	198.8	192.4	6.4	1
Concrete ingredients	331.0	339.0	−8.0	3
Lumber	343.0	329.6	13.4	4
Gas fuels	1073.0	1034.3	38.7	6
Drugs and pharm.	247.4	265.9	−18.5	5
Synthetic fibers	157.6	151.1	6.5	2
				$T_- = 8$

$$\bar{d} = \frac{\sum d_i}{n} = \frac{38.5}{6} = 6.4167$$

$$s_d^2 = \frac{\sum d_i^2 - \frac{\left(\sum d_i\right)^2}{n}}{n - 1} = \frac{2166.71 - \frac{38.5^2}{6}}{6 - 1} = 383.9337$$

$$s_d = \sqrt{383.9337} = 19.5942$$

To determine whether the mean number of these indexes differ, we test:

H_0: $\mu_d = 0$
H_a: $\mu_d \neq 0$

The test statistic is $t = \dfrac{\bar{d} - 0}{s_d/\sqrt{n}} = \dfrac{6.4167}{\dfrac{19.5942}{\sqrt{6}}} = .802$

The rejection region requires $\alpha/2 = .05/2 = .025$ in each tail of the t distribution with df $= n - 1 = 6 - 1 = 5$. From Table VI, Appendix B, $t_{.025} = 2.571$. The rejection region is $t < -2.571$ or $t < 2.571$.

Since the observed value of the test statistic does not fall in the rejection region ($t = .802 \not> 2.571$), H_0 is not rejected. There is insufficient evidence to indicate the mean values of these indexes differ at $\alpha = .05$.

We must assume that the populations of differences is normal and that we have a random sample from the population of differences.

b. To determine whether the probability distribution of the economic indexes has changed, we test:

H_0: The two sampled populations have identical probability distributions
H_a: The probability distribution for the economic indexes has changed

The test statistic is $T_- = 8$.

The rejection region is $T_- \leq 1$ from Table XII, Appendix B, with $n = 6$ and $\alpha = .05$.

Since the observed value of the test statistic does not fall in the rejection region ($T_- = 8 \not\leq 1$), H_0 is not rejected. There is insufficient evidence to indicate the probability distribution of the economic indexes has changed at $\alpha = .05$.

We must assume we have a random sample from the population of differences and that the population of differences is continuous.

17.31   Using Table XIII, Appendix B,

a. $P(\chi^2 \geq 3.07382) = .995$

b. $P(\chi^2 \leq 24.4331) = 1 - .975 = .025$

c. $P(\chi^2 \geq 14.6837) = .10$

d. $P(\chi^2 < 34.1696) = 1 - .025 = .975$

e. $P(\chi^2 < 6.26214) = 1 - .975 = .025$

f. $P(\chi^2 \leq .584375) = 1 - .90 = .10$

17.33   a.   A completely randomized design was used.

b.   The hypotheses are:

H_0: The three probability distributions are identical
H_a: At least two of the three probability distributions differ in location

c.   The rejection region requires $\alpha = .01$ in the upper tail of the χ^2 distribution with df = p $- 1 = 3 - 1 = 2$. From Table XIII, Appendix B, $\chi^2_{.01} = 9.21034$. The rejection region is $H > 9.21034$.

d.   Some preliminary calculations are:

I		II		III	
Observation	Rank	Observation	Rank	Observation	Rank
66	13	19	2	75	14.5
23	3	31	6	96	19
55	10	16	1	102	21
88	18	29	4	75	14.5
58	11	30	5	98	20
62	12	33	7	78	16
79	17	40	8		
49	9				
	$R_A = 93$		$R_B = 33$		$R_C = 105$

The test statistic is

$$H = \frac{12}{n(n+1)} \sum \frac{R_j^2}{n_j} - 3(n+1)$$

$$= \frac{12}{21(21+1)} \left[\frac{93^2}{8} + \frac{33^2}{7} + \frac{105^2}{6} \right] - 3(21+1) = 79.85 - 66 = 13.85$$

Since the observed value of the test statistic falls in the rejection region ($H = 13.85 >$ 9.21034), H_0 is rejected. There is sufficient evidence to indicate at least two of the three probability distributions differ in location at $\alpha = .01$.

17.35 a. Some preliminary calculations are:

Income		Growth		Maximum Growth	
Rate	Rank	Rate	Rank	Rate	Rank
8.3%	5	20.2%	17	2.2%	3
19.0	14	1.9	2	16.9	12
15.6	11	11.2	7	23.3	19
26.8	20	4.5	4	12.1	8
17.1	13	19.3	15	10.4	6
20.1	16	13.3	9	59.9	21
14.9	10	23.1	18	−9.3	1
	$R_A = 89$		$R_B = 72$		$R_C = 70$

To determine if the rate-of-return distributions differ among the three types of mutual funds, we test:

H_0: The three probability distributions are identical

H_a: At least two of the rate-of-return distributions differ

The test statistic is

$$H = \frac{12}{n(n + 1)} \sum \frac{R_j^2}{n_j} - 3(n + 1)$$

$$= \frac{12}{21(21 + 1)} \left[\frac{89^2}{7} + \frac{72^2}{7} + \frac{70^2}{7} \right] - 3(21 + 1) = 66.8089 - 66 = .8089$$

The rejection region requires $\alpha = .05$ in the upper tail of the χ^2 distribution with df $= p - 1 = 3 - 1 = 2$. From Table XIII, Appendix B, $\chi_{.05}^2 = 5.99147$. The rejection region is $H > 5.99147$.

Since the observed value of the test statistic does not fall in the rejection region ($H = .8089 \not> 5.99147$), H_0 is not rejected. There is insufficient evidence to indicate the rate-of-return distributions differ among the three types of mutual funds at $\alpha = .05$.

b. The necessary assumptions are:

1. The 3 samples are random and independent.
2. There are 5 or more measurements in each sample.
3. The 3 probability distributions from which the samples are drawn are continuous.

c. A Type I error would be concluding at least two of the rate-of-return distributions differ when they do not.

A Type II error would be concluding the three rate-of-return distributions are identical when they are not.

d. The F test could be used if the three distributions were normal with equal variances.

17.37 Some preliminary calculations are:

Urban	Rank	Suburban	Rank	Rural	Rank
4.3	4.5	5.9	14	5.1	9
5.2	10.5	6.7	17	4.8	7
6.2	15.5	7.6	19	3.9	2
5.6	12	4.9	8	6.2	15.5
3.8	1	5.2	10.5	4.2	3
5.8	13	6.8	18	4.3	4.5
4.7	6				
	$R_A = 62.5$		$R_B = 86.5$		$R_C = 41$

To determine if the level of property taxes among the three types of school districts, we test:

H_0: The three probability distributions are identical
H_a: At least two of the three probability distributions differ in location

The test statistic is $H = \dfrac{12}{n(n+1)} \sum \dfrac{R_j^2}{n_j} - 3(n+1)$

$$= \frac{12}{19(20)} \left[\frac{62.5^2}{7} + \frac{86.5^2}{6} + \frac{41^2}{6} \right] - 3(20) = 65.8498 - 60$$
$$= 5.8498$$

The rejection region requires $\alpha = .05$ in the upper tail of the χ^2 distribution with df $= p - 1$ $= 3 - 1 = 2$. From Table XIII, Appendix B, $\chi^2_{.05} = 5.99147$. The rejection region is $H > 5.99147$.

Since the observed value of the test statistic does not fall in the rejection region ($H = 5.8498$ $\not> 5.99147$), H_0 is not rejected. There is insufficient evidence to indicate the level of property taxes differ among the three types of school districts at $\alpha = .05$.

17.39 a. The assumptions for the F test are:

1. All p population probability distributions are normal.
2. The p population variances are equal.
3. Samples are selected randomly and independently from the respective populations.

The assumptions for the Kruskal-Wallis H test are:

1. The k samples are random and independent.
2. There are 5 or more measurements in each sample.
3. The observations can be ranked.

The assumptions for the Kruskal-Wallis H test are less restrictive than those for the F test.

b. Some preliminary calculations are:

Insurance	Rank	Publishing	Rank	Electric Utilities	Rank	Banking	Rank
.24	9	.03	2	.84	19	.54	14
.09	4.5	.32	11	1.00	20	.29	10
.12	6	.51	13	1.03	21	.82	17
.09	4.5	.15	8	1.11	22	.13	7
.00	1	.73	16	.83	18	.41	12
.07	3			1.16	23		
				.70	15		
$R_A = 28$		$R_B = 50$		$R_C = 138$		$R_D = 60$	

To determine whether debt-to-equity ratios differ among the four industries, we test:

H_0: The four probability distributions are identical
H_a: At least two of the four probability distributions differ in location

The test statistic is $H = \dfrac{12}{n(n+1)} \sum \dfrac{R_j^2}{n_j} - 3(n+1)$

$$= \frac{12}{23(24)} \left[\frac{23^2}{6} + \frac{50^2}{5} + \frac{138^2}{7} + \frac{60^2}{5} \right] - 3(24)$$

$$= 88.5052 - 72 = 16.5052$$

The rejection region requires $\alpha = .05$ in the upper tail of the χ^2 distribution with df $= p - 1 = 4 - 1 = 3$. From Table XIII, Appendix B, $\chi^2_{.05} = 7.81473$. The rejection region is $H > 7.81473$.

Since the observed value of the test statistic falls in the rejection region ($H = 16.5052 > 7.81473$), H_0 is rejected. There is sufficient evidence to indicate the debt-to-equity ratios differ among the four industries at $\alpha = .05$.

c. The distributions of debt-to-ratios for the electric utility and banking industries could be compared using the Wilcoxon rank sum test.

17.41 Since there are no ties, we will use the shortcut formula.

a. Some preliminary calculations are:

x Rank (u_i)	y Rank (v_i)	$d_i = u_i - v_i$	d_i^2
2	2	0	0
4	4	0	0
5	5	0	0
1	1	0	0
3	3	0	$\underline{0}$
			Total = 0

$$r_s = 1 - \frac{6\sum d_i^2}{n(n^2 - 1)} = 1 - \frac{6(0)}{5(5^2 - 1)} = 1$$

b.

x Rank (u_i)	y Rank (v_i)	$d_i = u_i - v_i$	d_i^2
2	3	-1	1
3	4	-1	1
4	2	2	4
5	1	4	16
1	5	-4	$\underline{16}$
			Total = 38

$$r_s = 1 - \frac{6\sum d_i^2}{n(n^2 - 1)} = 1 - \frac{6(38)}{5(5^2 - 1)} = 1 - 1.9 = -.9$$

c.

x Rank (u_i)	y Rank (v_i)	$d_i = u_i - v_i$	d_i^2
1	1	0	0
4	4	0	0
2	2	0	0
3	3	0	$\underline{0}$
			Total = 0

$$r_s = 1 - \frac{6\sum d_i^2}{n(n^2 - 1)} = 1 - \frac{6(0)}{4(4^2 - 1)} = 1 - 0 = 1$$

d.

x Rank (u_i)	y Rank (v_i)	$d_i = u_i - v_i$	d_i^2
2	1	1	1
5	3	2	4
4	5	−1	1
3	2	1	1
1	4	−3	9
			Total = 16

$$r_s = 1 - \frac{6\sum d_i^2}{n(n^2 - 1)} = 1 - \frac{6(16)}{5(5^2 - 1)} = 1 - .8 = .2$$

17.43   a.   Since there are no ties, we can use the formula

$$r_s = 1 - \frac{6\sum d_i^2}{n(n^2 - 1)}$$

Differences

$d_i = u_i - v_i$	d_i^2
0	0
0	0
0	0
0	0
−1	1
1	1
−2	4
−3	9
1	1
0	0
−3	9
0	0
6	36
1	1
	62

$$r_s = 1 - \frac{6(62)}{14(14^2 - 1)}$$
$$= 1 - .136$$
$$= .864$$

$H_0: \rho_s = 0$
$H_a: \rho_s > 0$

The test statistic is $r_s = .864$

From Table XVI, Appendix B, $r_{s,.05} = .457$ for $n = 14$. The rejection region is $r_s > .457$.

Since the observed value of the test statistic falls in the rejection region ($r_s = .864 > .457$), H_0 is rejected. There is sufficient evidence to indicate the magnitude of brake pressure and the electrodermal response are positively correlated at $\alpha = .05$. This supports Helander's finding.

b. Type I error = concluding the variables are positively related when in fact they are not.

Type II error = concluding the variables are not related when in fact they are positively related.

c. p-value = $P(r_s \geq .864) < .005$ from Table XVI, Appendix B, with $n = 14$.

d. The necessary assumptions are:
1. The sample of experimental units on which the two variables are measured is randomly selected.
2. The probability distributions of the two variables are continuous.

17.45 a. Since there are no ties, we can use the formula

$$r_s = 1 - \frac{6\sum d_i^2}{n(n^2 - 1)}$$

Differences $d_i = u_i - v_i$	d_i^2
1	1
−3	9
−1	1
4	16
−1	1
1	1
0	0
0	0
−1	1
−3	9
−1	1
2	4
3	9
−4	16
3	9
$\sum d_i^2 = 78$	

$$r_s = 1 - \frac{6\sum d_i^2}{(n^2 - 1)}$$
$$= 1 - \frac{6(78)}{15(15^2 - 1)}$$
$$= 1 - .139 = .861$$

b. To determine if job satisfaction and income are positively correlated, we test:

H_0: $\rho_s = 0$
H_a: $\rho_s > 0$

The test statistic is $r_s = .861$

From Table XVI, Appendix B, $r_{s.,05} = .441$, with $n = 15$. The rejection region is $r_s > .441$.

Since the observed value of the test statistic falls in the rejection region ($r_s = .861 > .441$), H_0 is rejected. There is sufficient evidence to indicate job satisfaction and income are positively correlated at $\alpha = .05$.

17.47 Since there are no ties, we can use the formula:

$$r_s = 1 - \frac{6\sum d_i^2}{n(n^2 - 1)}$$

Differences

$d_i = u_i - v_i$	d_i^2
0	0
0	0
−1	1
−1	1
2	4
−1	1
−1	1
−2	4
3	9
1	1
−1	1
1	1
0	0
	$\sum d_i^2 = 24$

$$r_s = 1 - \frac{6\sum d_i^2}{n(n^2 - 1)}$$
$$= 1 - \frac{6(24)}{13(13^2 - 1)}$$
$$= 1 - .0659 = .9341$$

The rankings of the manufacturers locating in Arkansas and the ranking of the manufacturers locating elsewhere are positively related. Because $r_s = .9341$, this relationship is very strong.

17.49 a. Some preliminary calculations:

Before		After	
Observation	Rank	Observation	Rank
12	10	4	3
5	4	2	1
10	9	7	6
9	8	3	2
14	11	8	7
6	5		
	$T_A = 47$		$T_B = 19$

To determine if the traffic light aided in reducing the number of accidents, we test:

H_0: The two sampled populations have identical probability distributions
H_a: The probability distribution after the traffic light was installed is shifted to the left of that before

The test statistic is $T_B = 19$.

The rejection region is $T_B \le 19$ from Table XI, Appendix B, with $n_A = 6$, $n_B = 5$, and $\alpha = .025$.

Since the observed value of the test statistic falls in the rejection region ($T_B = 19 \le 19$), H_0 is rejected. There is sufficient evidence to indicate the traffic light aided in reducing the number of accidents at $\alpha = .025$.

b. In order to use the t test, the data must be normally distributed. It is doubtful that the number of accidents is normally distributed.

17.51 a. Some preliminary calculations are:

Truck	Static Weight of Truck (u_i)	Weigh-in-Motion Prior (v_i)	Weigh-in-Motion After (w_i)	$u_i v_i$	$u_i w_i$
1	3	3	3	9	9
2	4	4	4	16	16
3	10	9	10	90	100
4	1	1.5	2	1.5	2
5	6	6	6	36	36
6	8	8	8	64	64
7	2	1.5	1	3	2
8	5	5	5	25	25
9	7	7	7	49	49
10	9	10	9	90	81
	55	55	55	383.5	384

$$SS_{uv} = \sum u_i v_i - \frac{\sum u_i \sum v_i}{n} = 383.5 - \frac{55(55)}{10} = 81$$

$$SS_{uw} = \sum u_i w_i - \frac{\left(\sum u_i \sum w_i\right)}{n} = 384 - \frac{55(55)}{10} = 81.5$$

$$SS_{uu} = \sum u_i^2 - \frac{\left(\sum u_i\right)^2}{n} = 385 - \frac{55^2}{10} = 81.5$$

$$SS_{vv} = \sum v_i^2 - \frac{\left(\sum v_i\right)^2}{n} = 384.5 - \frac{55^2}{10} = 82$$

$$SS_{ww} = \sum w_i^2 - \frac{\left(\sum w_i\right)^2}{n} = 385 - \frac{55^2}{10} = 82.5$$

$$r_{s1} = \frac{SS_{uv}}{\sqrt{SS_{uu}SS_{vv}}} = \frac{81}{\sqrt{82.5(82)}} = .9848$$

$$r_{s2} = \frac{SS_{uw}}{\sqrt{SS_{uu}SS_{ww}}} = \frac{81.5}{\sqrt{82.5(82.5)}} = .9879$$

The correlation coefficient for x and y_1 is $r_{s1} = .9848$.

Since $r_{s1} > 0$, the relationship between static weight and weigh-in-motion prior to adjustment is positive. Because the value is close to 1, the relationship is very strong. It is larger than $r_1 = .965$ found in Exercise 10.47.

The correlation coefficient for x and y_2 is $r_{s2} = .9879$.

Since $r_{s2} > 0$, the relationship between static weight and weigh-in-motion after the adjustment is positive. Because the value is close to 1, the relationship is very strong. It is smaller than $r_2 = .996$ found in Exercise 10.47.

b.   In order for r_s to be exactly 1, the rankings for the static weight and the weigh-in-motion must be the same for each truck.

In order for r_s to be exactly 0, the rankings for one of the variables (static weight) must be equal to 11 minus ranking of the other variable (weigh-in-motion) for each truck.

17.53   To determine if the median of the Type A is greater than 100, we test:

H_0: $M = 100$
H_a: $M > 100$

The test statistic is

$$S = \{\text{Number of measurements greater than } 100\} = 3$$

The p-value $= P(x \geq 3)$ where x is a binomial random variable with $n = 5$ and $p = .5$. From Table II,

$$p\text{-value} = P(x \geq 3) = 1 - P(x \leq 2) = 1 - .500 = .500$$

Since the p-value $= .500 > \alpha = .05$, H_0 is not rejected. There is insufficient evidence to indicate the median of the Type A is greater than 100 at $\alpha = .05$.

To determine if the median of the Type B is greater than 100, we test:

$$H_0: \ M = 100$$
$$H_a: \ M > 100$$

The test statistic is $S = \{\text{Number of measurements greater than } 100\} = 5$.

The p-value $= P(x \geq 5)$ where x is a binomial random variable with $n = 5$ and $p = .5$. From Table II,

$$p\text{-value} = P(x \geq 5) = 1 - P(x \leq 4) = 1 - .969 = .031$$

Since the p-value $= .031 < \alpha = .05$, H_0 is rejected. There is sufficient evidence to indicate the median of the Type B is greater than 100 at $\alpha = .05$.

17.55   Some preliminary calculations are:

Store 1	Rank	Store 2	Rank
18.8	12	12.3	7
27.9	18	19.2	13
12.2	6	6.3	1
85.3	20	24.5	17
13.1	8	11.0	5
29.5	19	10.3	4
16.3	11	15.6	9
22.1	15	9.8	3
15.7	10	8.6	2
24.0	16	19.3	14
	$T_A = 135$		$T_B = 75$

To determine if the probability distributions of the customers' incomes differ in location for the 2 stores, we test:

H_0: The two sampled populations have identical probability distributions
H_a: The probability distribution for store 1 is shifted to the right or to the left of that for store 2

The test statistic is $T_A = 135$.

The rejection region is $T_A \leq 79$ or $T_A \geq 131$ from Table XI, Appendix B, with $n_A = n_B = 10$ and $\alpha = .05$.

Since the observed value of the test statistic falls in the rejection region ($T_A = 135 \geq 131$), H_0 is rejected. There is sufficient evidence to indicate the probability distributions of the customer's incomes differ in location for the two stores at $\alpha = .05$.

17.57   Some preliminary calculations are:

Before		After	
Observation	Rank	Observation	Rank
10	19	4	5.5
5	8.5	3	3.5
3	3.5	8	16.5
6	12	5	8.5
7	14.5	6	12
11	20	4	5.5
8	16.5	2	2
9	18	5	8.5
6	12	7	14.5
5	8.5	1	1
	$T_{Before} = 132.5$		$T_{After} = 77.5$

To determine if the situation has improved under the new policy, we test:

H_0: The two sampled population probability distributions are identical
H_a: The probability distribution associated with after the policy was instituted is shifted to the left of that before

The test statistic is $T_{Before} = 132.5$.

The rejection region is $T_{Before} \geq 127$ from Table XI, Appendix B, with $n_A = n_B = 10$ and $\alpha = .05$.

Since the observed value of the test statistic falls in the rejection region ($T_{Before} = 132.5 \geq 127$), H_0 is rejected. There is sufficient evidence to indicate the situation has improved under the new policy at $\alpha = .05$.

17.59 a.   The design utilized was a completely randomized design.

   b.   Some preliminary calculations are:

Site 1	Rank	Site 2	Rank	Site 3	Rank
34.3	6	39.3	17	34.5	7
35.5	11	45.5	25	29.3	2
32.1	3	50.2	28	37.2	14
28.3	1	72.1	29	33.2	5
40.5	19	48.6	27	32.6	4
36.2	12	42.2	21	38.3	16
43.5	23	103.5	30	43.3	22
34.7	8	47.9	26	36.7	13
38.0	15	41.2	20	40.0	18
35.1	9	44.0	24	35.2	10
	$R_A = 107$		$R_B = 247$		$R_C = 111$

To determine if the probability distributions for the three sites differ, we test:

H_0: The three sampled population probability distributions are identical
H_a: At least two of the three sampled population probability distributions differ in location

The test statistic is $H = \dfrac{12}{n(n+1)} \sum \dfrac{R_j^2}{n_j} - 3(n+1)$

$$= \dfrac{12}{30(31)} \left[\dfrac{107^2}{10} + \dfrac{247^2}{10} + \dfrac{111^2}{10} \right] - 3(31) = 109.3923 - 93$$

$$= 16.3923$$

The rejection region requires $\alpha = .05$ in the upper tail of the χ^2 distribution with df $= p - 1 = 3 - 1 = 2$. From Table XIII, Appendix B, $\chi^2_{.05} = 5.99147$. The rejection region is $H > 5.99147$.

Since the observed value of the test statistic falls in the rejection region ($H = 16.3923 > 5.99147$), H_0 is rejected. There is sufficient evidence to indicate the probability distributions for at least two of the three sites differ at $\alpha = .05$.

Since H_0 was rejected, we need to compare all pairs of sites.

c. Some preliminary calculations are:

Site 1	Rank	Site 2	Rank	Site 1	Rank	Site 3	Rank
34.3	3	39.3	9	34.3	6	34.3	7
35.5	6	45.5	15	35.5	11	29.3	2
32.1	2	50.2	18	32.1	3	37.2	14
28.3	1	72.1	19	28.3	1	33.2	5
40.5	10	48.6	17	40.5	18	32.6	4
36.2	7	42.2	12	36.2	12	38.3	16
43.5	13	103.5	20	43.5	20	43.3	19
34.7	4	47.9	16	34.7	8	36.7	13
38.0	8	41.2	11	38.0	15	40.0	17
35.1	5	44.0	14	35.1	9	35.2	10
$T_A = 59$		$T_B = 151$		$T_A = 103$		$T_C = 107$	

Site 2	Rank	Site 3	Rank
39.3	9	34.5	4
45.5	15	29.3	1
50.2	18	37.2	7
72.1	19	33.2	3
48.6	17	32.6	2
42.2	12	38.3	8
103.5	20	43.3	13
47.9	16	36.7	6
41.2	11	40.0	10
44.0	14	35.2	5
$T_B = 151$		$T_C = 59$	

For each pair, we test:

H_0: The two sampled population probability distributions are identical
H_a: The probability distribution for one site is shifted to the right or left of the other.

The rejection region for each pair is $T \leq 79$ or $T \geq 131$ from Table XI, Appendix B, with $n_1 = n_2 = 10$ and $\alpha = .05$.

For sites 1 and 2:

The test statistic is $T_A = 59$.

Since the observed value of the test statistic falls in the rejection region, ($T_A = 59 \leq 79$), H_0 is rejected. There is sufficient evidence to indicate the probability distribution for site 1 is shifted to the left of that for site 2 at $\alpha = .05$.

For sites 1 and 3:

The test statistic is $T_A = 103$.

Since the observed value of the test statistic does not fall in the rejection region ($T_A = 103 \not< 79$ and $103 \not> 131$), H_0 is not rejected. There is insufficient evidence to indicate the probability distribution for site 1 is shifted to the right or left of that for site 3 at $\alpha = .05$.

For sites 2 and 3:

The test statistic is $T_B = 151$.

Since the observed value of the test statistic falls in the rejection region ($T_B = 151 \geq 131$), H_0 is rejected. There is sufficient evidence to indicate the probability distribution for site 2 is shifted to the right of that for site 3 at $\alpha = .05$.

Thus, the income for those at site 2 is significantly higher than at the other two sites.

d.   The necessary assumptions are:

1.   The 3 samples are random and independent.
2.   There are 5 or more measurements in each sample.
3.   The 3 probability distributions from which the samples are drawn are continuous.

For parametric tests, the assumptions are:

1.   The 3 populations are normal.
2.   The samples are random and independent
3.   The 3 population variances are equal.

17.61   Some preliminary calculations are:

Door-to-Door	Rank	Telephone	Rank	Grocery Store Stand	Rank	Department Store Stand	Rank
47	14	63	18	113	22	25	5
93	21	19	2	50	15	36	10
58	16	29	7	68	19	21	3
37	11.5	24	4	37	11.5	27	6
62	17	33	9	39	13	18	1
				77	20	31	8
	$R_A = 79.5$		$R_B = 40$		$R_C = 100.5$		$R_D = 33$

To determine if the probability distributions of number of sales differ in location for at least 2 of the 4 techniques, we test:

H_0: The 4 population probability distributions are identical
H_a: At least 2 of the 4 probability distributions differ in location

The test statistic is $H = \dfrac{12}{n(n+1)} \sum \dfrac{R_j^2}{n_j} - 3(n+1)$

$$= \dfrac{12}{22(23)} \left[\dfrac{79.5^2}{5} + \dfrac{40^2}{5} + \dfrac{100.5^2}{6} + \dfrac{33^2}{6} \right] - 3(23) = 81.7927 - 69$$

$$= 12.7927$$

The rejection region requires $\alpha = .10$ in the upper tail of the χ^2 distribution with df $= p - 1$ $= 4 - 1 = 3$. From Table XIII, Appendix B, $\chi^2_{.10} = 6.25139$. The rejection region is $H > 6.25139$.

Since the observed value of the test statistic falls in the rejection region ($H = 12.7927 > 6.25139$), H_0 is rejected. There is sufficient evidence to indicate the probability distributions of number of sales differ for at least 2 of the 4 techniques at $\alpha = .10$.

17.63   Some preliminary calculations are:

Candidate	u_i	v_i	Difference $d_i = u_i - v_i$	d_i^2
1	6	4	2	4
2	4	5	−1	1
3	5	6	−1	1
4	1	2	−1	1
5	2	1	1	1
6	3	3	0	0
				$\sum d_i^2 = 8$

$$r_s = 1 - \dfrac{6 \sum d_i^2}{n(n^2 - 1)}$$

$$= 1 - \dfrac{6(8)}{6(36 - 1)} = .7714$$

To determine if the candidates' qualification scores are related to their interview performance, we test:

H_0: $\rho_s = 0$
H_a: $\rho_s \neq 0$

The test statistic is $r_s = .7714$

From Table XVI, Appendix B, for $\alpha/2 = .10/2 = .05$, $r_{s,.05} = .829$ for $n = 6$. The rejection region is $r_s < -.829$ or $r_s > .829$. Since the observed value of the test statistic does not fall in the rejection region ($r_s = .7714 \not> .829$), H_0 is not rejected. There is insufficient evidence to indicate the qualification scores are related to the interview performance at $\alpha = .10$.

17.65
$$SS_{uv} = \sum u_i v_i - \frac{\sum u_i v_i}{n} = 2774.75 - \frac{210(210)}{20} = 569.75$$

$$SS_{uu} = \sum u_i^2 - \frac{\left(\sum u_i\right)^2}{n} = 2869.5 - \frac{210^2}{20} = 664.5$$

$$SS_{vv} = \sum v_i^2 - \frac{\left(\sum v_i\right)^2}{n} = 2869.5 - \frac{210^2}{20} = 664.5$$

$$r_s = \frac{SS_{uv}}{\sqrt{SS_{uu}SS_{vv}}} = \frac{569.75}{\sqrt{664.5(664.5)}} = .8574$$

Since $r_s = .8574$ is greater than 0, the relationship between current importance and ideal importance is positive. The relationship is fairly strong since r_s is close to 1. This implies the views on current importance and ideal importance are very similar.

17.67 a. To determine if the median level differs from the target, we test:

H_0: $M = .75$
H_a: $M \neq .75$

b. S_1 = number of observations less than .75 and S_2 = number of observations greater than .75.

The test statistic is S = larger of S_1 and S_2.

The p-value $= 2P(x \geq S)$ where x is a binomial random variable with $n = 25$ and $p = .5$. If the p-value is less than $\alpha = .10$, reject H_0.

c. A Type I error would be concluding the median level is not .75 when it is. If a Type I error were committed, the supervisor would correct the fluoridation process when it was not necessary. A Type II error would be concluding the median level is .75 when it is not. If a Type II error were committed, the supervisor would not correct the fluoridation process when it was necessary.

d. S_1 = number of observations less than .75 = 7 and S_2 = number of observations greater than .75 = 18.

The test statistic is S = larger of S_1 and S_2 = 18.

The p-value $= 2P(x \geq 18)$ where x is a binomial random variable with $n = 25$ and $p = .5$. From Table II,

p-value $= 2P(x \geq 18) = 2(1 - P(x \leq 17)) = 2(1 - .978)$
$= 2(.022) = .044$

Since the p-value $= .044 < \alpha = .10$, H_0 is rejected. There is sufficient evidence to indicate the median level of fluoridation differs from the target of .75 at $\alpha = .10$.

e. A distribution heavily skewed to the right might look something like the following:

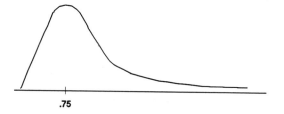

.75

One assumption necessary for the *t* test is that the distribution from which the sample is drawn is normal. A distribution which is heavily skewed in one direction is not normal. Thus, the sign test would be preferred.

CHAPTER EIGHTEEN

··

The Chi-Square Test and the Analysis of Contingency Tables

18.1   a.   For df = 17, $\chi^2_{.05} = 27.5871$

       b.   For df = 100, $\chi^2_{.990} = 70.0648$

       c.   For df = 15, $\chi^2_{.10} = 22.3072$

       d.   For df = 3, $\chi^2_{.005} = 12.8381$

18.3   a.   The rejection region requires $\alpha = .10$ in the upper tail of the χ^2 distribution with df = $k - 1 = 4 - 1 = 3$. From Table XIII, Appendix B, $\chi^2_{.10} = 6.25139$. The rejection region is $X^2 > 6.25139$.

       b.   The rejection region requires $\alpha = .01$ in the upper tail of the χ^2 distribution with df = $k - 1 = 6 - 1 = 5$. From Table XIII, Appendix B, $\chi^2_{.01} = 15.0863$. The rejection region is $X^2 > 15.0863$.

       c.   The rejection region requires $\alpha = .05$ in the upper tail of the χ^2 distribution with df = $k - 1 = 8 - 1 = 7$. From Table XIII, Appendix B, $\chi^2_{.05} = 14.0671$. The rejection region is $X^2 > 14.0671$.

18.5   The sample size n will be large enough so that, for every cell, the expected cell count, $E(n_i)$, will be equal to 5 or more.

18.7   a.   If all probabilities are equal, $p_1 = p_2 = p_3 = p_4 = .25$

          Some preliminary calculations are:

$$E(n_1) = np_{1,0} = 220(.25) = 55$$
$$E(n_2) = E(n_3) = E(n_4) = np_{i,0} = 220(.25) = 55$$

          To determine if the probabilities differ, we test:

$$H_0: p_1 = p_2 = p_3 = p_4 = .25$$
$$H_a: \text{At least one of the multinomial probabilities does not equal } .25$$

The test statistic is $X^2 = \sum \dfrac{[n_i - E(n_i)]^2}{E(n_i)}$

$$= \frac{(48 - 55)^2}{55} + \frac{(56 - 55)^2}{55} + \frac{(63 - 55)^2}{55} + \frac{(53 - 55)^2}{55} = 2.145$$

The rejection region requires $\alpha = .05$ in the upper tail of the χ^2 distribution with df = $k - 1 = 4 - 1 = 3$. From Table XIII, Appendix B, $\chi^2_{.05} = 7.81473$. The rejection region is $X^2 > 7.81473$.

Since the observed value of the test statistic does not fall in the rejection region $(X^2 = 2.145 \not> 7.81473)$, H_0 is not rejected. There is insufficient evidence to indicate the probabilities differ at $\alpha = .05$.

b. The Type I error is concluding the multinomial probabilities differ when, in fact, they do not.

The Type II error is concluding the multinomial probabilities do not differ when, in fact, they do.

18.9 a. To determine if a difference in the preference of entrepreneurs for the cars exists, we test:

H_0: $p_1 = p_2 = p_3 = 1/3$
H_a: At least one of the multinomial probabilities does not equal 1/3

where p_1 = proportion who prefer U.S. cars
p_2 = proportion who prefer European cars
p_3 = proportion who prefer Japanese cars

$E(n_1) = np_{1,0} = 100 \left(\dfrac{1}{3}\right) = 33\dfrac{1}{3}$

$E(n_2) = np_{2,0} = 100 \left(\dfrac{1}{3}\right) = 33\dfrac{1}{3}$

$E(n_3) = np_{3,0} = 100 \left(\dfrac{1}{3}\right) = 33\dfrac{1}{3}$

The test statistic is $X^2 = \sum \dfrac{[n_i - E(n_i)]^2}{E(n_i)}$

$$= \frac{\left[45 - 33\dfrac{1}{3}\right]^2}{33\dfrac{1}{3}} + \frac{\left[46 - 33\dfrac{1}{3}\right]^2}{33\dfrac{1}{3}} + \frac{\left[9 - 33\dfrac{1}{3}\right]^2}{33\dfrac{1}{3}} = 26.66$$

The rejection region requires $\alpha = .05$ in the upper tail of the χ^2 distribution with df = $k - 1 = 3 - 1 = 2$. From Table XIII, Appendix B, $\chi^2_{.05} = 5.99147$. The rejection region is $X^2 > 5.99147$.

Since the observed value of the test statistic falls in the rejection region ($X^2 = 26.66$ > 5.99147), H_0 is rejected. There is sufficient evidence to indicate a difference in the preference of entrepreneurs for the cars at $\alpha = .05$.

b.   To determine if there is a difference in the preference of entrepreneurs for domestic versus foreign cars, we test:

H_0: $p_1 = p_2 = .5$
H_a: At least one probability differs from .5

where p_1 = proportion who prefer U.S. cars
      p_2 = proportion who prefer European cars

$E(n_1) = np_{1,0} = 100(.5) = 50$
$E(n_2) = np_{2,0} = 100(.5) = 50$

The test statistic is $X^2 = \sum \dfrac{[n_i - E(n_i)]^2}{E(n_i)} = \dfrac{(45 - 50)^2}{50} + \dfrac{(55 - 50)^2}{50} = 1$

The rejection region requires $\alpha = .05$ in the upper tail of the χ^2 distribution with df = $k - 1 = 2 - 1 = 1$. From Table XIII, Appendix B, $\chi^2_{.05} = 3.84146$. The rejection region is $X^2 > 3.84146$.

Since the observed value of the test statistic does not fall in the rejection region ($X^2 = 1 \not> 3.84146$), H_0 is not rejected. There is insufficient evidence to indicate a difference in the preference of entrepreneurs for domestic and foreign cars at $\alpha = .05$.

c.   We must assume the sample size n is large enough so that the expected cell count, $E(n_i)$, will be equal to 5 or more for every cell.

18.11   a.   To determine if the number of overweight trucks per week is distributed over the 7 days of the week in direct proportion to the volume of truck traffic, we test:

H_0: $p_1 = .191, p_2 = .198, p_3 = .187, p_4 = .180, p_5 = .155, p_6 = .043,$
       $p_7 = .046$
H_a: At least one of the probabilities differs from the hypothesized value

$E(n_1) = np_{1,0} = 414(.191) = 79.074$
$E(n_2) = np_{2,0} = 414(.198) = 81.972$
$E(n_3) = np_{3,0} = 414(.187) = 77.418$
$E(n_4) = np_{4,0} = 414(.180) = 74.520$
$E(n_5) = np_{5,0} = 414(.155) = 64.170$
$E(n_6) = np_{6,0} = 414(.043) = 17.802$
$E(n_7) = np_{7,0} = 414(.046) = 19.044$

The test statistic is $X^2 = \sum \dfrac{[n_i - E(n_i)]^2}{E(n_i)} = \dfrac{(90 - 79.074)^2}{79.074} + \dfrac{(82 - 81.972)^2}{81.972}$

$+ \dfrac{(72 - 77.418)^2}{77.418} + \dfrac{(70 - 74.520)^2}{74.520} + \dfrac{(51 - 64.170)^2}{64.170} + \dfrac{(18 - 17.802)^2}{17.802}$

$+ \dfrac{(31 - 19.044)^2}{19.044} = 12.374$

The rejection region requires $\alpha = .05$ in the upper tail of the χ^2 distribution with df $= k - 1 = 7 - 1 = 6$. From Table XIII, Appendix B, $\chi^2_{.05} = 12.5916$. The rejection region is $X^2 > 12.5916$.

Since the observed value of the test statistic does not fall in the rejection region ($X^2 = 12.374 \not> 12.5916$), H_0 is not rejected. There is insufficient evidence to indicate the number of overweight trucks per week is distributed over the 7 days of the week is not in direct proportion to the volume of truck traffic at $\alpha = .05$.

b. The p-value is $P(\chi^2 \geq 12.374)$. From Table XIII, Appendix B, with df $= k - 1 = 7 - 1 = 6$, $.05 < P(\chi^2 \geq 12.374) < .10$.

18.13 If the die is balanced, $p_1 = p_2 = p_3 = p_4 = p_5 = p_6 = 1/6$

Some preliminary calculations are:

$$E(n_1) = E(n_2) = E(n_3) = E(n_4) = E(n_5) = E(n_6) = np_{i,0} = 120(1/6) = 20$$

To determine if the die is fair, we test:

$H_0: p_1 = p_2 = p_3 = p_4 = p_5 = p_6 = 1/6$
$H_a:$ At least one of the multinomial probabilities does not equal 1/6

The test statistic is $X^2 = \sum \dfrac{[n_i - E(n_i)]^2}{E(n_i)} = \dfrac{(28 - 20)^2}{20} + \dfrac{(27 - 20)^2}{20} + \dfrac{(20 - 20)^2}{20}$

$+ \dfrac{(18 - 20)^2}{20} + \dfrac{(15 - 20)^2}{20} + \dfrac{(12 - 20)^2}{20} = 10.3$

The rejection region requires $\alpha = .10$ in the upper tail of the χ^2 distribution with df $= k - 1 = 6 - 1 = 5$. From Table XIII, Appendix B, $\chi^2_{.10} = 9.23635$. The rejection region is $X^2 > 9.23635$.

Since the observed value of the test statistic falls in the rejection region ($X^2 = 10.3 > 9.23635$), H_0 is rejected. There is sufficient evidence to indicate the die is not fair at $\alpha = .10$.

18.15 a. df $= (r - 1)(c - 1) = (5 - 1)(4 - 1) = 12$. From Table XIII, Appendix B, $\chi^2_{.05} = 21.0261$. The rejection region is $X^2 > 21.0261$.

b. $df = (r - 1)(c - 1) = (4 - 1)(6 - 1) = 15$. From Table XIII, Appendix B, $\chi^2_{.10} = 22.3072$. The rejection region is $X^2 > 22.3072$.

c. $df = (r - 1)(c - 1) = (3 - 1)(3 - 1) = 4$. From Table XIII, Appendix B, $\chi^2_{.01} = 13.2767$. The rejection region is $X^2 > 13.2767$.

18.17   a. To convert the frequencies to percentages, divide the numbers in each column by the column total and multiply by 100. Also, divide the row totals by the overall total and multiply by 100. The column totals are 28, 61, and 70, while the row totals are 89 and 70. The overall sample size is 159. The table of percentages are:

Column

	1	2	3	
Row 1	$\dfrac{12}{28} \cdot 100 = 42.9\%$	$\dfrac{32}{61} \cdot 100 = 52.5\%$	$\dfrac{45}{70} \cdot 100 = 64.3\%$	$\dfrac{89}{159} \cdot 100 = 56.0\%$
Row 2	$\dfrac{16}{28} \cdot 100 = 57.1\%$	$\dfrac{29}{61} \cdot 100 = 47.5\%$	$\dfrac{25}{70} \cdot 100 = 35.7\%$	$\dfrac{70}{159} \cdot 100 = 44.0\%$

b.

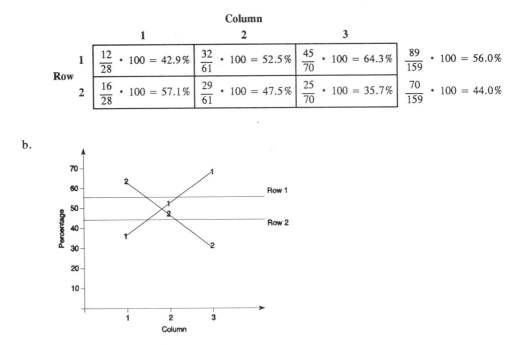

c. If the rows and columns are independent, the row percentages in each column would be close to the row total percentages. This pattern is not evident in the plot, implying the rows and columns are not independent. This does not agree with the results of 18.16 where we did not reject H_0. The percents are not parallel lines, but they are not different enough from parallel for the sample size to say the rows and columns are dependent.

18.19 To convert the frequencies to percentages, divide the numbers in each column by the column total and multiply by 100. Also, divide the row totals by the overall total and multiply by 100.

B

	B_1	B_2	B_3	
A_1	$\dfrac{39}{136} \cdot 100 = 28.7\%$	$\dfrac{70}{159} \cdot 100 = 44.0\%$	$\dfrac{42}{138} \cdot 100 = 30.4\%$	$\dfrac{151}{433} \cdot 100 = 34.9\%$
Row A_2	$\dfrac{67}{136} \cdot 100 = 49.3\%$	$\dfrac{51}{159} \cdot 100 = 32.1\%$	$\dfrac{70}{138} \cdot 100 = 50.7\%$	$\dfrac{188}{433} \cdot 100 = 43.4\%$
A_3	$\dfrac{30}{136} \cdot 100 = 22.1\%$	$\dfrac{38}{159} \cdot 100 = 23.9\%$	$\dfrac{26}{138} \cdot 100 = 18.8\%$	$\dfrac{94}{433} \cdot 100 = 21.7\%$

The graph supports the conclusion that the rows and columns are not independent. The percentages of category A at B_2 are switched for A_1 and A_2 from what is expected.

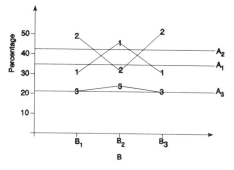

18.21 For each table, divide the numbers in each column by the column total and multiply by 100. Also, divide the row totals by the overall total and multiply by 100.

a.

B

	B_1	B_2		
A_1	$\dfrac{50}{60} \cdot 100 = 83.3\%$	$\dfrac{10}{50} \cdot 100 = 20\%$	$\dfrac{60}{110} \cdot 100 = 54.5\%$	
A A_2	$\dfrac{10}{60} \cdot 100 = 16.7\%$	$\dfrac{40}{50} \cdot 100 = 80\%$	$\dfrac{50}{110} \cdot 100 = 45.5\%$	

Since the row percentages in each column are not similar to the row total percentages, the graph implies the variables are dependent.

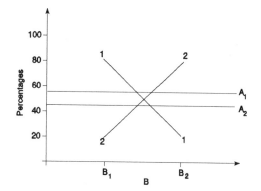

b.

	D_1	D_2	
C_1	$\dfrac{15}{95} \cdot 100 = 15.8\%$	$\dfrac{70}{85} \cdot 100 = 82.4\%$	$\dfrac{85}{180} \cdot 100 = 47.2\%$
C_2	$\dfrac{80}{95} \cdot 100 = 84.2\%$	$\dfrac{15}{85} \cdot 100 = 17.6\%$	$\dfrac{95}{180} \cdot 100 = 52.8\%$

C, D

Since the row percentages in each column are not similar to the row total percentages, the graph implies the variables are dependent.

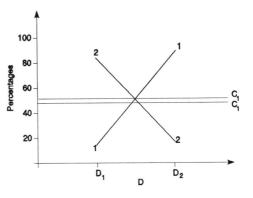

c.

	F_1	F_2	F_3	
E_1	$\dfrac{10}{40} \cdot 100 = 25\%$	$\dfrac{30}{40} \cdot 100 = 75\%$	$\dfrac{15}{30} \cdot 100 = 50\%$	$\dfrac{55}{110} \cdot 100 = 50\%$
E_2	$\dfrac{30}{40} \cdot 100 = 75\%$	$\dfrac{10}{40} \cdot 100 = 25\%$	$\dfrac{15}{30} \cdot 100 = 50\%$	$\dfrac{55}{110} \cdot 100 = 50\%$

E, F

Since the row percentages in each column are not similar to the row total percentages, the graph implies the variables are dependent.

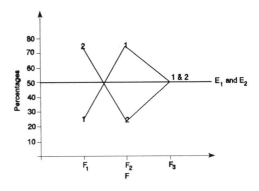

18.23 **a.** Some preliminary calculations are:

$$\hat{E}(n_{11}) = \frac{r_1 c_1}{n} = \frac{31(70)}{117} = 18.55 \qquad \hat{E}(n_{21}) = \frac{86(70)}{117} = 51.45$$

$$\hat{E}(n_{12}) = \frac{31(30)}{117} = 7.95 \qquad\qquad \hat{E}(n_{22}) = \frac{86(30)}{117} = 22.05$$

$$\hat{E}(n_{13}) = \frac{31(17)}{117} = 4.50 \qquad\qquad \hat{E}(n_{23}) = \frac{86(17)}{117} = 12.50$$

To determine if there is a relationship between an employee's age and the kind of accident, we test:

H_0: Age and kind of accident are independent
H_a: Age and kind of accident are dependent

The test statistic is $X^2 = \sum\sum \dfrac{\left[n_{ij} - \hat{E}(n_{ij})\right]^2}{\hat{E}(n_{ij})} = \dfrac{(9 - 18.55)^2}{18.55} + \dfrac{(61 - 51.45)^2}{51.45}$

$+ \dfrac{(17 - 7.95)^2}{7.95} + \dfrac{(13 - 22.05)^2}{22.05} + \dfrac{(5 - 4.50)^2}{4.50} + \dfrac{(12 - 12.5)^2}{12.5} = 20.78$

The rejection region requires $\alpha = .05$ in the upper tail of the χ^2 distribution with df $= (r - 1)(c - 1) = (2 - 1)(3 - 1) = 2$. From Table XIII, Appendix B, $\chi^2_{.05} = 5.99147$. The rejection region is $X^2 > 5.99147$.

Since the observed value of the test statistic falls in the rejection region ($X^2 = 20.78 > 5.99147$), H_0 is rejected. There is sufficient evidence to indicate a relationship exists between an employee's age and the kind of accident that the employee had at $\alpha = .05$.

b. The most frequent type of accident is sprain (total sprains is 70). Of the 70 people who suffered sprains, 61 are 25 or over while only 9 are under 25. Thus, older employees are more likely to have sprains.

Of the 30 people who suffered burns, 17 are under 25 while 13 are 25 and over. Thus, younger employees are more likely to suffer burns.

c. The necessary assumption is:

1. The sample size, n, will be large enough so that, for every cell, the expected cell count, $E(n_{ij})$, will be equal to 5 or more.

d. To find the percentages of those under 25 who are injured, divide the number under 25 in each column by the column total and multiply by 100.

The percentages are:

	Kind of Accident			
	Sprain	Burn	Cut	Total
Under 25	$\frac{9}{70} \cdot 100 = 12.9\%$	$\frac{17}{30} \cdot 100 = 56.7\%$	$\frac{5}{17} \cdot 100 = 29.4\%$	$\frac{31}{117} \cdot 100 = 26.50\%$

Since the row percentages in each column are not similar to the row total percentage, the graph indicates the variables are dependent.

18.25   a.   The hypotheses are:

H_0: Viewing and buying are independent
H_a: Viewing and buying are dependent

$$\hat{E}(n_{11}) = \frac{r_1 c_1}{n} = \frac{1175(797)}{2452} = 381.923 \quad \hat{E}(n_{12}) = \frac{1175(363)}{2452} = 173.950$$

$$\hat{E}(n_{13}) = \frac{1175(401)}{2452} = 192.159 \quad\quad\quad \hat{E}(n_{14}) = \frac{1175(891)}{2452} = 426.968$$

$$\hat{E}(n_{21}) = \frac{292(797)}{2452} = 94.912 \quad\quad\quad \hat{E}(n_{22}) = \frac{292(363)}{2452} = 43.228$$

$$\hat{E}(n_{23}) = \frac{292(401)}{2452} = 47.754 \quad\quad\quad \hat{E}(n_{24}) = \frac{292(891)}{2452} = 106.106$$

$$\hat{E}(n_{31}) = \frac{246(797)}{2452} = 79.960 \quad\quad\quad \hat{E}(n_{32}) = \frac{246(363)}{2452} = 36.418$$

$$\hat{E}(n_{33}) = \frac{246(401)}{2452} = 40.231 \quad\quad\quad \hat{E}(n_{34}) = \frac{246(891)}{2452} = 89.391$$

$$\hat{E}(n_{41}) = \frac{739(797)}{2452} = 240.205 \quad\quad\quad \hat{E}(n_{42}) = \frac{739(363)}{2452} = 109.403$$

$$\hat{E}(n_{43}) = \frac{739(401)}{2452} = 120.856 \quad\quad\quad \hat{E}(n_{44}) = \frac{739(891)}{2452} = 268.535$$

The test statistic is $X^2 = \sum\sum \dfrac{\left[n_{ij} - \hat{E}(n_{ij})\right]^2}{\hat{E}(n_{ij})} = \dfrac{(460 - 381.923)^2}{381.923} + \dfrac{(76 - 94.912)^2}{94.912}$

$$+ \frac{(86 - 79.96)^2}{79.96} + \ldots + \frac{(347 - 268.535)^2}{268.535} = 89.09$$

The rejection region requires $\alpha = .05$ in the upper tail of the χ^2 distribution with df = $(r - 1)(c - 1) = (4 - 1)(4 - 1) = 9$. From Table XIII, Appendix B, $\chi^2_{.05} = 16.9190$. The rejection region is $X^2 > 16.9190$.

Since the observed value of the test statistic falls in the rejection region ($X^2 = 89.09 > 16.9190$), H_0 is rejected. There is sufficient evidence to indicate viewing and buying are dependent at $\alpha = .05$.

c.   We must assume the same size, n, will be large enough so that, for every cell, the expected cell count, $E(n_{ij})$, will be 5 or more.

18.27 Some preliminary calculations are:

$$\hat{E}(n_{11}) = \frac{r_1 c_1}{n} = \frac{67(83)}{311} = 17.88 \qquad \hat{E}(n_{12}) = \frac{67(117)}{311} = 25.21$$

$$\hat{E}(n_{21}) = \frac{69(83)}{311} = 18.41 \qquad \hat{E}(n_{22}) = \frac{69(117)}{311} = 25.96$$

$$\hat{E}(n_{31}) = \frac{104(83)}{311} = 27.76 \qquad \hat{E}(n_{32}) = \frac{104(117)}{311} = 39.13$$

$$\hat{E}(n_{41}) = \frac{71(83)}{311} = 18.95 \qquad \hat{E}(n_{42}) = \frac{71(117)}{311} = 26.71$$

$$\hat{E}(n_{13}) = \frac{67(79)}{311} = 17.02 \qquad \hat{E}(n_{14}) = \frac{67(32)}{311} = 6.89$$

$$\hat{E}(n_{23}) = \frac{69(79)}{311} = 17.53 \qquad \hat{E}(n_{24}) = \frac{69(32)}{311} = 7.10$$

$$\hat{E}(n_{33}) = \frac{104(79)}{311} = 26.42 \qquad \hat{E}(n_{34}) = \frac{104(32)}{311} = 10.70$$

$$\hat{E}(n_{43}) = \frac{71(79)}{311} = 18.04 \qquad \hat{E}(n_{44}) = \frac{71(32)}{311} = 7.31$$

To determine if there is a relationship between length of stay and hospitalization coverage, we test:

H_0: Length of stay and hospitalization coverage are independent
H_a: Length of stay and hospitalization coverage are dependent

The test statistic is $X^2 = \sum\sum \dfrac{\left[n_{ij} - \hat{E}(n_{ij})\right]^2}{\hat{E}(n_{ij})} = \dfrac{(26 - 17.88)^2}{17.88} + \dfrac{(21 - 18.41)^2}{18.41}$

$$+ \frac{(25 - 27.76)^2}{27.76} + \cdots + \frac{(11 - 7.31)^2}{7.31} = 40.70$$

The rejection region requires $\alpha = .01$ in the upper tail of the χ^2 distribution with df $= (r - 1)(c - 1) = (4 - 1)(4 - 1) = 9$. From Table XIII, Appendix B, $\chi^2_{.01} = 21.666$. The rejection region is $X^2 > 21.666$.

Since the observed value of the test statistic falls in the rejection region ($X^2 = 40.70 > 21.666$), H_0 is rejected. There is sufficient evidence to indicate length of stay and hospitalization coverage are dependent at $\alpha = .01$.

18.29 a. No. If January change is down, half the next 11-month changes are up and half are down.

b. The percentages of years for which the 11-month movement is up based on January change are found by dividing the numbers in the first column by the corresponding row total and multiplying by 100. We also divide the first column total by the overall total and multiply by 100.

January Change:

Up $\quad \dfrac{25}{35} \cdot 100 = 71.4\%$

Down $\quad \dfrac{9}{18} \cdot 100 = 50\%$

Total $\quad \dfrac{34}{53} \cdot 100 = 64.2\%$

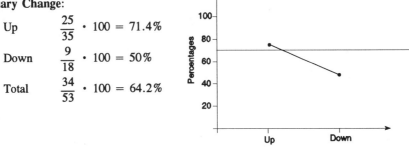

c. H_0: The January change and the next 11-month change are independent

$\quad H_a$: The January change and the next 11-month change are dependent

d. Some preliminary calculations are:

$$\hat{E}(n_{11}) = \frac{r_1 c_1}{n} = \frac{35(34)}{53} = 22.453 \qquad \hat{E}(n_{12}) = \frac{35(19)}{53} = 12.547$$

$$\hat{E}(n_{21}) = \frac{18(34)}{53} = 11.547 \qquad \hat{E}(n_{22}) = \frac{18(19)}{53} = 6.453$$

The test statistic is $X^2 = \displaystyle\sum\sum \frac{\left[n_{ij} - \hat{E}(n_{ij})\right]^2}{\hat{E}(n_{ij})} = \frac{(25 - 22.453)^2}{22.453} + \frac{(9 - 11.547)^2}{11.547}$

$$+ \frac{(10 - 12.547)^2}{12.547} + \frac{(9 - 6.453)^2}{6.453} = 2.373$$

The rejection region requires $\alpha = .05$ in the upper tail of the χ^2 distribution with df $= (r - 1)(c - 1) = (2 - 1)(2 - 1) = 1$. From Table XIII, Appendidix B, $\chi^2_{.05} = 3.84146$. The rejection region is $X^2 > 3.84146$.

Since the observed value of the test statistic does not fall in the rejection region ($X^2 = 2.373 \ngtr 3.84146$), H_0 is not rejected. There is insufficient evidence to indicate the January change and the next 11-month change are dependent at $\alpha = .05$.

e. Yes. For $\alpha = .10$, the rejection region is $X^2 > \chi^2_{.10} = 2.70554$, from Table XIII, Appendix B, with df $= 1$. Since the observed value of the test statistic does not fall in the rejection region ($X^2 = 2.373 \ngtr 2.70554$), H_0 is not rejected. The conclusion is the same.

18.31 a. Some preliminary calculations are:

$$E(n_1) = np_{1,0} = 237(.37) = 87.69$$
$$E(n_2) = np_{2,0} = 237(.533) = 126.321$$
$$E(n_3) = np_{3,0} = 237(.097) = 22.989$$

To determine if consumer attitudes toward warranties changed, we test:

H_0: $p_1 = .37$, $p_2 = .533$, $p_3 = .097$
H_a: At least one of the proportions differs from the hypothesized values

The test statistic is $X^2 = \sum \dfrac{[n_i - E(n_i)]^2}{E(n_i)}$

$$= \dfrac{(156 - 87.69)^2}{87.69} + \dfrac{(61 - 126.321)^2}{126.321} + \dfrac{(20 - 22.989)^2}{22.989} = 87.38$$

The rejection region requires $\alpha = .05$ in the upper tail of the χ^2 distribution with df $= k - 1 = 3 - 1 = 2$. From Table XIII, Appendix B, $\chi^2_{.05} = 5.99147$. The rejection region is $X^2 > 5.99147$.

Since the observed value of the test statistic falls in the rejection region ($X^2 = 87.38 > 5.99147$), H_0 is rejected. There is sufficient evidence to indicate the consumer attitudes toward warranties have changed at $\alpha = .05$.

b.   Post-Magnuson-Moss Warranty Act

Yes $\qquad \dfrac{156}{237} \cdot 100\% = 65.8\%$

Uncertain $\dfrac{61}{237} \cdot 100\% = 25.7\%$

No $\qquad \dfrac{20}{237} \cdot 100\% = 8.4\%$

The percentages that said "no" prior to the Warranty Act and after are about the same, 9.7% and 8.4%. However, the percentage that said "yes" increased from 37.0% to 65.8% and the percentage that said "uncertain" dropped from 53.3% to 25.7%.

18.33   First, we convert the percentages to frequencies. The contingency table is:

	Yes	No
Men	60	35
Women	160	47

Some preliminary calculations are:

$\hat{E}(n_{11}) = \dfrac{r_1 c_1}{n} = \dfrac{95(220)}{302} = 69.205 \qquad \hat{E}(n_{12}) = \dfrac{r_1 c_2}{n} = \dfrac{95(82)}{302} = 25.795$

$\hat{E}(n_{21}) = \dfrac{r_2 c_1}{n} = \dfrac{207(220)}{302} = 150.795 \qquad \hat{E}(n_{22}) = \dfrac{r_2 c_2}{n} = \dfrac{207(82)}{302} = 56.205$

To determine if the interviewee's sex and anxiety are related, we test:

H_0: Anxiety over hotel room interviewing and the interviewee's sex are independent
H_a: Anxiety over hotel room interviewing and the interviewee's sex are dependent

The test statistic is $X^2 = \sum\sum \dfrac{\left[n_{ij} - \hat{E}(n_{ij})\right]^2}{\hat{E}(n_{ij})} = \dfrac{(60 - 69.205)^2}{69.205} + \dfrac{(35 - 25.795)^2}{25.795}$

$+ \dfrac{(160 - 150.795)^2}{150.795} + \dfrac{(47 - 56.205)^2}{56.205} = 6.58$

The rejection region requires $\alpha = .05$ in the upper tail of the χ^2 distribution with df $=$ $(r - 1)(c - 1) = (2 - 1)(2 - 1) = 1$. From Table XIII, Appendix B, $\chi^2_{.05} = 3.84146$. The rejection region is $X^2 > 3.84146$.

Since the observed value of the test statistic falls in the rejection region ($X^2 = 6.58 >$ 3.84146), H_0 is rejected. There is sufficient evidence to indicate the anxiety over hotel room interviewing is related to the interviewee's sex at $\alpha = .05$.

18.35   Some preliminary calculations are:

If the transactions are randomly assigned to each of the locations, then $p_{i,0} = .20$ for $i = 1, 2, 3, 4,$ and 5.

$E(n_i) = np_{i,0} = 425(.2) = 85$

To determine if there is a difference in the proportions of transactions assigned to the five memory locations, we test:

H_0: $p_1 = p_2 = p_3 = p_4 = p_5 = .2$
H_a: At least one of the proportions differ from .2

The test statistic is $X^2 = \sum \dfrac{\left[n_i - E(n_i)\right]^2}{E(n_i)}$

$= \dfrac{(90 - 85)^2}{85} + \dfrac{(78 - 85)^2}{85} + \dfrac{(100 - 85)^2}{85} + \dfrac{(72 - 85)^2}{85} + \dfrac{(85 - 85)^2}{85}$

$= 5.506$

The rejection region requires $\alpha = .025$ in the upper tail of the χ^2 distribution with df $=$ $k - 1 = 5 - 1 = 4$. From Table XIII, Appendix B, $\chi^2_{.025} = 11.1433$. The rejection region is $X^2 > 11.1433$.

Since the observed value of the test statistic does not fall in the rejection region ($X^2 =$ 5.506 $\not>$ 11.1433), H_0 is not rejected. There is insufficient evidence to indicate the proportions of transactions assigned to the five memory locations differ at $\alpha = .025$.

18.37 Some preliminary calculations are:

$$\hat{E}(n_{11}) = \frac{r_1 c_1}{n} = \frac{75(152)}{300} = 38 = \hat{E}(n_{21}) = \hat{E}(n_{31}) = \hat{E}(n_{41})$$

$$\hat{E}(n_{12}) = \frac{r_1 c_2}{n} = \frac{75(122)}{300} = 30.5 = \hat{E}(n_{22}) = \hat{E}(n_{32}) = \hat{E}(n_{42})$$

$$\hat{E}(n_{13}) = \frac{r_1 c_3}{n} = \frac{75(26)}{300} = 6.5 = \hat{E}(n_{23}) = \hat{E}(n_{33}) = \hat{E}(n_{43})$$

To determine if customer preferences are different for the four restaurants, we test:

H_0: Customer preferences and restaurants are independent
H_a: Customer preferences and restaurants are dependent

The test statistic is $X^2 = \sum\sum \dfrac{\left[n_{ij} - \hat{E}(n_{ij})\right]^2}{\hat{E}(n_{ij})} = \dfrac{(38 - 38)^2}{38} + \dfrac{(32 - 30.5)^2}{30.5}$

$$+ \frac{(5 - 6.5)^2}{6.5} + \cdots + \frac{(8 - 6.5)^2}{6.5} = 2.60$$

The rejection region requires $\alpha = .10$ in the upper tail of the χ^2 distribution with df = $(r - 1)(c - 1) = (4 - 1)(3 - 1) = 6$. From Table XIII, Appendix B, $\chi^2_{.10} = 10.6446$. The rejection region is $X^2 > 10.6466$.

Since the observed value of the test statistic does not fall in the rejection region ($X^2 = 2.60 \not> 10.6446$), H_0 is not rejected. There is insufficient evidence to indicate the customer preferences are different for the four restaurants at $\alpha = .10$.

18.39 a. Some preliminary calculations are:

The contingency table is:

		Defectives	Non-Defectives	
	1	25	175	200
Shift	2	35	165	200
	3	80	120	200
		140	460	600

$$\hat{E}(n_{11}) = \frac{r_1 c_1}{n} = \frac{200(140)}{600} = 46.667$$

$$\hat{E}(n_{21}) = \hat{E}(n_{31}) = \frac{200(140)}{600} = 46.667$$

$$\hat{E}(n_{12}) = \hat{E}(n_{22}) = \hat{E}(n_{32}) = \frac{200(460)}{600} = 153.333$$

To determine if quality of the filters are related to shift, we test:

H_0: Quality of filters and shift are independent
H_a: Quality of filters and shift are dependent

The test statistic is $X^2 = \sum\sum \dfrac{\left[n_{ij} - \hat{E}(n_{ij})\right]^2}{\hat{E}(n_{ij})} = \dfrac{(25 - 46.667)^2}{46.667} + \dfrac{(35 - 46.667)^2}{46.667}$

$+ \dfrac{(80 - 46.667)^2}{46.667} + \dfrac{(175 - 153.333)^2}{153.333} + \dfrac{(165 - 153.333)^2}{153.333} + \dfrac{(120 - 153.333)^2}{153.333}$

$= 47.98$

The rejection region requires $\alpha = .05$ in the upper tail of the χ^2 distribution with df $= (r - 1)(c - 1) = (3 - 1)(2 - 1) = 2$. From Table XIII, Appendix B, $\chi^2_{.05} = 5.99147$. The rejection region is $X^2 > 5.99147$.

Since the observed value of the test statistic falls in the rejection region ($X^2 = 47.98 > 5.99147$), H_0 is rejected. There is sufficient evidence to indicate quality of filters and shift are related at $\alpha = .05$.

b. The form of the confidence interval for p is

$\hat{p} \pm z_{\alpha/2} \sqrt{\dfrac{\hat{p}\hat{q}}{n}}$ where $\hat{p} = \dfrac{25}{200} = .125$

For confidence coefficient .95, $\alpha = 1 - .95 = .05$ and $\alpha/2 = .05/2 = .025$. From Table IV, Appendix B, $z_{.025} = 1.96$. The 95% confidence interval is:

$.125 \pm 1.96 \sqrt{\dfrac{.125(.875)}{200}} \Rightarrow .125 \pm .046 \Rightarrow (.079, .171)$

18.41 a. Some preliminary calculations are:

$\hat{E}(n_{11}) = \dfrac{r_1 c_1}{n} = \dfrac{107(129)}{400} = 34.5075$ $\qquad \hat{E}(n_{13}) = \dfrac{107(103)}{400} = 27.5525$

$\hat{E}(n_{21}) = \dfrac{237(129)}{400} = 76.4325$ $\qquad \hat{E}(n_{23}) = \dfrac{237(103)}{400} = 61.0275$

$\hat{E}(n_{31}) = \dfrac{56(129)}{400} = 18.06$ $\qquad \hat{E}(n_{33}) = \dfrac{56(103)}{400} = 14.42$

$\hat{E}(n_{12}) = \dfrac{107(132)}{400} = 35.31$ $\qquad \hat{E}(n_{14}) = \dfrac{107(36)}{400} = 9.63$

$\hat{E}(n_{22}) = \dfrac{237(132)}{400} = 78.21$ $\qquad \hat{E}(n_{24}) = \dfrac{237(36)}{400} = 21.33$

$\hat{E}(n_{32}) = \dfrac{56(132)}{400} = 18.48$ $\qquad \hat{E}(n_{34}) = \dfrac{56(36)}{400} = 5.04$

To determine if a relationship exists between employee income and amount stolen, we test:

H_0: Employee income and amount stolen are independent
H_a: Employee income and amount stolen are dependent

The test statistic is $X^2 = \sum\sum \dfrac{\left[n_{ij} - \hat{E}(n_{ij})\right]^2}{\hat{E}(n_{ij})} = \dfrac{(46 - 34.5075)^2}{34.5075} + \dfrac{(78 - 76.4325)^2}{76.4325}$

$$+ \dfrac{(5 - 18.06)^2}{18.06} + \cdots + \dfrac{(12 - 5.04)^2}{5.04} = 38.68$$

The rejection region requires $\alpha = .05$ in the upper tail of the χ^2 distribution with df = $(r - 1)(c - 1) = (3 - 1)(4 - 1) = 6$. From Table XIII, Appendix B, $\chi^2_{.05} = 12.5916$. The rejection region is $X^2 > 12.5916$.

Since the observed value of the test statistic falls in the rejection region ($X^2 = 38.68 > 12.5916$), H_0 is rejected. There is sufficient evidence to indicate a relationship exists between employee income and amount stolen at $\alpha = .05$.

b. To convert the responses to percentages, divide the numbers in each column by the column total and multiply by 100. The percentages in each income category are found by dividing the row totals by the total sample size and multiplying by 100.

Amount Stolen

		Under 5,000	5000–9,999	10,000–19,999	20,000 or more	
Income of Employee	Under 15	$\frac{46}{129} \cdot 100 = 35.7\%$	$\frac{39}{132} \cdot 100 = 29.5\%$	$\frac{17}{103} \cdot 100 = 16.5\%$	$\frac{5}{36} \cdot 100 = 13.9\%$	$\frac{107}{400} \cdot 100 = 26.75\%$
	15–25	$\frac{78}{129} \cdot 100 = 60.5\%$	$\frac{79}{132} \cdot 100 = 59.8\%$	$\frac{61}{103} \cdot 100 = 59.2\%$	$\frac{19}{36} \cdot 100 = 52.8\%$	$\frac{237}{400} \cdot 100 = 59.25\%$
	Over 25	$\frac{5}{129} \cdot 100 = 3.9\%$	$\frac{14}{132} \cdot 100 = 10.6\%$	$\frac{25}{103} \cdot 100 = 24.3\%$	$\frac{12}{36} \cdot 100 = 33.3\%$	$\frac{56}{400} \cdot 100 = 14\%$

Since the lines are not close to the horizontal lines representing the row percentages, there is evidence to support the conclusion in part a.

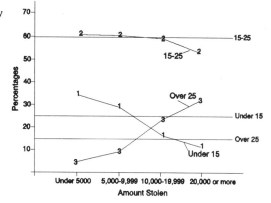

18.43 Some preliminary calculations are:

$$\hat{E}(n_{11}) = \frac{r_1 c_1}{n} = \frac{117(155)}{300} = 60.45 \qquad\qquad \hat{E}(n_{13}) = \frac{117(31)}{300} = 12.09$$

$$\hat{E}(n_{21}) = \frac{120(155)}{300} = 62 \qquad\qquad \hat{E}(n_{23}) = \frac{120(31)}{300} = 12.4$$

$$\hat{E}(n_{31}) = \frac{63(155)}{300} = 32.55 \qquad\qquad \hat{E}(n_{33}) = \frac{63(31)}{300} = 6.51$$

$$\hat{E}(n_{12}) = \frac{117(114)}{300} = 44.46$$

$$\hat{E}(n_{22}) = \frac{120(114)}{300} = 45.6$$

$$\hat{E}(n_{32}) = \frac{63(114)}{300} = 23.94$$

To determine if time and size of purchase are related, we test:

H_0: Time and size of purchase are independent
H_a: Time and size of purchase are dependent

The test statistic is $X^2 = \sum\sum \dfrac{\left[n_{ij} - \hat{E}(n_{ij})\right]^2}{\hat{E}(n_{ij})} = \dfrac{(65 - 60.45)^2}{60.45} + \dfrac{(61 - 62)^2}{62}$

$$+ \frac{(29 - 32.55)^2}{32.55} + \ldots + \frac{(7 - 6.51)^2}{6.51} = 3.13$$

The rejection region requires $\alpha = .05$ in the upper tail of the χ^2 distribution with df $= (r - 1)(c - 1) = (3 - 1)(3 - 1) = 4$. From Table XIII, Appendix B, $\chi^2_{.05} = 9.48773$. The rejection region is $X^2 > 9.48773$.

Since the observed value of the test statistic does not fall in the rejection region ($X^2 = 3.13 \not> 9.48773$), H_0 is not rejected. There is insufficient evidence to indicate time and size of purchase are related at $\alpha = .05$.

18.45 a. Some preliminary calculations are:

$$\begin{aligned}
E(n_1) &= np_{1,0} = 2000(.2) = 400 & E(n_4) &= np_{4,0} = 2000(.2) = 400 \\
E(n_2) &= np_{2,0} = 2000(.2) = 400 & E(n_5) &= np_{5,0} = 2000(.2) = 400 \\
E(n_3) &= np_{3,0} = 2000(.2) = 400
\end{aligned}$$

To determine if a difference in preference for the five candidates exists, we test:

H_0: $p_1 = p_2 = p_3 = p_4 = p_5 = .2$
H_a: At least one of the proportions differs from the hypothesized value

The test statistic is $X^2 = \sum \dfrac{\left[n_i - E(n_i)\right]^2}{E(n_i)} = \dfrac{(385 - 400)^2}{400} + \dfrac{(493 - 400)^2}{400}$

$$+ \frac{(628 - 400)^2}{400} + \frac{(235 - 400)^2}{400} + \frac{(259 - 400)^2}{400} = 269.91$$

The rejection region requires $\alpha = .01$ in the upper tail of the χ^2 distribution with df $= k - 1 = 5 - 1 = 4$. From Table XIII, Appendix B, $\chi^2_{.01} = 13.2767$. The rejection region is $X^2 > 13.2767$.

Since the observed value of the test statistic falls in the rejection region ($X^2 = 269.91 > 13.2767$), H_0 is rejected. There is sufficient evidence to indicate a difference in preference for the five candidates exists at $\alpha = .01$.

b. The observed significance is $P(\chi^2 > 269.91)$. From Table XIII, Appendix B, with df $= 4$, $P(\chi^2 > 269.91) < .005$.

18.47 a. Some preliminary calculations are:

$$\hat{E}(n_{11}) = \frac{r_1 c_1}{n} = \frac{819(331)}{1600} = 169.4306 \quad \hat{E}(n_{22}) = \frac{781(60)}{1600} = 29.2875$$

$$\hat{E}(n_{12}) = \frac{781(331)}{1600} = 161.5694 \quad \hat{E}(n_{31}) = \frac{819(1209)}{1600} = 618.8569$$

$$\hat{E}(n_{21}) = \frac{819(60)}{1600} = 30.7125 \quad \hat{E}(n_{32}) = \frac{781(1209)}{1600} = 590.1431$$

To determine if life insurance preference of students depends on their sex, we test:

H_0: Life insurance preference of students and students' sex are independent
H_a: Life insurance preference of students and students' sex are dependent

The test statistic is $X^2 = \sum\sum \dfrac{\left[n_{ij} - \hat{E}(n_{ij})\right]^2}{\hat{E}(n_{ij})}$

$$= \frac{(116 - 169.4306)^2}{169.4306} + \frac{(215 - 161.5694)^2}{161.5694} + \cdots + \frac{(533 - 590.1431)^2}{590.1431} = 46.25$$

The rejection region requires $\alpha = .05$ in the upper tail of the χ^2 distribution with df $= (r - 1)(c - 1) = (2 - 1)(3 - 1) = 2$. From Table XIII, Appendix B, $\chi^2_{.05} = 5.99147$. The rejection region is $X^2 > 5.99147$.

Since the observed value of the test statistic falls in the rejection region ($X^2 = 46.25 > 5.99147$), H_0 is rejected. There is sufficient evidence to indicate life insurance preferences of students depend on their sex at $\alpha = .05$.

b. The observed significance level is $P(\chi^2 > 46.25)$. From Table XIII, Appendix B, with df $= 2$, $P(\chi^2 > 46.25) < .005$.

c. To calculate the appropriate percentages, divide the numbers in each column by the total for that column and multiply by 100. Divide each row total by the total sample size and multiply by 100.

	Preferred a Term Policy	Preferred Whole Life	No Preference	
Females	$\dfrac{116}{331} \cdot 100 = 35.0\%$	$\dfrac{27}{60} \cdot 100 = 45\%$	$\dfrac{676}{1209} \cdot 100 = 55.9\%$	$\dfrac{819}{1600} \cdot 100 = 51.2\%$
Males	$\dfrac{215}{331} \cdot 100 = 65\%$	$\dfrac{33}{60} \cdot 100 = 55\%$	$\dfrac{533}{1209} \cdot 100 = 44.1\%$	$\dfrac{781}{1600} \cdot 100 = 48.8\%$

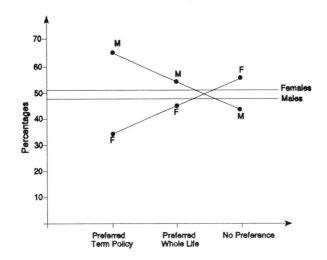

CHAPTER NINETEEN

. . . .

Decision Analysis

19.3 a. This problem is decision-making under uncertainty. The management does not know whether the applicant will default on the loan or not.

   b. This problem is decision-making under conflict. The manufacturer is "playing against" his competitors.

   c. This problem is decision-making under certainty. The company knows the outcomes of the possible decisions.

19.5 1. Actions: The set of two or more alternatives the decision-maker has chosen to consider. The decision-maker's problem is to choose one action from this set.
   2. States of Nature: The set of two or more mutually exclusive and collectively exhaustive chance events upon which the outcome of the decision-maker's chosen action depends.
   3. Outcomes: The set of consequences resulting from all possible action/state of nature combinations.
   4. Objective variables: The quantity used to measure and express the outcomes of a decision problem.

19.9 **Actions:**

   a_1: Add a new building to existing facilities.
   a_2: Do not add a new building.

States of Nature:

   S_1: Economy continues to expand.
   S_2: Economy remains stable or experiences a downward trend.

Outcomes: [The outcome is indicated for each action-state combination (a_i, S_j).]

   (a_1, S_1): \$650,000
   (a_1, S_2): \$475,000
   (a_2, S_1): \$550,000
   (a_2, S_2): \$550,000

Objective Variable:

   Profit per year

19.13 Action a_3 is inadmissible because it is dominated by action a_2. That is, for each state of nature, the payoff for a_2 is greater than or equal to the payoff for a_3.

. . . .

The opportunity losses are calculated as shown in the table below.

	State of Nature			
	S_1	S_2	S_3	S_4
a_1	0	0	$40 - 0 = 4$	$80 - 50 = 30$
Action a_2	$-58 - (-69) = 11$	$-2 - (-10) = 8$	0	$80 - 71 = 9$
a_3	$-58 - (-100) = 42$	$-2 - (-40) = 38$	$40 - (-10) = 50$	0

19.15 a. Action a_4 is dominated by action a_2 and thus is inadmissible.

b. For each action-state combination (a_i, S_j), the opportunity loss table displays the difference between the payoff received for action a and the maximum payoff attainable for state S_j. From this information, it is not possible to determine the original payoffs.

19.17 The payoff is $15 million if the minicomputer is not marketed in Europe. If it is marketed in Europe, and a 3% market share is attained, the company profits $3 million (or $18 million for the company), 2% market share gives $1 million profit ($16 million), 1% market share gives $1 million loss ($14 million). The payoff table is:

		State of Nature		
		1%	2%	3%
Action	Market	14	16	18
	Do Not Market	15	15	15

The opportunity losses are calculated as shown in the following table.

		State of Nature		
		1%	2%	3%
Action	Market	$15 - 14 = 1$	0	0
	Do Not Market	0	$16 - 15 = 1$	$18 - 15 = 3$

19.19 a. The increase in sales will be $200,000 if advertising is not increased. If it is increased by $1 million, sales will increase by $1.6 million, a profit of $600,000, if campaign is successful. If advertising is increased by $1 million, and the campaign fails, sales will increase by only $400,000, a loss of $600,000. The payoff table is as shown below.

		State of Nature	
		Successful	Unsuccessful
Action	Increase Advertising	$600,000	$-600,000
	Do Not Increase Advertising	$200,000	$200,000

b. Neither action is inadmissable. Depending on the state of nature, one action is more profitable than the other.

c.

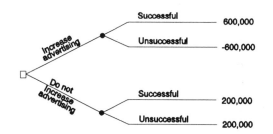

19.21 a. If the company settles out of court, they will lose $50 million. If they go to court and win, they will lose $1 million in court costs. If they go to court and lose, they will lose $100 million plus the $1 million in court fees. The payoff table is shown:

		State of Nature	
		Win	Lose
Action	Go to Court	$-$1 million	$-$101 million
	Settle Out of Court	$-$50 million	$-$50 million

b.

Go to court — Win — -$1 million

Go to court — Lose — -$101 million

Settle out of court — Win — -$50 million

Settle out of court — Lose — -$50 million

19.23 $E(x) = \sum_{\text{All } x} xp(x)$

$EP(a_1) = (-150)(.1) + (-500)(.3) + (10)(.4) + (300)(.2) = -101$

$EP(a_2) = (-200)(.1) + (-100)(.3) + (300)(.4) + (100)(.2) = 90$

$EP(a_3) = (-150)(.1) + (-40)(.3) + (-10)(.4) + (85)(.2) = -14$

We would choose a_2, the action that produced the largest expected payoff.

19.25

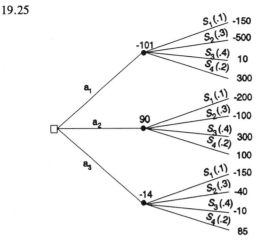

19.27 $EOL(a_1) = (56)(.6) + (0)(.4) = 33.6$
$EOL(a_2) = (0)(.6) + (70)(.4) = 28.0$

We would choose a_2, since it produces the smallest expected opportunity loss.

19.29 a. $EP(a_1) = .5(800,000) + .5(-200,000) = 300,000$
$EP(a_2) = .5(660,000) + .5(10,000) = 335,000$

We would choose action a_2: Design B, because the expected payoff is larger.

b. $EP(a_1) = 2/3(800,000) + 1/3(-200,000) = 466,666.67$
$EP(a_2) = 2/3(660,000) + 1/3(10,000) = 443,333.33$

We would choose action a_1: Design A, because the expected payoff is larger.

19.31 The payoff table, with state of nature probabilities shown in parentheses is given below.

		State (Percent Defective)		
		5% (.55)	10% (.24)	50% (.21)
Action	a_1: Purchase the Guarantee	−1250	−1500	−1500
	a_2: Do Not Purchase the Guarantee	−250	−500	−2500

$EP(a_1) = -1250(.55) - 1500(.24) - 1500(.21) = -1362.5$
$EP(a_2) = -250(.55) - 500(.24) - 2500(.21) = -782.5$

The greater expected payoff is associated with action a_2: Do not purchase the guarantee.

19.33 a. The payoff table for this decision problem is shown below.

	State	
	Successful (2/3)	Unsuccessful (1/3)
Action a_1: Increase Advertising Budget	$1,600,000 - 1,000,000$ $= 600,000$	$400,000 - 1,000,000$ $= -600,000$
a_2: Do Not Increase Budget	200,000	200,000

$EP(a_1) = 600,000(2/3) - 600,000(1/3) = 200,000$
$EP(a_2) = 200,000(2/3) + 200,000(1/3) = 200,000$

Thus, both actions have the same expected payoff and neither is preferred over the other.

b.

	State	
	Successful (2/3)	Unsuccessful (1/3)
Action a_1: Increase Advertising Budget	0	800,000
a_2: Do Not Increase Budget	400,000	0

c. $EOL(a_1) = 0(2/3) + 800,000(1/3) = 266,667$
$EOL(a_2) = 400,000(2/3) + 0(1/3) = 266,667$

Both actions have the same expected outcome. Thus, either action can be chosen.

19.35 a. Utilities were computed using the function $U(x) = x^2$, and are shown in the following table.

	State of Nature		
	S_1 (.25)	S_2 (.30)	S_3 (.45)
Action a_1	10,000	5,625	625
a_2	4,900	9,025	0

b. $EU(a_1) = 10000(.25) + 5625(.30) + 625(.45) = 4469.05$
$EU(a_2) = 4900(.25) + 9025(.30) + 0(.45) = 3932.5$

Therefore, the expected utility criterion selects action a_1.

19.37 a. Utilities were computed using the function $U(x) = -1 + .01x$, and are shown in the following table.

		State of Nature			
		S_1 (.30)	S_2 (.05)	S_3 (.45)	S_4 (.20)
	a_1	.50	.80	0	−.20
Action	a_2	.30	.65	.90	.40
	a_3	.15	.20	.60	1.10

b. $EU(a_1) = (.50)(.30) + (.80)(.05) + (0)(.45) + (-.20)(.20) = .1500$
$EU(a_2) = (.30)(.30) + (.65)(.05) + (.90)(.45) + (.40)(.20) = .6075$
$EU(a_3) = (.15)(.30) + (.20)(.05) + (.60)(.45) + (1.10)(.20) = .5450$

The expected utility criterion selects action a_2.

19.39 a.

		State of Nature		
		Dry (.5)	Moderate (.3)	Gusher (.2)
Action	Invest	−500,000	600,000	1,500,000
	Do Not Invest	0	0	0

b. $EP(\text{Invest}) = (-500,000)(.5) + (600,000)(.3) + (1,500,000)(.2) = 230,000$
$EP(\text{Do not invest}) = (0)(.5) + (0)(.3) + (0)(.2) = 0$

Using the expected payoff criterion, we would choose to invest in the oil well.

c. First, from Exercise 19.38, we compute the utility values for the table.

The maximum payoff is 1,500,000. The utility of 1,500,000 is $U(1,500,000) = 1$

The minimum payoff is −500,000. The utility of −500,000 is $U(-500,000) = 0$.

From part a, Exercise 19.38, $U(600,000) = .9$ and $U(0) = .8$.

The utility table is:

		State of Nature		
		Dry (.5)	Moderate (.3)	Gusher (.2)
Action	Invest	0	.9	1.0
	Do Not Invest	.8	.8	.8

EU(Invest) $= 0(.5) + .9(.3) + 1(.2) = .47$
EU(Do Not Invest) $= .8(.5) + .8(.3) + .8(.2) = .8$

The expected utility criterion prescribes action: Do not invest.

19.41   a.

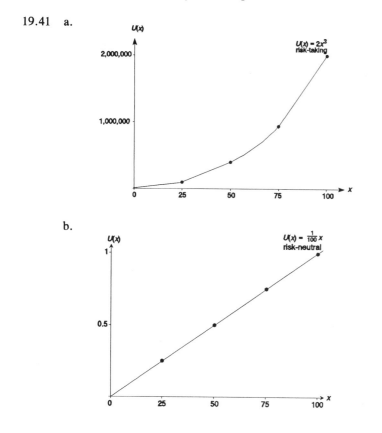

b.

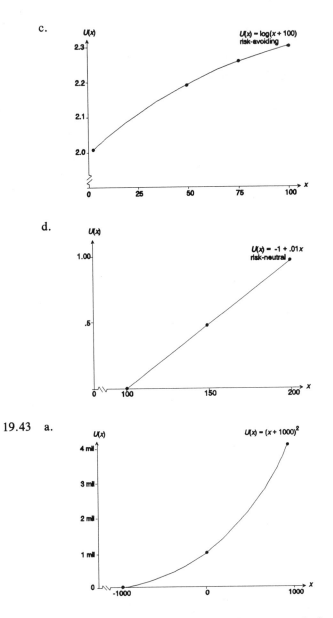

c.

$U(x) = \log(x + 100)$
risk-avoiding

d.

$U(x) = -1 + .01x$
risk-neutral

19.43   a.

$U(x) = (x + 1000)^2$

b.   The graph of the utility function is convex and, thus, represents a risk-taking attitude.

c.   The utilities were computed using the utility function

$$U(x) = (x + 1000)^2$$

and are shown in the table.

	State of Nature		
	S_1 (.1)	S_2 (.5)	S_3 (.4)
Action a_1	160,000	4,000,000	1,690,000
a_2	40,000	1,000,000	3,062,500
a_3	1,123,600	680,625	2,592,100

$EU(a_1) = (160,000)(.1) + (4,000,000)(.5) + (1,690,000)(.4) = 2,692,000$
$EU(a_2) = (40,000)(.1) + (1,000,000)(.5) + (3,062,500)(.4) = 1,729,000$
$EU(a_3) = (1,123,600)(.1) + (680,625)(.5) + (2,592,100)(.4) = 1,489,512.5$

The expected utility criterion identifies action a_1 as the optimal action.

19.45 a. The plot of the utility function is:

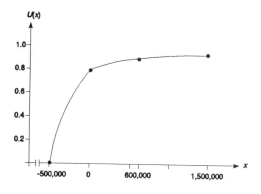

b. Since the graph is concave, the investor is a risk-avoider.

19.49

(1)	(2)	(3)	(4)	(5)
State of Nature	Prior Probability	Conditional Probability of Sample Information	$(2) \times (3)$	Posterior Probability $(4) \div$ Total of (4)
S_1	.10	.90	.0900	.1925
S_2	.25	.10	.0250	.0535
S_3	.45	.65	.2925	.6257
S_4	.20	.30	.0600	.1283
	1.00		.4675	1.0000

19.51 Prior information is information available before any sample information is obtained. Posterior information is additional information obtained from sampling.

19.53

(1)	(2)	(3)	(4)	(5)
	Prior	Conditional	Probability of Intersection	Posterior Probability
State of Nature	Probability	Probability	(2) × (3)	(4) ÷ Total of (4)
S_1: Correctly Adjusted	.9	.05	.045	.474
S_2: Incorrectly Adjusted	.1	.50	.050	.526
	1.00		.095	1.000

The probability that the machine is incorrectly adjusted is .526.

19.55 a.

p	Prior Probability
.01	$p = .25$
.05	$2p = .50$
.10	$p = .25$
	$4p = 1$
	$\Rightarrow p = .25$

b. $P(I \mid S_1) = P(x = 3 \mid p = .01) = \binom{5}{3}(.01)^3(.99)^2 = .0000098$

$P(I \mid S_2) = P(x = 3 \mid p = .05) = \binom{5}{3}(.05)^3(.95)^2 = .0011281$

$P(I \mid S_3) = P(x = 3 \mid p = .10) = \binom{5}{3}(.10)^3(.90)^2 = .0081000$

(1)	(2)	(3)	(4)	(5)
	Prior	Conditional	Probability of Intersection	Posterior Probability
State of Nature	Probability	Probability	(2) × (3)	(4) ÷ Total of (4)
S_1: $p = .01$	.25	.0000098	.0000025	.00096
S_2: $p = .05$	.50	.0011281	.0005641	.21766
S_3: $p = .10$	.25	.0081000	.0020250	.78137
	1.00		.0025916	1.00000

19.57 $EP(a_1) = \sum_{\text{All states}} (\text{Payoff})(\text{Posterior probability of state})$

$\qquad = (30)(.1925) + (20)(.0535) + (0)(.6257) + (15)(.1283) = 8.7695$

$EP(a_2) = (40)(.1925) + (20)(.0535) + (5)(.6257) + (0)(.1283) = 11.8985$

$EP(a_3) = (9)(.1925) + (15)(.0535) + (-21)(.6257) + (25)(.1283) = -7.3972$

The action with the highest expected payoff is a_2.

19.59 a. $EP(a_1) = (-150)(.5) + (225)(.3) + (120)(.2) = 16.5$
$EP(a_2) = (200)(.5) + (-172)(.3) + (100)(.2) = 68.4$
$EP(a_3) = (300)(.5) + (80)(.3) + (-100)(.2) = 154$

The action with the highest expected payoff is a_3.

b.

(1) State	(2) Prior Probability	(3) Conditional Probability	(4) Probability of Intersection (2) × (3)	(5) Posterior Probability (4) ÷ Total of (4)
S_1	.5	.2	.10	.3125
S_2	.3	.6	.18	.5625
S_3	.2	.2	.04	.1250
			.32	1.0000

c. $EP(a_1) = (-150)(.3125) + 225(.5625) + 120(.1250) = 94.6875$
$EP(a_2) = (200)(.3125) + (-172)(.5625) + 100(.1250) = -21.75$
$EP(a_3) = (300)(.3125) + (80)(.5625) + (-100)(.1250) = 126.25$

The action with the highest expected payoff is a_3.

19.61 Payoffs are in millions of dollars.

a.

Action	State of Nature (Prior Probabilities in Parentheses)	
	Pass Bill (.6)	Do Not Pass Bill (.4)
a_1: Stay in Business	2.5	-1.5
a_2: Lease Facility	.5	.5

b.

(1) State of Nature	(2) Prior Probability	(3) Conditional Probability	(4) Probability of Intersection (2) × (3)	(5) Posterior Probability (4) ÷ Total of (4)
S_1	.6	.8	.48	.923
S_2	.4	.1	.04	.077
			.52	1.000

d. $EP(a_1) = 2.5(.923) + (-1.5)(.077) = 2.192$ million dollars
$EP(a_2) = .5(.923) + .5(.077) = .5$ million dollars

The action having the highest expected payoff is a_1: Stay in business

Decision Analysis

19.63 a. First we construct the probability revision tables to derive posterior probabilities for each of the two sample outcomes:

(1)	(2)	(3)	(4)	(5)
			Probability of	
State of Nature	Prior Probability	Conditional Probability	Intersection (2) × (3)	Posterior Probability (4) ÷ Total of (4)
		Sample Outcome: S_1 True		
S_1	.4	.4	.16	.4706
S_2	.6	.3	.18	.5294
			.34	1.0000
		Sample Outcome: S_2 True		
S_1	.4	.6	.24	.3636
S_2	.6	.7	.42	.6364
			.66	1.0000

From the upper portion of the previous table,

$$EP(a_1 \mid \text{Sample information indicates } S_1 \text{ True}) = 600(.4706) - 100(.5294)$$
$$= \$229.42$$
$$EP(a_2 \mid \text{Sample information indicates } S_1 \text{ True}) = -200(.4706) + 300(.5294)$$
$$= \$64.70$$

The expected payoff criterion selects action a_1, when the sample information indicates S_1 is the true state of nature.

From the lower portion of the table,

$$EP(a_1 \mid \text{Sample information indicates } S_2 \text{ True}) = 600(.3636) - 100(.6364)$$
$$= \$154.52$$
$$EP(a_2 \mid \text{Sample information indicates } S_2 \text{ True}) = -200(.3636) + 300(.6364)$$
$$= \$118.20$$

The expected payoff criterion selects action a_1, when the sample information indicates S_2 is the true state of nature.

The expected payoff of sampling (EPS) is computed as follows:

$$EPS = 229.42(.34) + 154.52(.66) = \$179.986$$
$$EVSI = EPS - EPNS = 179.986 - 180 = -.014$$
$$(EPNS = 180 \text{ is from Exercise 19.62})$$
$$ENGS = EVSI - CS = -.014 - 10 = -10.014$$

b. Since the ENGS for the second source is smaller than for the first source, the decision maker should choose the first source.

19.65 a. We now construct probability revision tables to derive posterior probabilities for each of the four possible sample outcomes.

(1)	(2)	(3)	(4)	(5)
	Prior	**Conditional**	**Probability of Intersection**	**Posterior Probability**
State of Nature	**Probability**	**Probability**	**(2) × (3)**	**(4) ÷ Total of (4)**
Sample Outcome: Product is "Very Successful"				
Failure	.6	.1	.06	1/3
Successful	.3	.2	.06	1/3
Very Successful	.1	.6	.06	1/3
			.18	1
Sample Outcome: Product is "Successful"				
Failure	.6	.1	.06	.300
Successful	.3	.4	.12	.600
Very Successful	.1	.2	.02	.100
			.20	1.000
Sample Outcome: Product is "Failure"				
Failure	.6	.5	.30	.811
Successful	.3	.2	.06	.612
Very Successful	.1	.1	.01	.027
			.37	1.000
Sample Outcome: Product is "Uncertain"				
Failure	.6	.3	.18	.72
Successful	.3	.2	.06	.24
Very Successful	.1	.1	.01	.04
			.25	1.00

From the first portion of the table, Product is "Very Successful,"

$$EP(a_1) = -200,000(1/3) + 300,000(1/3) + 600,000(1/3) = \$233,333.33$$
$$EP(a_2) = 0(1/3) + 0(1/3) + 0(1/3) = 0$$

The expected payoff criterion selects action a_1.

From the second portion of the table, Product is "Successful,"

$$EP(a_1) = -200,000(.3) + 300,000(.6) + 600,000(.1) = \$180,000$$
$$EP(a_2) = 0(.3) + 0(.6) + 0(.1) = 0$$

The expected payoff criterion selects action a_1.

From the third portion of the table, Product is "Failure,"

$$EP(a_1) = -200,000(.811) + 300,000(.162) + 600,000(.027) = \$-97,400$$
$$EP(a_2) = 0(.811) + 0(.162) + 0(.027) = 0$$

The expected payoff criterion selects action a_2.

From the fourth portion of the table, Product is "Uncertain,"

$$EP(a_1) = -200,000(.72) + 300,000(.24) + 600,000(.04) = \$-48,000$$
$$EP(a_2) = 0(.72) + 0(.24) + 0(.04) = 0$$

The expected payoff criterion selects action a_2.

The expected payoff of sampling is:

$$EPS = 233,333.33(.18) + 180,000(.20) + 0(.37) + 0(.25) = \$78,000.00$$

The expected payoff with no sampling is:

$$EP(a_1) = -200,000(.6) + 300,000(.3) + 600,000(.1) = \$30,000$$
$$EP(a_2) = 0(.6) + 0(.3) + 0(.1) = 0$$

The EPNS = \$30,000

Thus, EVSI = EPS − EPNS = 78,000 − 30,000 = \$48,000

The most the company should pay for sampling information is \$48,000.

c.   ENGS = EVSI − CS(Cost of Sampling) = \$48,000 − \$30,000 = \$18,000.

Since this is a positive, the company should undertake the proposed market survey.

19.67   From Exercise 19.60, EPS = .06(−45,332) + .94(−42,210) = −42,397.32

EVSI = EPS − EPNS = −42,397.32 − (−42,400) = 2.68

Thus, it is worth \$2.68 for the hospital to be able to test one of the 100 television sets.

(If we round off EPS to the nearest \$1000, EVSI = 0).

19.69 a.

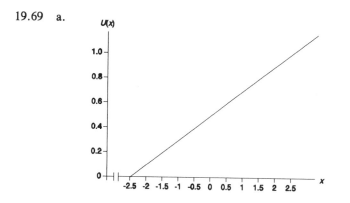

b.   The utility function is a straight-line, so the decision-maker is risk-neutral.

c.   Using the payoff table from Exercise 19.68 and the utility function $U(x) = .5 + .20x$, the utility table is:

		State of Nature	
		Bill Passes (.6)	Bill Does Not Pass (.4)
Action	a_1: Do Not Lease	1	.2
	a_2: Lease	.6	.6

$EU(a_1) = .6(1) + .4(.2) = .68$
$EU(a_2) = .6(.6) + .4(.6) = .60$

The expected utility criterion selects action a_1, do not lease.

19.71 a.   We would use relative frequency. We could obtain the information from previous insurance claims.

b.   Since we have no access to prior information, we would have to use subjective values. These could be obtained by your knowledge of market shares from similar new products.

c.   Since no prior information is available, we would have to use subjective values. These values could be obtained by your suspicions of how successful the research will be.

d.   Since previous bank records are available, we would use relative frequency to obtain probabilities.

19.73 a. $P(\text{repaid}) = \dfrac{1104}{1280} = .8625$

$P(\text{repaid with difficulty}) = \dfrac{120}{1280} = .09375$

$P(\text{defaulted}) = \dfrac{56}{1280} = .04375$

b. Using the payoff table in Exercise 19.72, the expected payoffs are:

$EP(\text{loan}) = .8625(7500) + .09375(6500) + .04375(-30,000) = 5765.625$
$EP(\text{no loan}) = .8625(6000) + .09375(6000) + .04375(6,000) = 6000$

The expected payoff criterion selects action, do not make loan.

19.75 a. We will first determine the quantity q of stockings demanded using the equations

$q = 10 - 2p$ for Economist 1

and

$q = 16 - 4p$ for Economist 2

These **quantities** are shown in the following table.

		Economist 1	Economist 2
	.99	8.02	12.04
Price	1.98	6.04	8.08
($)	2.75	4.50	5.00
	3.50	3.00	2.00

These quantities can now be used to determine **payoffs**, as illustrated in the following table.

		State of Nature	
		Economist 1 (.5)	Economist 2 (.5)
	a_1: .99	.99(8.02) = 7.94	.99(12.04) = 11.92
Action	a_2: 1.98	1.98(6.04) = 11.96	1.98(8.08) = 16.00
(Price, $)	a_3: 2.75	2.75(4.50) = 12.38	2.75(5.00) = 13.75
	a_4: 3.50	3.50(3.00) = 10.50	3.50(2.00) = 7.00

b.

		State of Nature	
		Economist 1 (.5)	Economist 2 (.5)
	a_1: .99	4.44	4.08
Action	a_2: 1.98	.42	0
(Price, $)	a_3: 2.75	0	2.25
	a_4: 3.50	1.88	9.00

c. $EP(a_1) = 7.94(.5) + 11.92(.5) = 9.93$
$EP(a_2) = 11.96(.5) + 16.00(.5) = 13.98$
$EP(a_3) = 12.38(.5) + 13.75(.5) = 13.07$
$EP(a_4) = 10.50(.5) + 7.00(.5) = 8.75$

According to the expected payoff criterion, the company should choose action a_2: Set the price at $1.98.

19.77 Using the utility funciton, $U(x) = \dfrac{\sqrt{1,000,000 + x}}{1400}$ and the payoff table from Exercise 19.76, the utility table is:

		State of Nature	
		Loses (.8)	Wins (.2)
Action	a_1: Goes to Court	0	.7143
	a_2: Out of Court	.6186	.6186

$EU(a_1) = .8(0) + .2(.7143) = .1429$
$EU(a_2) = .8(.6186) + .2(.6186) = .6186$

The expected utility criterion selects action a_2, settle out of court.

19.79 a.

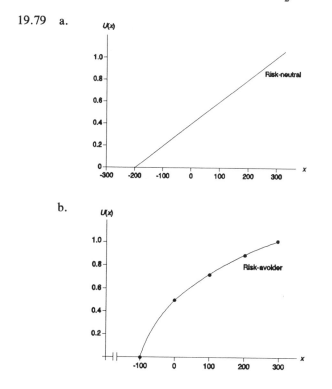

b.

c.

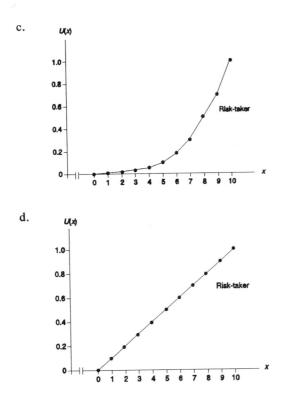

d.

19.81 a.

	(1)	(2)	(3)	(4)	(5)
		Prior	**Conditional**	**Probability of Intersection**	**Posterior Probability**
	State of Nature	**Probability**	**Probability**	**(2) × (3)**	**(4) ÷ Total of (4)**
S_1: Guilty		.1	1*	.1	.9911
S_2: Not Guilty		.9	.001**	.0009	
				.1009	
	S_1	.2	1	.2	.9960
	S_2	.8	.001	.0008	
				.2008	
	S_1	.3	1	.3	.9977
	S_2	.7	.001	.0007	
				.3007	
	S_1	.4	1	.4	.9985
	S_2	.6	.001	.0006	
				.4006	
	S_1	.5	1	.5	.9990
	S_2	.5	.001	.0005	
				.5005	
	S_1	.6	1	.6	.9993
	S_2	.4	.001	.0004	
				.6004	
	S_1	.7	1	.7	.9996
	S_2	.3	.001	.0003	
				.7003	
	S_1	.8	1	.8	.9998
	S_2	.2	.001	.0002	
				.8002	
	S_1	.9	1	.9	.9999
	S_2	.1	.001	.0001	
				.9001	

*P(Print | Guilty)
**P(Print | Not Guilty)

b.

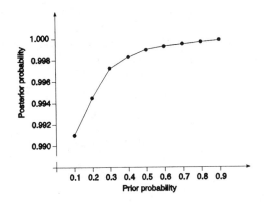

19.83   a.

		State of Nature (Prior Probabilities in Parentheses)	
		$S_1 = 5\%$ (.80)	$S_2 = 10\%$ (.20)
Action	Reject	-1000	0
	Accept	0	-4100

b. $EP(\text{Reject}) = (-1000)(.80) + (0)(.20) = -800$
$EP(\text{Accept}) = (0)(.80) + (-4100)(.20) = -820$

According to the expected payoff criterion, reject the lot.

c. $P(I \mid S_1) = P(x = 1 \mid p = .05) = \begin{pmatrix} 1 \\ 1 \end{pmatrix} (.05)^1 (.95)^0 = .05$

$P(I \mid S_2) = P(x = 1 \mid p = .10) = \begin{pmatrix} 1 \\ 1 \end{pmatrix} (.10)^1 (.90)^0 = .10$

(1)	(2)	(3)	(4)	(5)
			Probability of	
State of Nature	Prior Probability	Conditional Probability	Intersection (2) × (3)	Posterior Probability (4) ÷ Total of (4)
S_1	.8	.05	.04	.6667
S_2	.2	.10	.02	.3333
			.06	

$EP(\text{Reject}) = (-1000)(.6667) + (0)(.3333) = -666.7$
$EP(\text{Accept}) = (0)(.6667) + (-4100)(.3333) = -1366.53$

Thus, we would reject the lot.

d. $P(I \mid S_1) = P(x = 0 \mid p = .05) = \binom{1}{0}(.05)^0(.95)^1 = .95$

$P(I \mid S_2) = P(x = 0 \mid p = .10) = \binom{1}{0}(.10)^0(.90)^1 = .90$

(1) State of Nature	(2) Prior Probability	(3) Conditional Probability	(4) Probability of Intersection (2) × (3)	(5) Posterior Probability (4) ÷ Total of (4)
S_1	.8	.95	.76	.8085
S_2	.2	.90	.18	.1915
			.94	

$EP(\text{Reject}) = (-1000)(.8085) + (0)(.1915) = -808.5$
$EP(\text{Accept}) = (0)(.8085) + (-4100)(.1915) = -785.15$

Thus, we would accept the lot.

e. EVSI = EPS − EPNS where EPNS is the maximum EP from part b.

$$EPS = \sum \binom{\text{Maximum expected payoff}}{\text{for sample outcome}}\binom{\text{Marginal probability}}{\text{of sample outcome}}$$
$$= (-666.7)(.06) + (-785.15)(.94) = -778.043$$

EVSI = −778.043 − (−800) = 21.957

f. ENGS = 21.957 − 20 = 1.957

Since the expected net gain from sampling is positive, management should use a sample size equal to 1.

19.85 First, we find the probabilities of observing 1, 2, 3, 4, or 5 defectives given each of the states of nature.

$P(x = 0 \mid p = .05) = \binom{5}{0}.05^0.95^5 = .7738$

$P(x = 0 \mid p = .10) = \binom{5}{0}.10^0.90^5 = .5905$

$P(x = 1 \mid p = .05) = \binom{5}{1}.05^1.95^4 = .2026$

$P(x = 1 \mid p = .10) = \binom{5}{1}.10^1.90^4 = .3281$

$P(x = 2 \mid p = .05) = \binom{5}{2}.05^2.95^3 = .0214$

$P(x = 2 \mid p = .10) = \binom{5}{2}.10^2.90^3 = .0729$

$$P(x = 3 \mid p = .05) = \begin{bmatrix} 5 \\ 3 \end{bmatrix} .05^3 .95^2 = .0011$$

$$P(x = 3 \mid p = .10) = \begin{bmatrix} 5 \\ 3 \end{bmatrix} .10^3 .90^2 = .0081$$

$$P(x = 4 \mid p = .05) = \begin{bmatrix} 5 \\ 4 \end{bmatrix} .05^4 .95^1 = .00003$$

$$P(x = 4 \mid p = .10) = \begin{bmatrix} 5 \\ 4 \end{bmatrix} .10^4 .90^1 = .0005$$

$$P(x = 5 \mid p = .05) = \begin{bmatrix} 5 \\ 5 \end{bmatrix} .05^5 .95^0 = .0000003$$

$$P(x = 5 \mid p = .10) = \begin{bmatrix} 5 \\ 5 \end{bmatrix} .10^5 .90^0 = .00001$$

(1)	(2)	(3)	(4)	(5)
			Probability of	
State of	Prior	Conditional	Intersection	Posterior Probability
Nature	Probability	Probability	(2) × (3)	(4) ÷ Total of (4)
Sample Outcome: $x = 0$ defective				
S_1	.8	.7738	.6190	.8398
S_2	.2	.5905	.1181	.1602
			.7371	1.0000
Sample Outcome: $x = 1$ defective				
S_1	.8	.2036	.1629	.7129
S_2	.2	.3281	.0656	.2871
			.2285	1.0000
Sample Outcome: $x = 2$ defectives				
S_1	.8	.0214	.0171	.5394
S_2	.2	.0729	.0146	.4606
			.0317	1.0000
Sample Outcome: $x = 3$ defectives				
S_1	.8	.0011	.0009	.3600
S_2	.2	.0081	.0016	.6400
			.0025	1.0000
Sample Outcome: $x = 4$ defectives				
S_1	.8	.00003	.000024	.1935
S_2	.2	.0005	.0001	.8065
			.000124	1.0000
Sample Outcome: $x = 5$ defectives				
S_1	.8	.0000003	.00000024	.1071
S_2	.2	.00001	.000002	.8929
			.00000224	1.0000

$$EP(a_1 \mid x = 0) = .8398(-1000) + .1602(0) = -839.8$$

$$EP(a_2 \mid x = 0) = .8398(0) + .1602(-4100) = -656.82$$

The expected payoff criterion selects action a_2.

$$EP(a_1 \mid x = 1) = .7129(-1000) + .2871(0) = -712.9$$
$$EP(a_2 \mid x = 1) = .7129(0) + .2871(-4100) = -1177.11$$

The expected payoff criterion selects action a_1.

$$EP(a_1 \mid x = 2) = .5394(-1000) + .4606(0) = -539.4$$
$$EP(a_2 \mid x = 2) = .5394(0) + .4606(-4100) = -1888.46$$

The expected payoff criterion selects action a_1.

$$EP(a_1 \mid x = 3) = .3600(-1000) + .6400(0) = -360.0$$
$$EP(a_2 \mid x = 3) = .3600(0) + .6400(-4100) = -2624.00$$

The expected payoff criterion selects action a_1.

$$EP(a_1 \mid x = 4) = .1935(-1000) + .8065(0) = -193.5$$
$$EP(a_2 \mid x = 4) = .1935(0) + .8065(-4100) = -3306.65$$

The expected payoff criterion selects action a_1.

$$EP(a_1 \mid x = 5) = .1071(-1000) + .8929(0) = -107.1$$
$$EP(a_2 \mid x = 5) = .1071(0) + .8929(-4100) = -3660.89$$

The expected payoff criterion selects action a_1.

$$EPS = .7371(-656.82) + .2285(-712.9) + .0317(-539.4) + .0025(-360.0)$$
$$+ .000124(-193.5) + .00000224(-107.1) = -665.06$$

$$EVSI = EPS - EPNS = -665.05 - (-800) = 134.95$$

$$ENGS = EVSI - CS = 134.95 - (10 + 5(10)) = \$74.95$$

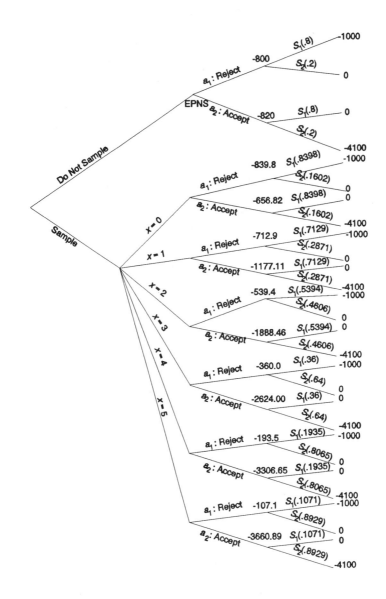

19.87 We will construct probability tables to derive posterior probabilities for each of the 7 possible outcomes.

(1) State of Nature	(2) Prior Probability	(3) Conditional Probability	(4) Probability of Intersection (2) × (3)	(5) Posterior Probability (4) ÷ Total of (4)
Sample Outcome: 0 − 5,000				
S_1	.05	.5	.025	.217
S_2	.20	.3	.06	.522
S_3	.30	.1	.03	.261
S_4	.20	0	0	0
S_5	.10	0	0	0
S_6	.08	0	0	0
S_7	.07	0	0	0
			.115	1.000
Sample Outcome: 5,001 − 10,000				
S_1	.05	.35	.0175	.099
S_2	.20	.4	.0800	.451
S_3	.30	.2	.0600	.338
S_4	.20	.1	.0200	.113
S_5	.10	0	0	0
S_6	.08	0	0	0
S_7	.07	0	0	0
			.1775	1.001
Sample Outcome: 10,001 − 20,000				
S_1	.05	.15	.0075	.030
S_2	.20	.2	.0400	.162
S_3	.30	.5	.1500	.606
S_4	.20	.2	.0400	.162
S_5	.10	.1	.0100	.040
S_6	.08	0	0	0
S_7	.07	0	0	0
			.2475	1.000
Sample Outcome: 20,001 − 30,000				
S_1	.05	0	0	0
S_2	.20	.1	.02	.1
S_3	.30	.2	.06	.3
S_4	.20	.5	.10	.5
S_5	.10	.2	.02	.1
S_6	.08	0	0	0
S_7	.07	0	0	0
			.2	1.0

(1)	(2)	(3)	(4)	(5)
			Probability of	
State of	**Prior**	**Conditional**	**Intersection**	**Posterior Probability**
Nature	**Probability**	**Probability**	**(2) × (3)**	**(4) ÷ Total of (4)**
		Sample Outcome:	30,001 − 40,000	
S_1	.05	0	0	0
S_2	.20	0	0	0
S_3	.30	0	0	0
S_4	.20	.2	.040	.381
S_5	.10	.5	.050	.476
S_6	.08	.1	.008	.076
S_7	.07	.1	.007	.067
			.105	1.000
		Sample Outcome:	40,001 − 50,000	
S_1	.05	0	0	0
S_2	.20	0	0	0
S_3	.30	0	0	0
S_4	.20	0	0	0
S_5	.10	.2	.020	.227
S_6	.08	.5	.040	.455
S_7	.07	.4	.028	.318
			.088	1.000
		Sample Outcome:	50,001 − 60,000	
S_1	.05	0	0	0
S_2	.20	0	0	0
S_3	.30	0	0	0
S_4	.20	0	0	0
S_5	.10	0	0	0
S_6	.08	.4	.032	.478
S_7	.07	.5	.035	.522
			.067	1.000

Using the payoff table in Exercise 19.86:

$$E(a_1 \mid 0 - 5000) = 0(.217) + 0(.522) + 0(.261) + (-3000)(0) + (-6000)(0)$$
$$+ (-9000)(0) + (-12,000)(0) = 0$$
$$E(a_2 \mid 0 - 5000) = -2(.217) - 2(.522) - 2(.261) - 2702(0) - 5702(0) - 8702(0)$$
$$- 11,702(0) = -2$$

Using similar procedures,

$$E(a_3 \mid 0 - 5000) = -4$$
$$E(a_4 \mid 0 - 5000) = -6$$
$$E(a_5 \mid 0 - 5000) = -8$$
$$E(a_6 \mid 0 - 5000) = -10$$
$$E(a_7 \mid 0 - 5000) = -12$$
$$E(a_8 \mid 0 - 5000) = -14$$

The expected payoff criterion selects action a_1 when the sample information indicates the demand is $0 - 5000$.

Similarly, given each state is the true state of nature, the expected payoff criterion selects:

True State of Nature	Selected by Expected Payoff Criterion	Expected Payoff
S_1	a_1	0
S_2	a_3	-286.5
S_3	a_5	-613.0
S_4	a_5	-1758.0
S_5	a_8	-4586.5
S_6	a_8	-7741.5
S_7	a_8	-8819.0

$$\begin{aligned} \text{EPS} = {} & .115(0) + .1775(-286.5) + .2475(-613.0) + .2(-1758) \\ & + .105(-4586.5) + .088(-7741.5) + .067(-8819.0) = -2307.9 \end{aligned}$$

$$\text{EVSI} = \text{EPS} - \text{EPNS} = -2307.9 - (-2314) = 6.1 \text{ or } \$6,100$$

where EPNS is found in Exercise 19.86.

$$\text{ENGS} = \text{EVSI} - \text{CS} = 6,100 - 5,000,000 = \$-4,993,900$$

Since this value is less than 0, the model should not be developed.

CHAPTER TWENTY

· ·

Survey Sampling

20.1　An object upon which a measurement is made is called an element; a sampling unit is a collection of elements; and a sample is a collection of sampling units selected from a list of sampling units or frame.

20.3　The frame was constructed from sources such as telephone directories, club membership lists, magazine subscriber lists, and lists of car owners. The sample was probably not representative of the whole population of voters in 1936; it was biased in favor of Republican voters. Also, it seemed to be affected by nonresponse bias. Only the proportion of the population with intense interest answered the questionnaire.

20.5　a.　Percentage sampled $= \dfrac{n}{N}(100\%) = \dfrac{1000}{2500}(100\%) = 40\%$

Finite population correction factor:

$$\sqrt{\frac{N-n}{N}} = \sqrt{\frac{2500 - 1000}{2500}} = \sqrt{.6} = .7746$$

　　　b.　Percentage sampled $= \dfrac{n}{N}(100\%) = \dfrac{1000}{5000}(100\%) = 20\%$

Finite population correction factor:

$$\sqrt{\frac{N-n}{N}} = \sqrt{\frac{5000 - 1000}{5000}} = \sqrt{.8} = .8944$$

　　　c.　Percentage sampled $= \dfrac{n}{N}(100\%) = \dfrac{1000}{10,000}(100\%) = 10\%$

Finite population correction factor:

$$\sqrt{\frac{N-n}{N}} = \sqrt{\frac{10,000 - 1000}{10,000}} = \sqrt{.9} = .9487$$

　　　d.　Percentage sampled $= \dfrac{n}{N}(100\%) = \dfrac{1000}{100,000}(100\%) = 1\%$

Finite population correction factor:

$$\sqrt{\frac{N-n}{N}} = \sqrt{\frac{100,000 - 1000}{100,000}} = \sqrt{.99} = .995$$

20.7 **a.** For $n = 36$, with the finite population correction factor:

$$\hat{\sigma}_{\bar{x}} = s/\sqrt{n}\left[\sqrt{\frac{N-n}{N}}\right] = \frac{24}{\sqrt{64}}\left[\sqrt{\frac{5000-64}{5000}}\right] = 3\sqrt{.9872} = 2.9807$$

without the finite population correction factor:

$$\hat{\sigma}_{\bar{x}} = s/\sqrt{n} = \frac{24}{\sqrt{64}} = 3$$

$\hat{\sigma}_{\bar{x}}$ without the finite population correction factor is slightly larger.

b. For $n = 400$, with the finite population correction factor:

$$\hat{\sigma}_{\bar{x}} = s/\sqrt{n}\left[\sqrt{\frac{N-n}{N}}\right] = \frac{24}{\sqrt{400}}\left[\sqrt{\frac{5000-400}{5000}}\right] = 1.2\sqrt{.92} = 1.1510$$

without the finite population correction factor:

$$\hat{\sigma}_{\bar{x}} = s/\sqrt{n} = \frac{24}{\sqrt{400}} = 1.2$$

c. In part **a**, n is smaller relative to N than in part **b**. Therefore, the finite population correction factor did not make as much difference in the answer in part **a** as in part **b**.

20.9 An approximate 95% confidence interval for μ is:

$$\bar{x} \pm 2\hat{\sigma}_{\bar{x}} \Rightarrow \bar{x} \pm 2\frac{s}{\sqrt{n}}\sqrt{\frac{N-n}{N}}$$

$$\Rightarrow 422 \pm 2\frac{14}{\sqrt{40}}\sqrt{\frac{375-40}{375}}$$

$$\Rightarrow 422 \pm 4.184 \Rightarrow (417.816, 426.184)$$

20.11 An approximate 95% confidence interval for τ is:

$$\hat{\tau} \pm 2\hat{\sigma}_{\hat{\tau}} \Rightarrow N\bar{x} \pm 2\sqrt{N^2\frac{s^2}{n}\left[\frac{N-n}{N}\right]}$$

$$\Rightarrow 3500(39.4) \pm 2\sqrt{3500^2\frac{4^2}{100}\left[\frac{3500-100}{3500}\right]}$$

$$\Rightarrow 137,900 \pm 2759.7101 \Rightarrow (135,140.2899, 140,659.7101)$$

20.13 a.   An approximate 95% confidence interval for μ is:

$$\bar{x} \pm 2\hat{\sigma}_{\bar{x}} \Rightarrow \bar{x} \pm 2\frac{s}{\sqrt{n}}\sqrt{\frac{N-n}{N}}$$

$$\Rightarrow 779{,}030 \pm 2\frac{1{,}083{,}162}{\sqrt{32}}\sqrt{\frac{1500-32}{1500}}$$

$$\Rightarrow 779{,}030 \pm 378{,}848.7164 \Rightarrow (400{,}181.2836,\ 1{,}157{,}878.7164)$$

b.   An approximate 95% confidence interval for τ is:

$$\hat{\tau} \pm 2\hat{\sigma}_{\hat{\tau}} \Rightarrow N\bar{x} \pm 2\sqrt{N^2\frac{s^2}{n}\left(\frac{N-n}{N}\right)}$$

$$\Rightarrow 1500(779{,}030) \pm 2\sqrt{1500^2\frac{1{,}083{,}162^2}{32}\left(\frac{1500-32}{1500}\right)}$$

$$\Rightarrow 1{,}168{,}545{,}000 \pm 568{,}273{,}074.8 \Rightarrow (600{,}271{,}925.2,\ 1{,}736{,}818{,}074.8)$$

c.   Approximately 95% confident the total amount spent by all firms in the population on external audits in 1981 is between \$600,271,925 and \$1,736,818,075.

d.   The necessary assumption is:

1. Sample size is sufficiently large.

20.15 a. p is the proportion of corn-related products having EDB residues above the safe level in a particular state.

b.   The estimated bound on the error of estimation:

$$2\hat{\sigma}_{\hat{p}} = 2\sqrt{\frac{\hat{p}(1-\hat{p})}{n}\left(\frac{N-n}{N}\right)} = 2\sqrt{\frac{.086(1-.086)}{175}\left(\frac{3000-175}{3000}\right)} = .041$$

c.   An approximate 95% confidence interval for p is:

$$\hat{p} \pm 2\hat{\sigma}_{\hat{p}} \Rightarrow .086 \pm .041 \Rightarrow (.045,\ .127)$$

d. $H_0\colon p = .07$
     $H_a\colon p > .07$

The test statistic is $z = \dfrac{\hat{p} - p_0}{\sigma_{\hat{p}}} = \dfrac{\hat{p} - p_0}{\sqrt{\dfrac{p_0(1-p_0)}{n}\left(\dfrac{N-n}{N}\right)}}$

$$= \frac{.086 - .07}{\sqrt{\dfrac{.07(.93)}{175}\left(\dfrac{3000-175}{3000}\right)}} = .85$$

The rejection region requires $\alpha = .05$ in the upper tail of the z distribution. From Table IV, Appendix B, $z_{.05} = 1.645$. The rejection region is $z > 1.645$.

Since the observed value of the test statistic does not fall in the rejection region ($z = .85$ $\not> 1.645$), H_0 is not rejected. There is insufficient evidence to indicate that more than 7% of the corn-related products in this state would have to be removed from shelves and warehouses at $\alpha = .05$.

20.17 a. An approximate 95% confidence interval for μ is:

$$\bar{x} \pm 2\hat{\sigma}_{\bar{x}} \Rightarrow \bar{x} \pm 2\frac{s}{\sqrt{n}}\sqrt{\frac{N-n}{N}}$$

$$\Rightarrow 330 \pm 2\frac{546}{\sqrt{60}}\sqrt{\frac{410-60}{410}}$$

$$\Rightarrow 330 \pm 130.25 \Rightarrow (199.75, 460.25)$$

b. An approximate 95% confidence interval for τ is:

$$\hat{\tau} \pm 2\hat{\sigma}_{\hat{\tau}} \Rightarrow N\bar{x} \pm 2\sqrt{N^2\frac{s^2}{n}\left[\frac{N-n}{N}\right]}$$

$$\Rightarrow 410(330) \pm 2\sqrt{410^2\frac{546^2}{60}\left[\frac{410-60}{410}\right]}$$

$$\Rightarrow 135,300 \pm 53,403.8987 \Rightarrow (81,896.1013, 188,703.8987)$$

20.19 a. $k = 3$
$N_1 = 25,000$
$N_2 = 10,000$
$N_3 = 5000$
$N = N_1 + N_2 + N_3 = 25,000 + 10,000 + 5000 = 40,000$

b. $n_1 = 20, n_2 = 15, n_3 = 10, n = n_1 + n_2 + n_3 = 45$

c. $\bar{x}_1 = \dfrac{\sum x_1}{n_1} = \dfrac{1513}{20} = 75.65$ $\bar{x}_2 = \dfrac{\sum x_2}{n_2} = \dfrac{776}{15} = 51.7333$

$\bar{x}_3 = \dfrac{\sum x_3}{n_3} = \dfrac{268}{10} = 26.8$

d. $s_1^2 = \dfrac{\sum x_1^2 - \dfrac{\left(\sum x_1\right)^2}{n_1}}{n_1 - 1} = \dfrac{118{,}873 - \dfrac{1513^2}{20}}{20 - 1} = 232.3447$

$s_2^2 = \dfrac{\sum x_2^2 - \dfrac{\left(\sum x_2\right)^2}{n_2}}{n_2 - 1} = \dfrac{42{,}178 - \dfrac{776^2}{15}}{15 - 1} = 145.2095$

$s_3^2 = \dfrac{\sum x_3^2 - \dfrac{\left(\sum x_3\right)^2}{n_3}}{n_3 - 1} = \dfrac{7880 - \dfrac{268^2}{10}}{10 - 1} = 77.5111$

e. $\hat{\mu} = \bar{x}_{\text{st}} = \dfrac{1}{N}\left(N_1\bar{x}_1 + N_2\bar{x}_2 + N_3\bar{x}_3\right)$

$= \dfrac{1}{40{,}000}\left[25{,}000(75.65) + 10{,}000(51.7333) + 5000(26.8)\right]$

$= \dfrac{1}{40{,}000}(2{,}542{,}583) = 63.5646$

$\text{Bound} = 2\hat{\sigma}_{\bar{x}_{\text{st}}} = 2\sqrt{\dfrac{1}{N^2}\sum N_i^2\left(\dfrac{N_i - n_i}{N_i}\right)\dfrac{s_i^2}{n_i}}$

$= 2\sqrt{\dfrac{1}{40{,}000^2}\left[25{,}000^2\left(\dfrac{25{,}000 - 20}{25{,}000}\right)\left(\dfrac{232.3447}{20}\right) + 10{,}000\left(\dfrac{10{,}000 - 15}{10{,}000}\right)\left(\dfrac{145.2095}{15}\right) + 5{,}000^2\left(\dfrac{5000 - 10}{5000}\right)\left(\dfrac{77.5111}{10}\right)\right]}$

$= 2\sqrt{\dfrac{1}{40{,}000^2}(8{,}414{,}964{,}691)} = 4.5867$

f. $\hat{\tau}_{\text{st}} = N\bar{x}_{\text{st}} = N_1\bar{x}_1 + N_2\bar{x}_2 + N_3\bar{x}_3 = 2{,}542{,}583$ (see part e.)

$\text{Bound} = 2\hat{\sigma}_{\hat{\tau}_{\text{st}}} = 2\sqrt{\sum N_i^2\left(\dfrac{N_i - n_i}{N_i}\right)\dfrac{s_i^2}{n_i}}$

$= 2\sqrt{8{,}414{,}964{,}691} = 183{,}466.23$ (see part e.)

g. $\hat{p}_1 = \dfrac{18}{20} = .90 \qquad \hat{p}_2 = \dfrac{9}{15} = .6 \qquad \hat{p}_3 = \dfrac{0}{10} = 0$

$\hat{p}_{\text{st}} = \dfrac{1}{N}\left(N_1\hat{p}_1 + N_2\hat{p}_2 + N_3\hat{p}_3\right)$

$= \dfrac{1}{40{,}000}\left[25{,}000(.9) + 10{,}000(.6) + 5000(0)\right] = .7125$

$$\text{Bound} = 2\hat{\sigma}_{\hat{p}_{st}} = 2\sqrt{\frac{1}{N^2}\sum N_i^2\left[\frac{N_i - n_i}{N_i}\right]\frac{\hat{p}_i(1 - \hat{p}_i)}{n_i - 1}}$$

$$= 2\sqrt{\frac{1}{40,000^2}\left[25,000^2\left(\frac{25,000 - 20}{25,000}\right)\left(\frac{.9(.1)}{19}\right) + 10,000^2\left(\frac{10,000 - 15}{10,000}\right)\left(\frac{.6(.4)}{14}\right)\right.}$$
$$\overline{\left. + 5,000^2\left[\frac{5000 - 10}{5000}\right](0)\right]}$$

$$= 2\sqrt{\frac{1}{40,000^2}(4,669,872.181)} = .1080$$

20.21   From Exercise 20.20,

$$\overline{x}_4 = \frac{\sum x_4}{n_4} = \frac{1246}{25} = 49.84$$

$$s_4^2 = \frac{\sum x_4^2 - \frac{\left(\sum x_4\right)^2}{n_4}}{n_4 - 1} = \frac{76,498 - \frac{1246^2}{25}}{24} = 599.89$$

$$\hat{\mu}_4 = \overline{x}_4 = 49.84$$

The bound is $2\hat{\sigma}_{\overline{x}_4} = 2\sqrt{\frac{s_4^2}{n_4}\left[\frac{N_4 - n_4}{N_4}\right]} = 2\sqrt{\frac{599.89}{25}\left[\frac{15,000 - 25}{15,000}\right]} = 9.7889$

20.23   a. $\overline{x}_1 = \frac{\sum x_1}{n_1} = \frac{181.6}{5} = 36.32$ $\overline{x}_2 = \frac{\sum x_2}{n_2} = \frac{443.8}{10} = 44.38$

$$\overline{x}_3 = \frac{\sum x_3}{n_3} = \frac{916.6}{15} = 61.1067$$

$$\hat{\mu} = \overline{x}_{st} = \frac{1}{N}\left(N_1\overline{x}_1 + N_2\overline{x}_2 + N_3\overline{x}_3\right)$$

$$= \frac{1}{6965}\left[3210(36.32) + 2015(44.38) + 1740(61.1067)\right]$$

$$= \frac{1}{6965}\left[3,212,338.558\right] = 44.844 \text{ thousand or } 44,844$$

b. $s_1^2 = \dfrac{\sum x_1^2 - \dfrac{\left(\sum x_1\right)^2}{n_1}}{n_1 - 1} = \dfrac{6628.3 - \dfrac{181.6^2}{5}}{4} = 8.147$

$s_2^2 = \dfrac{\sum x_2^2 - \dfrac{\left(\sum x_2\right)^2}{n_2}}{n_2 - 1} = \dfrac{20{,}272.52 - \dfrac{443.8^2}{10}}{9} = 64.0751$

$s_3^2 = \dfrac{\sum x_3^2 - \dfrac{\left(\sum x_3\right)^2}{n_3}}{n_3 - 1} = \dfrac{56{,}870.08 - \dfrac{916.6^2}{15}}{14} = 61.4078$

The bound is $2\hat{\sigma}_{\bar{x}_{st}} = 2\sqrt{\dfrac{1}{N^2}\sum N_i^2 \left[\dfrac{N_i - n_i}{N_i}\right]\dfrac{s_i^2}{n_i}}$

$= 2\sqrt{\dfrac{1}{6965^2}\left[3210^2\left(\dfrac{3210 - 5}{3210}\right)\dfrac{8.147}{5} + 2015^2\left(\dfrac{2015 - 10}{2015}\right)\dfrac{64.0751}{10}\right. \left. + 1740^2\left(\dfrac{1740 - 15}{1740}\right)\dfrac{61.4078}{15}\right]}$

$= 2\sqrt{\dfrac{1}{6965^2}(54{,}937{,}870.41)} = 2.128 \text{ thousand or } 2128$

The approximate 95% confidence interval is $44{,}844 \pm 2128 \Rightarrow (42{,}716, 46{,}972)$

We are 95% confident the mean 1980 income of the head administrators of the 6965 hospitals is between \$42,716 and \$46,972.

c. The 95% confidence interval is

$\bar{x}_3 \pm 2\sqrt{\dfrac{s_3^2}{n_3}\left(\dfrac{N_3 - n_3}{N_3}\right)} \Rightarrow 61.1067 \pm 2\sqrt{\dfrac{61.4078}{15}\left(\dfrac{1740 - 15}{1740}\right)}$

$\Rightarrow 61.1067 \pm 4.0292$

$\Rightarrow (57.0775, 65.1359) \text{ (in thousands)}$

20.25 $N = 360 + 74 + 95 = 529$

$\hat{\mu} = \bar{x}_{st} = \dfrac{1}{N}(N_1\bar{x}_1 + N_2\bar{x}_2 + N_3\bar{x}_3)$

$= \dfrac{1}{529}\left[360(14{,}900) + 74(39{,}250) + 95(23{,}800)\right] = 19{,}904.54$

$$\text{Bound} = 2\sqrt{\frac{1}{N^2}\sum N_i^2\left[\frac{N_i - n_i}{N_i}\right]\frac{s_i^2}{n_i}}$$

$$= 2\sqrt{\frac{1}{529^2}\left[360^2\left(\frac{360-30}{360}\right)\left(\frac{9,150,500}{30}\right) + 74^2\left(\frac{74-30}{74}\right)\left(\frac{25,003,000}{30}\right) + 95^2\left(\frac{95-30}{95}\right)\left(\frac{16,801,100}{30}\right)\right]}$$

$$= 2\sqrt{\frac{1}{529^2}(42,407,865,350)} = 778.57$$

20.27 For a bound of no more than \$200,000, solve the following equation:

$$200,000 = 2\frac{s}{\sqrt{n}}\sqrt{\frac{N-n}{N}}$$

For a first approximation, $\dfrac{N-n}{N} \approx 1$

From Exercise 20.13, we know $s = 1,083,162$.

$$200,000 = 2\left[\frac{1,083,162}{\sqrt{n}}\right]\sqrt{1} \Rightarrow n = 117.32$$

Now resolve the equation, using $n = 118$ for the correction factor (from Exercise 20.13, we know $N = 1500$):

$$200,000 = 2\left[\frac{1,083,162}{\sqrt{n}}\right]\sqrt{\frac{1500-118}{1500}} \Rightarrow n = 108.09$$

Now resolve the equation, using $n = 109$ for the correction factor:

$$200,000 = 2\left[\frac{1,083,162}{\sqrt{n}}\right]\sqrt{\frac{1500-109}{1500}} \Rightarrow n = 108.798$$

Thus, 109 large, diverse companies should be questioned concerning expenditures for external audits.

Since she already sampled 32, $109 - 32 = 77$ additional firms should be sampled.

20.29 From Exercise 20.23,

$$N_1 = 3210,\ N_2 = 2015,\ N_3 = 1740,\ N = 6965$$
$$s_1^2 = 8.147,\ s_2^2 = 64.075,\ s_3^2 = 61.408$$

a. To estimate the mean 1980 income of the head administrators to within $1000, solve the following equation (the data are in thousands of dollars):

$$1 = 2\sqrt{\frac{1}{N^2}\sum N_i^2\left[\frac{N_i - n_i}{N_i}\right]\frac{s_i^2}{n_i}}$$

For a first approximation, set $n_1 = n_2 = n_3 = n_s$ and $\dfrac{N_i - n_i}{N_i} \approx 1$

$$1 = 2\sqrt{\frac{1}{6965^2}\left[\frac{3210^2(8.147) + 2015^2(64.075) + 1740^2(61.408)}{n_s}\right]}$$

$$\Rightarrow n_s = 43.70 \approx 44$$

Now resolve the equation with $n_s = 44$ for the correction factor:

$$1 = 2\sqrt{\frac{1}{6965^2}\left[\frac{3210^2\left[\frac{3210 - 44}{3210}\right](8.147) + 2015^2\left[\frac{2015 - 44}{2015}\right](64.075) + 1740^2\left[\frac{1740 - 44}{1740}\right](61.408)}{n_s}\right]}$$

$$\Rightarrow n_s = 42.75 \approx 43$$

Now resolve the equation with $n_s = 43$ for the correction factor:

$$1 = 2\sqrt{\frac{1}{6965^2}\left[\frac{3210^2\left[\frac{3210 - 43}{3210}\right](8.147) + 2015^2\left[\frac{2015 - 43}{2015}\right](64.075) + 1740^2\left[\frac{1740 - 43}{1740}\right](61.408)}{n_s}\right]}$$

$$\Rightarrow n_s = 42.77 \approx 43$$

Thus, 43 administrators should be sampled from each stratum, so the total number of administrators sampled should be $43 \times 3 = 129$.

b. To estimate the mean 1980 income of the lead administrators to within $1000, solve the following equation (the data are in thousands of dollars):

$$1 = 2\sqrt{\frac{1}{N^2}\sum N_i^2\left[\frac{N_i - n_i}{N_i}\right]\frac{s_i^2}{n_i}}$$

For a first approximation, set $n_1 = 30$, $n_2 = 30$, and $\dfrac{N_3 - n_3}{N_3} \approx 1$.

$$1 = 2\sqrt{\frac{1}{6965^2}\left[3210^2\left[\frac{3210 - 30}{3210}\right]\frac{8.147}{30} + 2015^2\left[\frac{2015 - 30}{2015}\right]\frac{64.075}{30} + 1740^2\left[\frac{61.408}{n_3}\right]\right]}$$

$$\Rightarrow n_3 = 228.72 \approx 229$$

Now resolve the equation with $n_3 = 229$ for the correction factor:

$$1 = 2 \sqrt{\frac{1}{6965^2} \left[3210^2 \left(\frac{3210 - 30}{3210} \right) \frac{8.147}{30} + 2015^2 \left(\frac{2015 - 30}{2015} \right) \frac{64.075}{30} + 1740^2 \left(\frac{1740 - 229}{1740} \right) \frac{61.408}{n_3} \right]}$$

$$\Rightarrow n_3 = 198.62 \approx 199$$

Now resolve the equation with $n_3 = 199$ for the correction factor:

$$1 = 2 \sqrt{\frac{1}{6965^2} \left[3210^2 \left(\frac{3210 - 30}{3210} \right) \frac{8.147}{30} + 2015^2 \left(\frac{2015 - 30}{2015} \right) \frac{64.075}{30} + 1740^2 \left(\frac{1740 - 199}{1740} \right) \frac{61.408}{n_3} \right]}$$

$$\Rightarrow n_3 = 202.56 \approx 203$$

Thus, 203 administrators should be sampled from stratum 3.

20.31 $2\hat{\sigma}_{\hat{p}} = 2 \sqrt{\frac{\hat{p}(1 - \hat{p})}{n}} \sqrt{\frac{N - n}{N}} = .01$ \quad Using $\hat{p} = .62$ from Exercise 20.16.

$$\Rightarrow 2 \sqrt{\frac{.62(.38)}{n}} \sqrt{\frac{50,840 - n}{50,840}} = .01$$

$$\Rightarrow \sqrt{.62(.38) \left[\frac{1}{n} - \frac{1}{50,840} \right]} = .005$$

$$\Rightarrow \frac{1}{n} - \frac{1}{50,840} = \frac{.005^2}{.62(.38)} = .000106$$

$$\Rightarrow \frac{1}{n} = .000106 + \frac{1}{50,840} = .0001258$$

$$\Rightarrow n = \frac{1}{.0001258} = 7950.29 \approx 7951$$

20.33 **a.** n represents the number of accounts receivable selected ($n = 100$).

N represents the total number of accounts receivable at the department store in Boston ($N = 15,887$).

b. $\bar{x}$ represents the average age of the accounts receivable selected in the sample of $n = 100$.

μ represents the average age of all $N = 15,887$ accounts receivable at the department store in Boston.

c. $\hat{\tau}$ represents the estimated total monetary value of all $N = 15,887$ accounts receivable.

τ represents the actual total monetary value of all $N = 15,887$ accounts receivable at the department store in Boston.

d. $\hat{p}$ represents the estimated proportion of accounts receivable out of all $N = 15,887$ at the store that are more than 90 days old.

p represents the proportion of accounts receivable out of all $N = 15,887$ at the store that are more than 90 days old.

e. $\hat{\tau} \pm 2\hat{\sigma}_{\hat{\tau}}$ represents an approximate 95% confidence interval for the total monetary value of all $N = 15{,}887$ accounts receivable at the department store in Boston.

20.35 $N = 3500,\ n = 30$

$$\sum x_i = 10 + 0 + 5 + \cdots + 15 + 10 = 633$$

$$\sum x_i^2 = 10^2 + 0^2 + 5^2 + \cdots + 15^2 + 10^2 = 39{,}007$$

$$s^2 = \frac{\sum x_i^2 - \dfrac{\left(\sum x_i\right)^2}{n}}{n - 1} = \frac{39{,}007 - \dfrac{(633)^2}{30}}{30 - 1} = 884.5069$$

$$\bar{x} = \frac{\sum x_i}{n} = \frac{633}{30} = 21.10$$

a.   Estimator of τ:

$$\hat{\tau} = N\bar{x} = 3500(21.10) = 73{,}850$$

Estimated bound on the error of estimation:

$$2\hat{\sigma}_{\hat{\tau}} = 2\sqrt{N^2 \frac{s^2}{n}\left(\frac{N - n}{N}\right)} = 2\sqrt{3500^2\left[\frac{884.5069}{30}\right]\left[\frac{3500 - 30}{3500}\right]} = 37{,}845.89$$

b. p represents the proportion of PAC's that planned to support Reagan.

Let x represent the number of PAC's in the sample that expected to spend money in support of Reagan.

$$\hat{p} = \frac{x}{n} = \frac{21}{30} = .70$$

$$2\hat{\sigma}_{\hat{p}} = 2\sqrt{\frac{\hat{p}(1 - \hat{p})}{n}}\sqrt{\frac{N - n}{N}}$$

$$= 2\sqrt{\frac{.70(1 - .70)}{30}}\sqrt{\frac{3500 - 30}{3500}} = .167$$

An approximate 95% confidence interval for p is:

$$\hat{p} \pm 2\hat{\sigma}_{\hat{p}} = .70 \pm .167 \Rightarrow (.533, .867)$$

20.37 a. $\hat{p} = \dfrac{x}{n} = \dfrac{9296}{10{,}000} = .9296$

$$\text{Bound} = 2\hat{\sigma}_{\hat{p}} = 2\sqrt{\frac{\hat{p}(1 - \hat{p})}{n}}\sqrt{\frac{N - n}{N}}$$

$$= 2\sqrt{\frac{.9296(.0704)}{10{,}000}}\sqrt{\frac{500{,}000 - 10{,}000}{500{,}000}}$$

$$= 2\sqrt{.000006413} = .0051$$

b. Only $\frac{10,000}{500,000} \times 100\% = 2\%$ of the subscribers returned the questionnaire. Often in mail surveys, those that respond are those with strong views. Thus, the 10,000 that responded may not be representative. I would question the estimate in part **a**.

20.39 $k = 2, N = 535 + 366 = 901$

$$\hat{\mu} = \bar{x}_{st} = \frac{1}{N}(N_1\bar{x}_1 + N_2\bar{x}_2)$$

$$= \frac{1}{901}[535(8.5) + 366(5.2)] = 7.1595$$

$$\text{Bound} = 2\hat{\sigma}_{\bar{x}_{st}} = 2\sqrt{\frac{1}{N^2}\sum N_i^2 \left[\frac{N_i - n_i}{N_i}\right]\frac{s_i^2}{n_i}}$$

$$= 2\sqrt{\frac{1}{901^2}\left[535^2\left(\frac{535 - 50}{535}\right)\frac{16.8}{50} + 366^2\left(\frac{366 - 40}{366}\right)\frac{21.2}{40}\right]}$$

$$= 2\sqrt{.185293046} = .8609$$

The approximate 95% confidence interval is

$$\bar{x}_{st} \pm 2\hat{\sigma}_{\bar{x}_{st}} \Rightarrow 7.1595 \pm .8609 \Rightarrow (6.2986, 8.0204)$$